FUJITSU

shaping tomorrow with you

D1829328

MEF-CECP 2.0 Exam Study Guide

For Carrier Ethernet 2.0 (CE 2.0) Professionals

First Edition - December 2013

Jon Kieffer & Ralph Santitoro

Fujitsu Network Communications Inc.

https://partners.fnc.fujitsu.com

ISBN 10: 1494744244

ISBN 13: 9781494744243

This MEF-CECP 2.0 Exam Study Guide provides a broad range of Carrier Ethernet information useful for those who wish to learn more about MEF-standardized Carrier Ethernet and specifically for those planning to take the MEF Carrier Ethernet Certified Professional 2.0 (MEF-CECP 2.0) exam. Fujitsu has made every effort to ensure the accuracy of the MEF-CECP 2.0 Exam Study Guide, but editorial errors do occur from time to time.

Fujitsu welcomes your feedback and suggestions for this guide or future Carrier Ethernet educational material. Please use the email address below for all feedback. Finally, Fujitsu does not guarantee, either implicitly or explicitly, that an individual using this MEF-CECP 2.0 Exam Study Guide will pass the MEF-CECP 2.0 certification exam.

For comments or questions, contact us at ReaderFeedback@fnc.fujitsu.com.

Contents

1

Introduction

In this chapter:

1.1 Getting Started
1.2 Review

The MEF awards the Carrier Ethernet Certified Professional 2.0 (MEF-CECP 2.0) certification to individuals who pass the MEF-CECP 2.0 exam.

This guide is designed to help you prepare for the MEF-CECP 2.0 exam. This guide assumes that you have a basic understanding of Ethernet, but no prior knowledge of Carrier Ethernet. The guide covers all topics that are included in the MEF-CECP 2.0 exam. Material is presented systematically, beginning with MEF service definitions (the core content that accounts for most exam questions) and building outward. Each lesson ends with a set of multiple-choice review questions, similar to those appearing in the MEF-CECP 2.0 exam, to help you evaluate your learning. The guide also includes strategic guidance to help you focus your efforts and efficiently prepare for the exam.

This study guide was created by identifying essential content from MEF specifications, white papers, and reference presentations; reorganizing it into a systematic, logical progression; and adding context as necessary for clarity and continuity. It is intended to be used like a text book. Early lessons introduce fundamental concepts that later lessons build upon.

The guide is meant to be complete and comprehensive, but it is not meant to be used in isolation. In places, the guide directs you to visit the MEF Web site and familiarize yourself with its wealth of valuable content. The study guide includes numerous references to MEF specifications that you should consult to verify or enhance your understanding.

Fujitsu is an MEF-Accredited Training Provider

MEF ACCREDITED TRAINING PROVIDER

Comments

Fujitsu welcomes your feedback and suggestions for this guide or future Carrier Ethernet educational material. If you have comments or questions, please contact us at ***ReaderFeedback@fnc.fujitsu.com***.

Companion Material

Fujitsu publishes information related to this study guide at fujitsu.com/us/services/telecom/training/Ethernet-training.html.

- **Publication formats**–Retail sellers of this study guide for both print and a variety of electronic publication formats.
- **Errata** – Errors found in this study guide after publication will be listed along with corrections.
- **In-classroom training**–Fujitsu offers MEF-accredited in-classroom training for the MEF-CECP 2.0 exam.
- **Exam Trainer app**–Fujitsu offers a companion app for use in conjunction with this study guide. It contains hundreds of practice questions similar to those on the MEF-CECP 2.0 exam and is available for a variety of mobile devices.

Note: *A companion text book to this study guide is planned to help individuals who want more background in basic networking and Ethernet in order to make efficient use of this study guide. Information about that book, when it is published, will be available at* fujitsu.com/us/services/telecom/training/Ethernet-training.html.

1.1
Getting Started

In this section:

1.1.1
About the MEF

The Metro Ethernet Forum (MEF) is a global industry alliance comprising more than 220 organizations, including telecommunications service providers, cable MSOs, network equipment/software manufacturers, semiconductor vendors, and testing organizations. The MEF's mission is to accelerate the worldwide adoption of Carrier-class Ethernet networks and services. The MEF develops Carrier Ethernet technical specifications and implementation agreements to promote interoperability and deployment of Carrier Ethernet worldwide.

For more information about the MEF, including a complete listing of all current MEF members, visit MetroEthernetForum.org.

1.1.2
About Carrier Ethernet

The use of Ethernet for wide area networks (WANs) and services is motivated by the ubiquitous prevalence of Ethernet in local area networks (LANs) and by the pervasive use of Ethernet as the standard interface on most networked devices (PCs, servers, laptops, and so on).

Initially, TDM networks, which support T1/E1 private lines and SONET/SDH, were widely deployed because circuit-based voice traffic accounted for the vast majority of WAN traffic. At that time, data traffic was limited to low-speed, dial-up modem connections over voice circuits and fractional T1/E1 circuits for data transmission.

Today, we experience the opposite scenario where packet-based data traffic is dominant and circuit-based voice traffic accounts for only a tiny percentage of overall network traffic. High-speed, broadband data traffic using the Internet Protocol (IP) is most prevalent and continues to increase as more applications migrate towards IP-based implementations. Interestingly, circuit-based voice traffic, the original driver for TDM networks, is rapidly migrating to Voice over IP (VoIP), which makes voice traffic appear as packet-based data traffic to the network. Video, which originally required a dedicated network, is also migrating to video over IP, thus further driving the need for high-bandwidth packet-based WANs.

Most networked applications are IP-based and run over networking devices that connect to wireline networks over Ethernet. Even applications that run over a wireless network, such as smartphones and tablets, eventually reach a networking device that connects to an Ethernet network. Considering that most applications begin and end on an Ethernet network, the use of Carrier Ethernet as the WAN technology or service enables a common Ethernet technology to be used end-to-end.

Furthermore, compared to legacy, circuit-based TDM networks, Carrier Ethernet, being a packet-based technology, can much more efficiently accommodate the now-dominant packet-based traffic.

The MEF distinguishes Carrier Ethernet from familiar LAN-based Ethernet based on five attributes:

1. Standardized Services
2. Scalability
3. Reliability
4. Quality of Service
5. Service Management

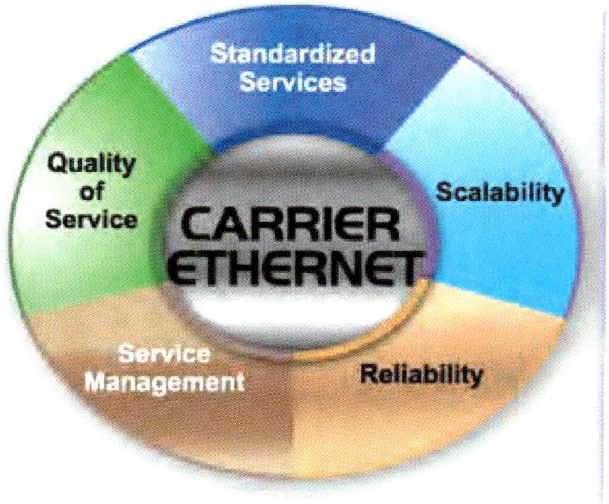

Reproduced with permission of the Metro Ethernet Forum

These five attributes are meant to be interpreted broadly as guiding principles that can adapt with technological evolution.

1.1.3 Explore the MEF Web Site and the Carrier Ethernet Reference Wiki

The MEF Web site is MetroEthernetForum.org.

Go ahead and explore this Web site now. It is a great way to start preparing for the MEF-CECP 2.0 exam.

In particular, discover how to navigate from the main page to each of the following pages:

- MEF Technical Specifications
- MEF Reference Presentations
- MEF White Papers
- MEF Certification Programs
- MEF Professional Certification

Look around. See what's there.

The MEF also maintains the *Carrier Ethernet Reference Wiki* at wiki.metroethernetforum.com/x/zIBP .

Again, look around. See what's there.

The MEF maintains the *Carrier Ethernet Reference Wiki* to serve as a tool to make it easier for the general public to access information on Carrier Ethernet as defined by the MEF. Much of its content overlaps with content in this study guide. However, its focus is broader: Carrier Ethernet in general, rather than Carrier Ethernet for the MEF-CECP 2.0 exam. Also, its content is generated collaboratively and is subject to continuous updating. The general public cannot update its pages directly. However, anyone can write a comment, using the comment dialog at the bottom of each page. Comments are reviewed by MEF authorities and may lead to revision of wiki page content.

This study guide includes many references to MEF material. Use those references. Know where to find them.

1.1.4 About the MEF-CECP 2.0 Exam

The MEF-CECP 2.0 exam targets individuals who perform job roles such as product manager, product planner, product developer, system architect, network engineer, network architect, systems engineer, sales engineer, or technical marketing. MEF-CECP 2.0 certifies that the individual possesses the knowledge required to define, design, develop, or market Carrier Ethernet products, services, or applications.

Note: *The MEF's first professional certification, the MEF-CECP certification, was updated and replaced in December 2013 by the MEF-CECP 2.0 certification. The MEF no longer offers MEF-CECP certification. If you are an MEF-CECP, your certification does not expire. However, to receive MEF-CECP 2.0 certification, you must pass the MEF-CECP 2.0 exam.*

The MEF-CECP 2.0 certification exam covers MEF Carrier Ethernet 2.0 (CE 2.0) concepts, capabilities, and specifications that were unavailable at the time the MEF-CECP exam was completed in late 2011. Such material includes, but is not limited to, E-Access services, new service performance metrics, MEF CE 2.0 equipment and services, and standardized classes of service as explained in this study guide. Current MEF-CECPs will want to study for and pass the new MEF-CECP 2.0 exam to demonstrate proficiency in MEF CE 2.0, the latest generation of Carrier Ethernet equipment and services certifications for all MEF services.

The exam is open to anyone and may be taken online from your home or office, at an MEF Approved Testing Center, or from an MEF Accredited Training Provider. Refer to EthernetAcademy.net for more details.

The MEF-CECP 2.0 exam is a 1-hour 45-minute exam consisting of 80 multiple-choice questions: 70 questions are scored, and 10 are unscored. The 10 unscored questions are included for statistical purposes (they are considered for use in future exams). To pass the exam, you need to correctly answer 46 of the 70 scored questions (66%).

Note: *Some MEF CECP 2.0 exam questions ask you to select more than one multiple-choice answer. For example:* ***Which three of the following options are required for ...? (Choose three)****. To correctly answer such questions, you must select all of the correct multiple-choice answers. Partial credit is not given for selecting some (but not all) correct answers.*

Individuals who pass the MEF-CECP 2.0 exam are awarded:

- An official MEF-CECP 2.0 certificate
- A listing, as being MEF-CECP 2.0 certified, on the online Directory of MEF Certified Professionals posted at the widely visited EthernetAcademy.net Web site
- Qualification to use the MEF-CECP 2.0 logo (for example, in e-mail signatures, business cards, and resumes) for instant recognition of their status

Figure 1: MEF-CECP 2.0 Logo

About one third of all questions are asked in the context of an application scenario.

Caution: *The MEF Web site and MEF standards are continually evolving and already includes material that supersedes the material covered by the MEF-CECP 2.0 exam (developed in 2013). For exam preparation purposes, you may ignore any MEF standards published after April 2013. For example, you may ignore MEF 10.3, published October 2013.*

1.1.5
MEF-CECP 2.0 Certification Blueprint

The *Carrier Ethernet Reference Wiki* at wiki.metroethernetforum.com/x/zIBP provides a link to the MEF-CECP 2.0 Certification Blueprint.

This blueprint lists the MEF-CECP 2.0 exam objectives and provides links to related references.

This study guide covers all of the objectives listed in the blueprint, but in a different structure, starting with fundamentals and gradually expanding outward.

Hint: *For now, you may want to glance over the MEF-CECP 2.0 Certification Blueprint to get a rough sense of the work ahead of you. If you are new to Carrier Ethernet, you should see many unfamiliar topics.*

1.1.6 Assumed Knowledge

The following tables list topics of assumed knowledge. Many resources are available for learning this material, including Firewall.cx and Wikipedia.org.

7-Layer OSI Reference Model
Focus is on the bottom three layers: • Layer 3 (Network Layer)—Routes IP packets based on network address (IP address) • Layer 2 (Data Link Layer)—Forwards Ethernet frames based on MAC address • Layer 1 (Physical Layer)—Transmits data over a link between two interfaces Each layer provides services to the next-higher layer.

Encapsulation
Refers to the process of putting headers and trailers around some data. A lower layer encapsulates the higher layer's data behind a header.

Protocol Data Unit (PDU)
Generic term for the set of bits that is recognized as a fundamental unit (structured group of bits) at any layer (Layer 2 and above) A lower-layer PDU is comprised of data (the PDU of the next-higher layer) plus a header, and possibly a trailer. A Layer 3 PDU is called a **packet** (IP packet, for example). A Layer 2 PDU is called a **frame** (Ethernet frame, for example).

Ethernet
Ethernet Frame Components: • MAC destination address–Address of destination interface • MAC source address–Address of source interface • Ethertype–Indicates which protocol is encapsulated in the payload • VLAN tag–Optional component containing VLAN ID and PCP bits • Frame check sequence (FCS)–Extra checksum characters added to an Ethernet frame for error detection
MAC Destination Addressing: Unicast, Multicast, Broadcast, L2CP (Layer 2 Control Protocol)
Collision Domain–The set of network interface cards (NICs) for which a frame sent by one NIC could result in a collision with a frame sent by any other NIC in the same collision domain
Broadcast Domain–The set of NICs for which a broadcast frame sent by one NIC is received by all other NICs in the same broadcast domain
Bridge–The Layer 2 network device that performs bridging (also commonly called an Ethernet switch or Layer 2 switch). This study guide uses the term bridge.
Bridging–Ethernet frame forwarding based on MAC address, MAC learning, and STP (Spanning Tree Protocol); the process by which bridges learn to forward unicast frames without broadcasting
Spanning Tree Protocol (STP)–The network protocol that ensures a loop-free topology for a bridged Ethernet LAN There are several types of STP: • STP (Spanning Tree Protocol) • RSTP (Rapid Spanning Tree Protocol) • MSTP (Multiple Spanning Tree Protocol)
Bridge Protocol Data Unit (BPDU)–Ethernet frames exchanged by bridges to support STP
IEEE 802.1Q (2005)–Basic standard for interconnecting LANs using forwarding based on MAC addresses, STP, and a single VLAN tag (C-Tag)

Router
Layer 3 device that routes packets based on IP address Also separates broadcast domains

1.2
Review

1. What does MEF stand for?
   a. Metro Ethernet Framework
   b. Metro Ethernet Forum
   c. Model Ethernet Forum
2. According to the MEF, how many attributes distinguish Carrier Ethernet from familiar LAN-based Ethernet?
   a. 4
   b. 5
   c. 6
3. Which one of the following statements is not true?
   a. This study guide assumes you have a basic knowledge of Ethernet.
   b. The bulk of MEF-CECP Exam questions relate to MEF specifications 6.1 and 10.2.
   c. MEF standards published after April 2013 are not included in the MEF-CECP 2.0 exam.
   d. This study guide is meant to be used in complete isolation.
4. If a bridge receives an Ethernet frame with an unknown unicast destination address, the frame is:
   a. Flooded
   b. Buffered until the address can be determined
   c. Dropped
5. When a bridge learns a MAC address, that address is:
   a. The destination address within a frame received by the bridge
   b. Flooded to all peer bridges
   c. The source address within a frame received by the bridge

Related Links

Answers: Introduction and Overview

2

MEF Services

In this chapter:

To date, the MEF has defined eight Carrier Ethernet services. Six of them are retail services (sold to subscribers):

1. EPL (Ethernet Private Line)
2. EP-LAN (Ethernet Private LAN)
3. EP-Tree (Ethernet Private Tree)
4. EVPL (Ethernet Virtual Private Line)
5. EVP-LAN (Ethernet Virtual Private LAN)
6. EVP-Tree (Ethernet Virtual Private Tree)

The other two are E-Access services (sold to service providers for use in supporting retail service to subscribers):

1. Access EPL (Access Ethernet Private Line)
2. Access EVPL (Access Ethernet Virtual Private Line)

This lesson only describes the six retail services. The two E-Access services are described in a later lesson.

Related Links

E-Access Services

2.1 Fundamental Components and Reference Models

In this section:

The MEF defines a framework for delivering Ethernet services over carrier grade networks. The business model for this framework involves two principle stakeholders:

- **Subscriber**–The organization purchasing the Carrier Ethernet service
- **Service Provider**–The organization providing the Carrier Ethernet service

In general terms, the service itself is an Ethernet connection between two or more sites. The sites belong to the subscriber. The network connecting the sites belongs to the service provider.

2.1.1 Basic Reference Model

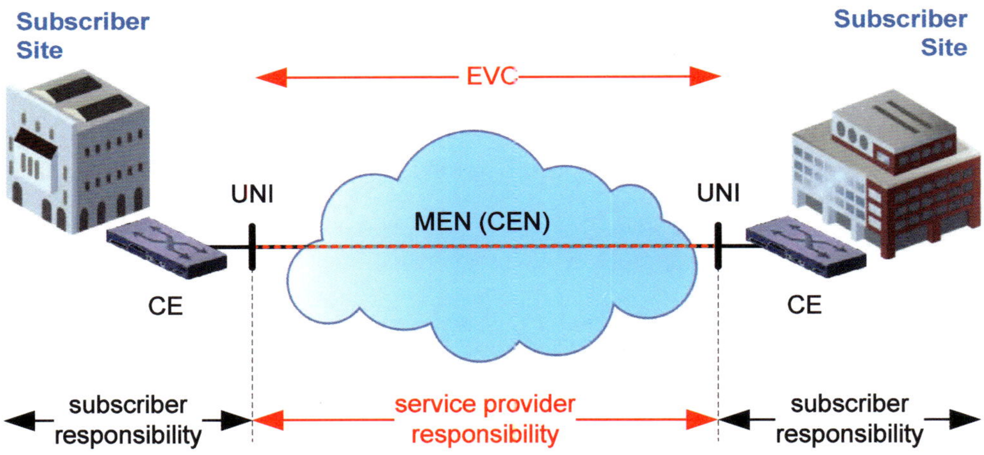

The basic reference model includes two functional components:

- **MEN (Metro Ethernet Network)** or **CEN (Carrier Ethernet Network)**–The service provider network used to transport Carrier Ethernet services.
- **CE (Customer Edge equipment)**–The equipment at the subscriber site that connects to the MEN. The CE can be a router or bridge/switch that is compliant with IEEE 802.1.

Note: *The term MEN originally implied that the network is metro-level, but it is now interpreted more generally. A MEN can be an access network, a metro network, a long-haul network, or any combination of networks required to support the Carrier Ethernet service. In more recent MEF specifications (and in all new MEF specifications), the term CEN is used in place of MEN.*

The other elements in the preceding figure are:

- **UNI (User-to-Network Interface)**–A physical demarcation point between the subscriber domain and the service provider domain

- **EVC (Ethernet Virtual Connection)**–The logical representation of a service connection between two or more UNIs

Every MEF service has exactly one EVC, but that EVC can connect more than two UNIs. The following figure shows an EVC connecting three UNIs.

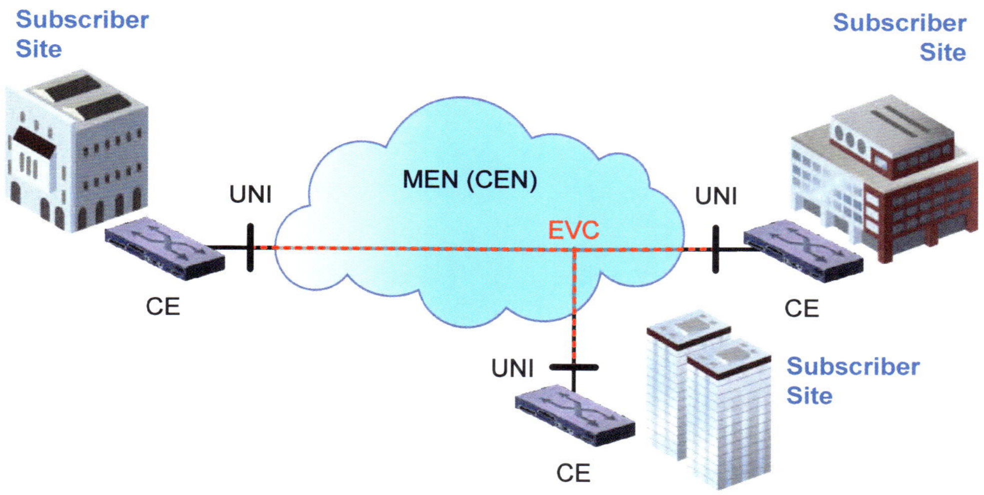

Figure 2: EVC for a Service Connecting Three Subscriber Sites

2.1.2 UNI Functionality

The UNI demarcation point is located at a specific place between the CE and the MEN (CEN). Typically, it is a port on the CE or a port on the MEN (CEN) edge device that connects to the CE.

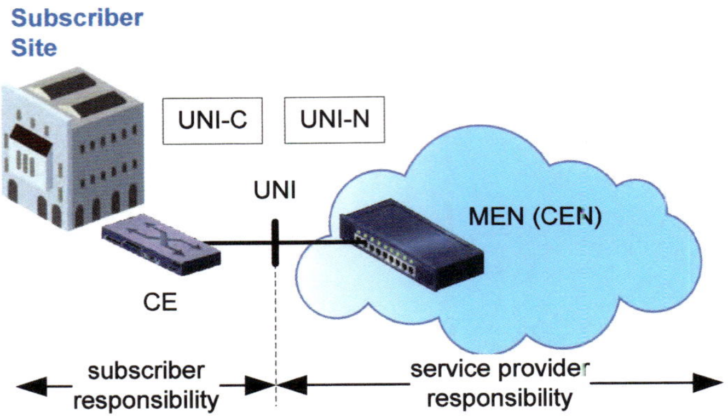

Figure 3: UNI-C and UNI-N

On either side of the UNI demarcation point, the MEF defines two functional elements:

- **UNI-C (UNI Client)**—Represents all of the subscriber-side functions required to connect the CE to the MEN (CEN). Individual functions in a UNI-C are entirely in the subscriber domain, but may in some cases be managed by the service provider.
- **UNI-N (UNI Network)**—Represents all of the MEN-side (CEN-side) functions required to connect the MEN (CEN) to the CE. The individual functions in a UNI-N are entirely in the service provider/network operator domain.

Together the UNI-C and UNI-N support a IEEE 802.3 Ethernet PHY/MAC interface, plus additional data, management, and control plane functions defined by the MEF. Details of UNI-C and UNI-N functionality are covered later in this study guide.

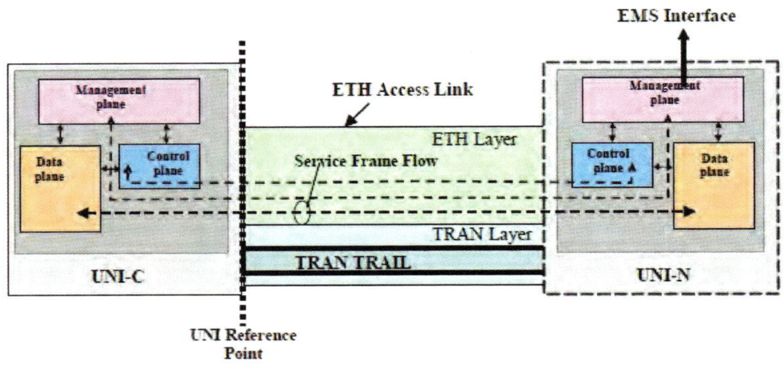

Figure 4: UNI Functionality

Related Links

UNI Requirements

2.1.3
General Reference Model

MEF 26.1 extends the basic reference model to define MEF services across multiple networks, with each MEN (CEN) independently owned and operated.

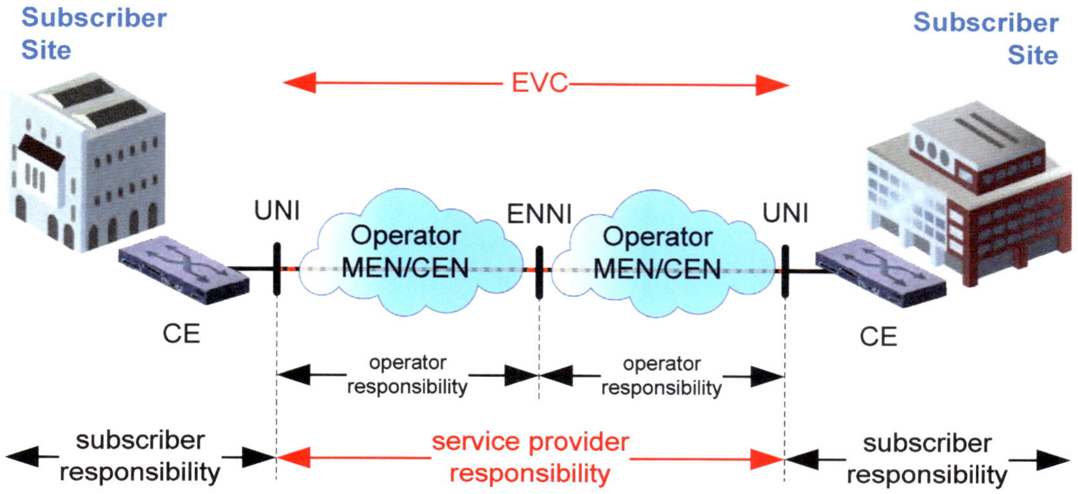

Figure 5: General Reference Model

In this model, the independent networks are called **Operator MENs** or **Operator CENs** (not MENs or CENs), and an interface between two Operator MEN/CENs is called an **ENNI (External Network-to-Network Interface)**.

This model is covered in detail later in this study guide. For now, simply recognize that the Carrier Ethernet service is unchanged from the subscriber's perspective:

- There is still only one Carrier Ethernet service provider responsible for the service as a whole.
- The Carrier Ethernet service is still implemented with exactly one EVC connecting two or more UNIs.
- The basic reference model is sufficient to define the service from the subscriber's perspective.
- The general reference model adds details of implementation that are only important to the service provider and subcontracting operators.

Related Links

Extending MEF Services over Multiple Operator MENs

2.2
Service Multiplexing

A UNI, in general, can support more than one EVC.

The MEF defines **Service Multiplexing** as the association of multiple EVCs to a UNI.

In the following figure, services are multiplexed on two of the UNIs.

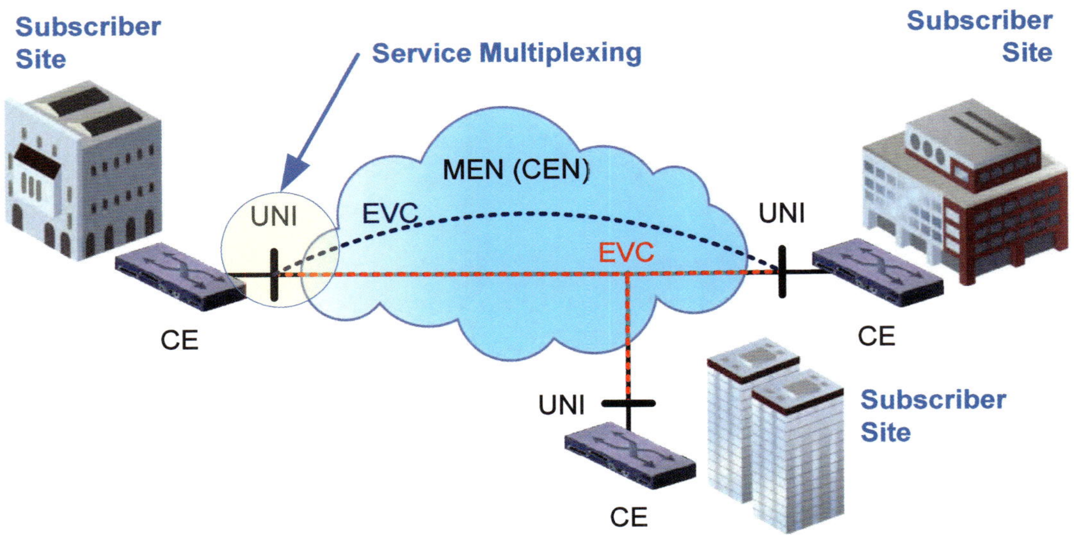

Figure 6: UNI Supporting Two EVCs

Notice that service multiplexing is a property of a UNI and is not a property of the Carrier Ethernet service or EVC.

The key benefit of service multiplexing is that it allows services to share UNIs, saving ports (physical connections) on subscriber and service provider edge equipment.

Related Links

Service Multiplexing, Bundling, and VLAN Tag Preservation Service Attributes

2.3 Assigning Ethernet Frames to EVCs

In this section:

Ethernet frames are assigned to EVCs at UNIs based on VLAN ID.

An Ethernet frame cannot be assigned to more than one EVC.

Note: *Sometimes people use the term* ***Ethernet packet*** *in place of* ***Ethernet frame****. The term* ***Ethernet frame*** *is preferred because it complies with terminology defined in the TCP/IP networking reference model: Layer 2 protocol data unit (PDUs) are called* ***frames****, and Layer 3 PDUs are called* ***packets****. Because Ethernet PDUs are at Layer 2, they are more properly called frames.*

2.3.1 Service Frames

The MEF defines a **service frame** as an Ethernet frame transmitted across the UNI toward the service provider or an Ethernet frame transmitted across the UNI toward the subscriber.

All service frames are either untagged or include an IEEE 802.1Q customer tag in the Ethernet header as shown in the following figure.

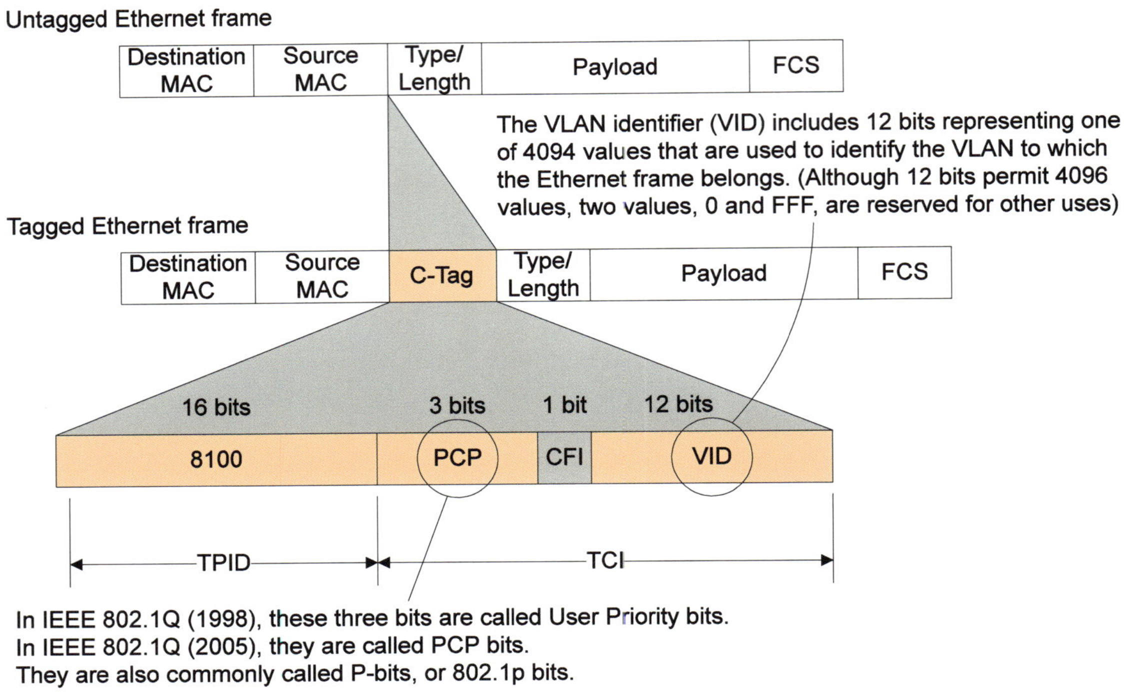

Figure 7: IEEE 802.1Q-2005 Ethernet Frame

About Ethernet Frame Terminology:

Prior to release of IEEE 802.1Q-2011, the frames depicted in the previous figure were commonly called *IEEE 802.1Q Ethernet frames* because they conform to IEEE 802.1Q-2005. However, with the release of IEEE 802.1Q-2011, the term *IEEE 802.1Q Ethernet frame* now properly refers to an Ethernet frame that conforms to IEEE 802.1Q-2011, which is far more general.[1]

In this study guide, we avoid the term *IEEE 802.1Q Ethernet frame*. The following terms are used as required:

- **IEEE 802.1Q-2005 Ethernet frame** – An Ethernet frame that is either untagged or includes a single C-Tag (VLAN tag with TPID=8100), as shown in the previous figure.
- **IEEE 802.1Q Customer Tag** – A C-Tag (VLAN tag with TPID=8100), as shown in the previous figure.

1 IEEE 802.1Q-2011 incorporates many amendments to IEEE 802.1Q-2005, including IEEE 802.1ad (Provider Bridging), IEEE 802.1ah (Provider Backbone Bridging), and IEEE 802.1Qay (Provider Backbone Bridging with Traffic Engineering Extensions).

Size of Service Frame Compared to Payload

An untagged Ethernet frame has 18 more bytes than its payload (for example, IP packet).

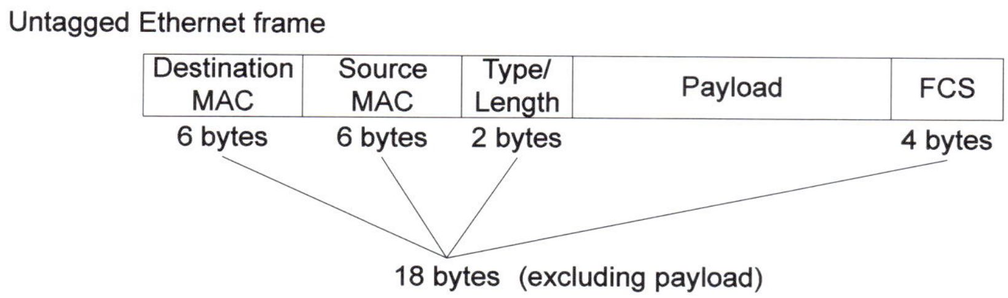

The C-Tag, if used, adds another 4 bytes to the Ethernet frame.

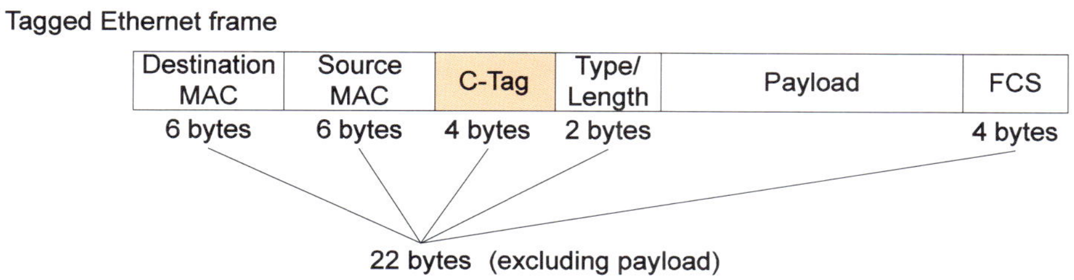

Note: *A byte consists of 8 bits.*

The payload for a standard 802.3 Ethernet frame is limited to 1500 bytes, which limits untagged Ethernet frames to 1518 bytes and C-Tagged Ethernet frames to 1522 bytes.

Note: *Some implementations of high-speed Ethernet (Gigabit Ethernet and faster) allow larger Ethernet frames known as **jumbo frames**.*

Layer 1 versus Layer 2 Ethernet Frames

Ethernet is both a Layer 1 technology and a Layer 2 technology.

In this study guide and in most MEF specifications, an *Ethernet frame* usually refers to the Layer 2 Ethernet frame. However, it is understood that at Layer 1 the Ethernet frame includes 20 more bytes (interframe gap and preamble bits) to support transmission over the physical (Layer 1) transmission medium.

As shown in the following example, the Layer 2 Ethernet frame starts with the Destination MAC address and ends with the Frame Check Sequence (FCS).

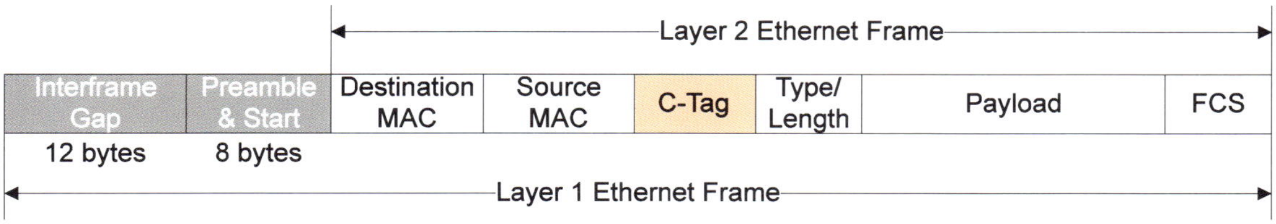

The Layer 1 Ethernet frame includes interframe gap and preamble bits.

2.3.2
CE-VLAN ID

In MEF terminology:

- **CE-VLAN tag**—Refers to the IEEE 802.1Q customer tag in the Ethernet header of a service frame

  Note: *The CE-VLAN tag is also commonly called the C-VLAN tag or C-Tag.*

- **CE-VLAN ID**—Refers to the VLAN identifier (VID) in the CE-VLAN tag

  Note: *The 12-bit VID field allows 4096 values, but only 4094 values can be used for VLAN identification because, per IEEE 802.1Q, values 0 and FFF are reserved for other uses.*

The subscriber can use the CE-VLAN tag to assign a service frame to a service (to an EVC) as follows:

1. The subscriber marks the service frame with a CE-VLAN ID before sending it across the UNI to the service provider.
2. When the service frame arrives at the UNI, the service provider associates it to an EVC based on CE-VLAN ID.

Of course, this requires coordination between the subscriber and the service provider because they must have an agreement about which CE-VLAN IDs are mapped to each EVC (per UNI).

2.3.3
Bundling

In a general sense, *bundling* implies that multiple CE-VLAN IDs are assigned to an EVC as shown in the following figure.

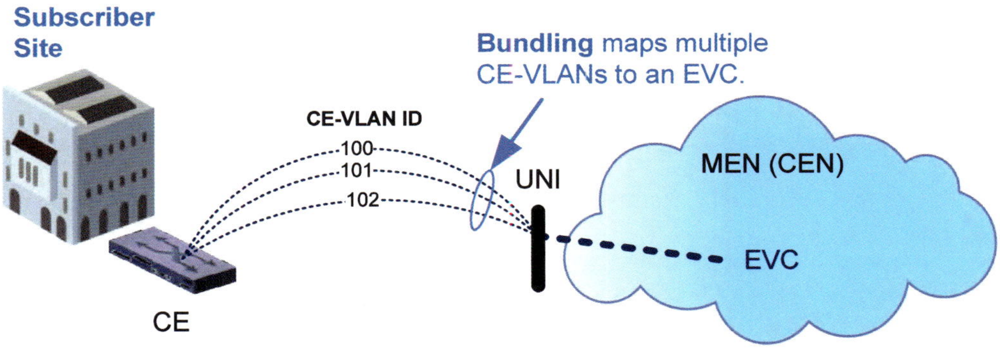

However, the MEF also uses the term more carefully. In MEF service definitions, **bundling** describes a UNI's ability to support *bundling* (the assignment of multiple CE-VLAN IDs to an EVC). MEF service definitions also make a distinction between **bundling** and **all-to-one bundling** as follows:

- **Bundling**–The association of multiple CE-VLAN IDs to an EVC at a UNI
- **All-to-one Bundling**–The association of all service frames to one EVC at the UNI

In MEF service definitions, **bundling** and **all-to-one bundling** are mutually exclusive. A UNI can support bundling or all-to-one bundling, not both.

Related Links

Service Multiplexing, Bundling, and VLAN Tag Preservation Service Attributes

2.3.4 Bundling versus Service Multiplexing

The following figure shows bundling together with service multiplexing to help you distinguish between the two.

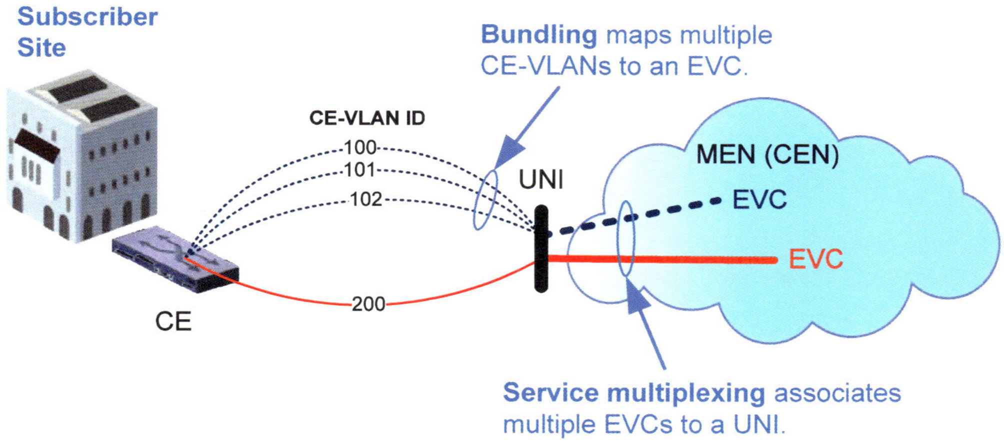

Test your understanding: *Can a UNI support service multiplexing together with all-to-one bundling? Explain why or why not.*

Answer: *No. All-to-one bundling implies that all service frames map to one EVC. A service frame cannot map to more than one EVC.*

2.4
Taxonomy of Carrier Ethernet Services

In this section:

MEF 6.1 defines three main Carrier Ethernet service **types**: E-Line, E-LAN, and E-Tree. Then, for each main service type, the MEF defines two services:

- A *port-based*, or *private*, service
- A *VLAN-based*, or *virtual private*, service

This results in six Carrier Ethernet services, as shown in the following table.

Service Type	Carrier Ethernet Service	
	Port-Based (All-to-One Bundling)	VLAN-Based (Ethernet frames mapped by VLAN ID)
E-Line (point-to-point EVC)	EPL (Ethernet Private Line)	EVPL (Ethernet Virtual Private Line)
E-LAN (multipoint-to-multipoint EVC)	EP-LAN (Ethernet Private LAN)	EVP-LAN (Ethernet Virtual Private LAN)
E-Tree (rooted-multipoint EVC)	EP-Tree (Ethernet Private Tree)	EVP-Tree (Ethernet Virtual Private Tree)

Hint: *If you are familiar with the terms **private** and **virtual private** in another context, try not to carry that meaning into this context because you may become confused. Just think of **private** and **virtual private** as names (like "orange" and "blue") that are yet to be defined. Focus more on the alternative terms **port-based** and **VLAN-based**, which are less likely to be confusing.*

Attention: *In addition to these six services, the MEF has defined two E-Access services (the **Access EPL** and the **Access EVPL**), which are described later in this study guide.*

Related Links

E-Access Services

2.4.1
Port-Based versus VLAN-Based Services

In a port-based (or private) service, all UNIs are configured for all-to-one bundling, and all service frames are mapped to the EVC, regardless of CE-VLAN ID.

In a VLAN-based (or virtual private) service, CE-VLAN IDs are explicitly mapped to the EVC at each UNI.

The key advantage of a port-based service is that the subscriber and the service provider do not have to coordinate VLAN IDs.

The key advantage of a VLAN-based service is that VLAN-based services can share UNIs, saving ports (physical connections) on subscriber and service provider edge equipment. To permit sharing, a UNI must be configured to support service multiplexing. The EVCs that share a UNI must use different CE-VLAN IDs.

Test your understanding: *With respect to VLAN-based service, classify each of the following UNI attributes as either mandatory, supported, or not supported: (1) Bundling, (2) All-to-one bundling, and (3) Service multiplexing.*

Answer: *(1) supported, (2) not supported, (3) supported*

Test your understanding: *With respect to port-based service, classify each of the following UNI attributes as either mandatory, supported, or not supported: (1) Bundling, (2) All-to-one bundling, and (3) Service multiplexing.*

Answer: *(1) not supported, (2) mandatory, (3) not supported*

2.4.2 The Three Service Types

Service type determines EVC topology and UNI connectivity.

E-Line Service Type

The E-Line service type is for creating a point-to-point service between two UNIs.

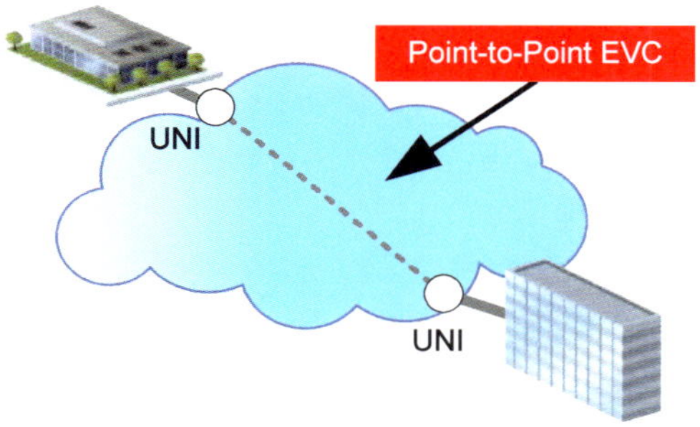

Example applications: *private line, virtual private line, Internet access*

Related Links

Support for TDM Private Lines
Internet Access

E-LAN Service Type

The E-LAN service type is for creating a multipoint-to-multipoint service that allows any UNI to forward Ethernet frames to any other UNI.

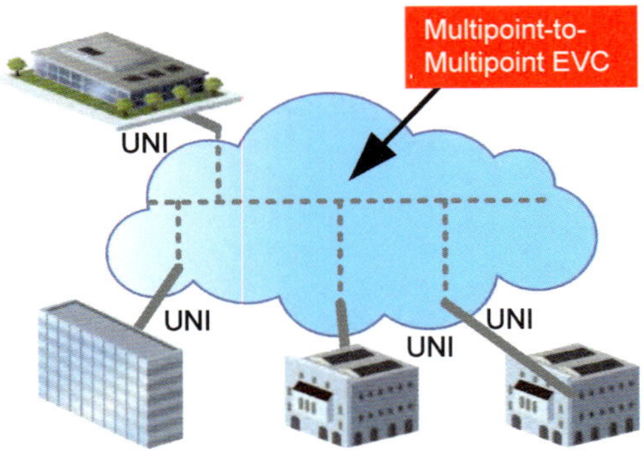

This service type supports bridging (forwards frames based on MAC address learning).

Example applications: *Layer 2 virtual private network (VPN), Layer 3 VPN, multicast network*

Related Links

Virtual Private Networks (VPNs)

E-Tree Service Type

The E-Tree service type is for creating a rooted-multipoint service that prevents some UNIs from forwarding Ethernet frames to other UNIs. Each UNI is declared to be either a **root** or a **leaf**. Roots are allowed to forward Ethernet frames to leaves and roots. A leaf can only forward frames to a root (no leaf-to-leaf forwarding).

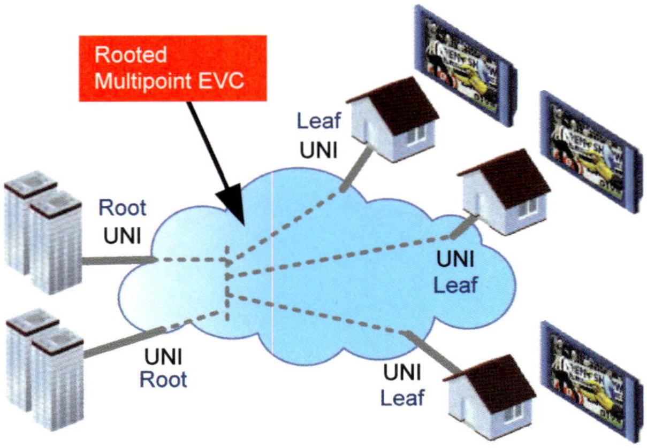

Like the E-LAN service type, this service type supports bridging (forwards frames based on MAC address learning). However, unlike the E-LAN service type, frame forwarding is also subject to leaf/root constraints.

Example applications: *Internet access, broadcast network, mobile backhaul, residential broadband backhaul*

Related Links

Ethernet Access to IP Services
Mobile Backhaul Services

2.5
EPL Service

An Ethernet Private Line (EPL) is a port-based service of type E-Line. It transports all service frames bidirectionally between two UNIs.

In the following figure, the enterprise on the left uses two EPLs: one to connect to a storage service provider and another to connect to the Internet. The two EPLs share no UNIs.

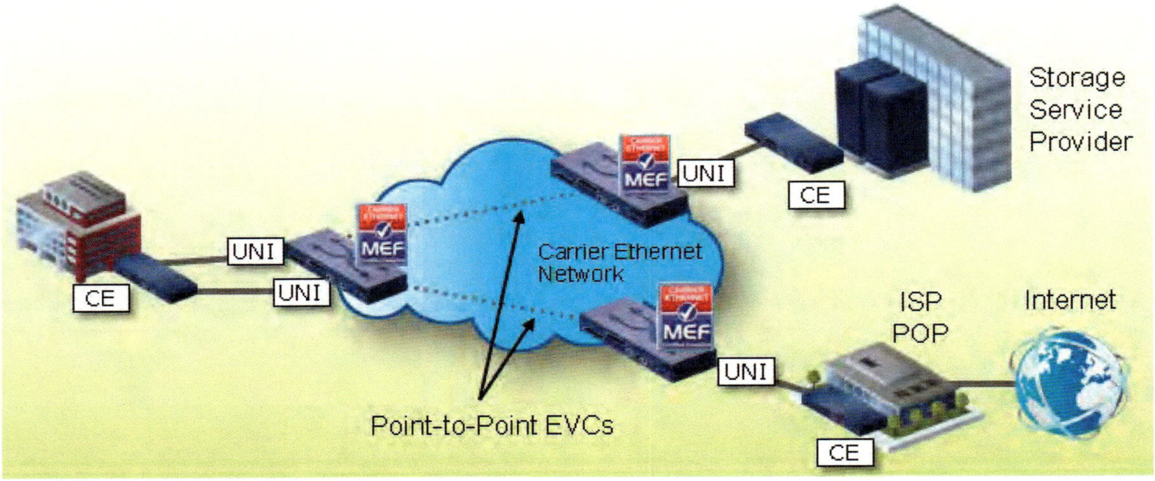

Reproduced with permission of the Metro Ethernet Forum

Distinctions of EPL service include:

- EPL service is the most popular Carrier Ethernet service due to its simplicity.
- EPL service is a replacement for TDM private line service.
- EPL service requires a dedicated physical connection (UNI) at each end.
- EPL service requires no VLAN coordination between the subscriber and the service provider.
- EPL service provides a high degree of transparency, transporting all service frames and preserving the content of all service frames, including Layer 2 control protocol (L2CP) frames.

Related Links

Support for TDM Private Lines

2.6
EVPL Service

An Ethernet Virtual Private Line (EVPL) is a VLAN-based service of type E-Line.

Two EVPLs are shown in the following figure. Both EVPLs connect to the site on the left through the same UNI.

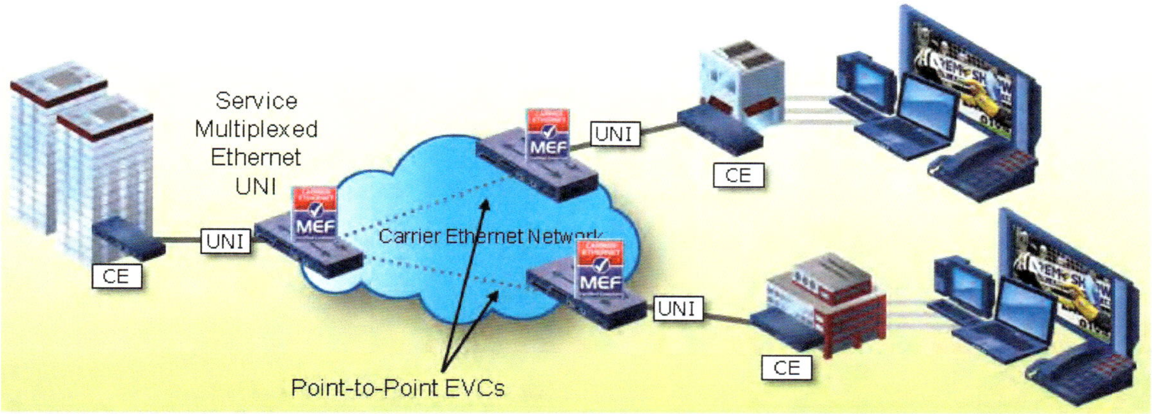

Reproduced with permission of the Metro Ethernet Forum

Distinctions of EVPL service include:

- EVPL service is a replacement for TDM private line service and for Frame Relay or ATM Layer 2 VPN services and also offers higher bandwidth capability.
- EVPL service allows service multiplexing at UNIs (multiple services can be delivered over a single physical connection).
- EVPL service requires VLAN coordination between the subscriber and the service provider.
- EVPL service provides less transparency compared to EPL service because tunneling of certain L2CP service frames, such as Bridge Protocol Data Unit (BPDU) frames to support Spanning Tree Protocol (STP), is not supported.

Note: *Tunneling is the process by which an L2CP service frame is passed through the service provider network without being processed and is delivered unchanged to the proper UNI(s).*

Related Links

Support for TDM Private Lines
Frame Relay Replacement by EVPLs

2.7
EP-LAN Service

An Ethernet Private LAN (EP-LAN) service is a port-based service of type E-LAN. It allows any UNI to forward Ethernet frames to any other UNI.

In the following figure, the EP-LAN service connects three separate locations into one EVC.

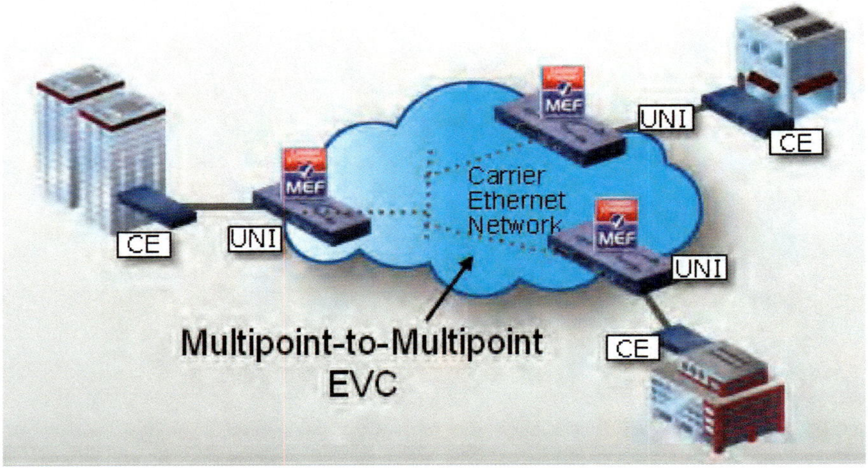

Reproduced with permission of the Metro Ethernet Forum

Distinctions of EP-LAN service include:

- Ethernet frames are forwarded based on MAC address (EVC must support MAC address learning).
- EP-LAN service requires a dedicated physical connection (UNI) to support each EVC end point.
- EP-LAN service requires no VLAN coordination between the subscriber and the service provider.
- EP-LAN service provides a high degree of transparency:
    - Preserves CE-VLAN tags, including VLAN IDs and PCP bits
    - Supports tunneling of Layer 2 Control Protocol (L2CP) frames, including Bridge Protocol Data Unit (BPDU) frames to support Spanning Tree Protocol (STP)

Note: *Tunneling is the process by which an L2CP service frame is passed through the service provider network without being processed and is delivered unchanged to the proper UNI(s).*

2.8
EVP-LAN Service

An Ethernet Virtual Private LAN (EVP-LAN) service is a VLAN-based service of type E-LAN.

In the following figure, an EVP-LAN service shares a UNI with a EVPL service. The same UNI supports both Internet access and access to the corporate LAN.

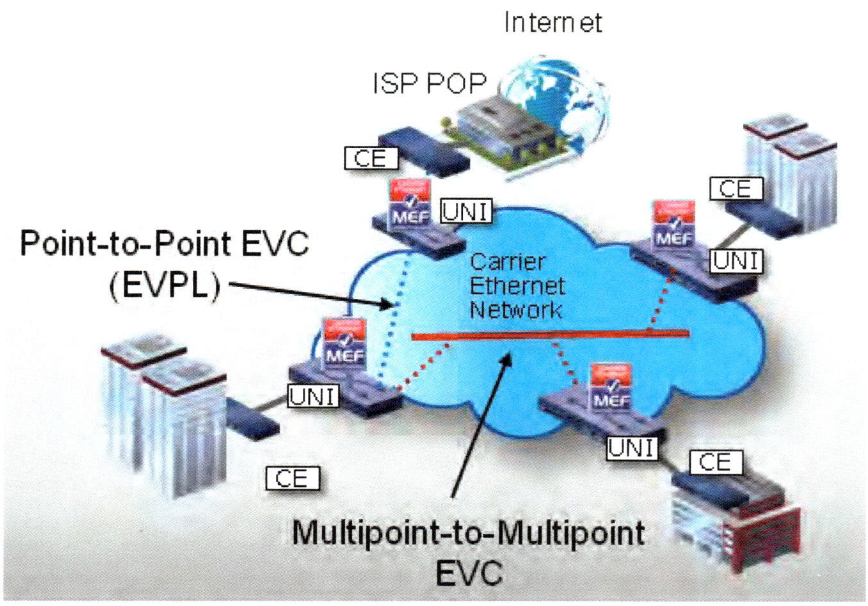

Reproduced with permission of the Metro Ethernet Forum

Distinctions of EVP-LAN service include:

- Ethernet frames are forwarded based on MAC address (EVC must support MAC address learning).
- EVP-LAN service allows service multiplexing at UNIs (multiple services can be delivered over a single physical connection).
- EVP-LAN service requires VLAN coordination between the subscriber and the service provider.
- EVP-LAN service provides less transparency compared to EP-LAN service because tunneling of certain L2CP service frames, such as Bridge Protocol Data Unit (BPDU) frames to support Spanning Tree Protocol (STP), is not supported.

2.9
EP-Tree Service

An Ethernet Private Tree (EP-Tree) service is a port-based service of type E-Tree.

Recall: *An E-Tree service allows roots to forward Ethernet frames to roots or leaves, but leaves can only forward Ethernet frames to roots.*

In the following figure, the EP-Tree service connects three franchise sites to a common corporate LAN.

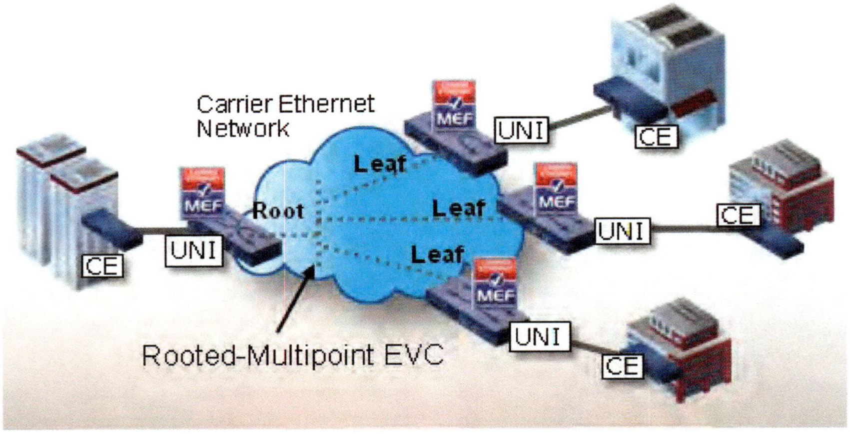

Reproduced with permission of the Metro Ethernet Forum

Distinctions of EP-Tree service include:

- Ethernet frames are forwarded based on MAC address (EVC must support MAC address learning).
- EP-Tree service enforces security, preventing interaction between leaves.
- EP-Tree service requires a dedicated physical connection (UNI) to support each leaf and each root.
- EP-Tree service requires no VLAN coordination between the subscriber and the service provider.
- EP-Tree service provides a high degree of transparency:
    - Preserves CE-VLAN tags, including VLAN IDs and PCP bits
    - Supports tunneling of Layer 2 Control Protocol (L2CP) frames, including Bridge Protocol Data Unit (BPDU) frames to support Spanning Tree Protocol (STP)

Note: *Tunneling is the process by which an L2CP service frame is passed through the service provider network without being processed and is delivered unchanged to the proper UNI(s).*

2.10
EVP-Tree Service

An Ethernet Virtual Private Tree (EVP-Tree) service is a VLAN-based service of type E-Tree.

Recall: *An E-Tree service allows roots to forward Ethernet frames to roots or leaves, but leaves can only forward Ethernet frames to roots.*

In the following figure, the EP-Tree service connects a distribution site, at the top, to three leaf sites. Each leaf of the EVP-Tree service shares a UNI with an EVP-LAN service. At each leaf, the same UNI supports the EVP-Tree service and the EVP-LAN service.[2]

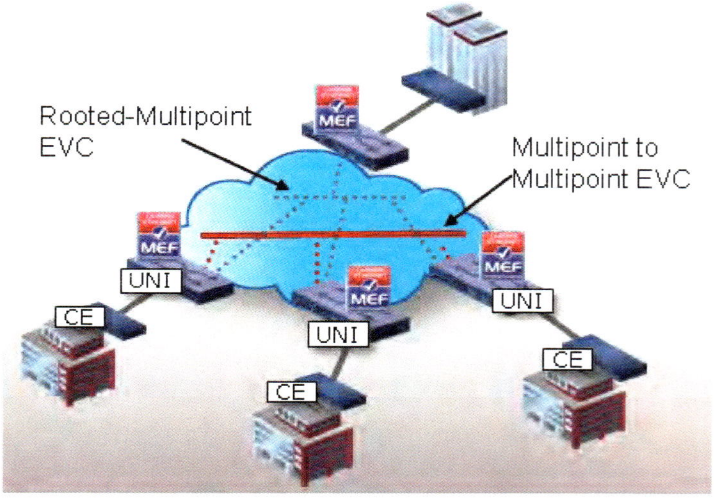

Reproduced with permission of the Metro Ethernet Forum

Distinctions of EVP-Tree service include:

- Ethernet frames are forwarded based on MAC address (EVC must support MAC address learning).
- EVP-Tree service enforces security, preventing interaction between leaves.
- EVP-Tree service allows service multiplexing at UNIs (multiple services can be delivered over a single physical connection).
- EVP-Tree service requires VLAN coordination between the subscriber and the service provider.
- EVP-Tree service provides less transparency compared to EP-Tree service because tunneling of certain L2CP service frames, such as Bridge Protocol Data Unit (BPDU) frames to support Spanning Tree Protocol (STP), is not supported.

2 This example (taken from an MEF power point presentation) is not a typical application because the E-Tree service prevents leaf UNIs from forwarding frames to each other, but the E-LAN service exactly undermines that property. Together the two services offer the same functionality of one type E-LAN service connecting all four sites.

2.11
Matrix of Service Distinctions

The following table lists key distinctions and indicates the Carrier Ethernet services to which each applies.

Distinction	Service					
	Type E-Line		Type E-LAN		Type E-Tree	
	EPL	EVPL	EP-LAN	EVP-LAN	EP-Tree	EVP-Tree
Is the most popular Carrier Ethernet service due to its simplicity	×					
Is a replacement for TDM private line service	×	×				
Is a replacement for Frame Relay or ATM Layer 2 VPN services and offers higher bandwidth capability		×				
Requires a dedicated physical connection (UNI) to support each EVC end point	×		×		×	
Allows service multiplexing at UNIs (multiple services can be delivered over a single physical connection)		×		×		×
Requires VLAN coordination between the subscriber and the service provider		×		×		×
Provides a high degree of transparency	×		×		×	
Supports tunneling of Bridge Protocol Data Unit (BPDU) frames to support Spanning Tree Protocol (STP)	×		×		×	
Prevents interaction between leaf UNIs					×	×
Supports bridging (forwards Ethernet frames based on MAC address learning)			×	×	×	×

2.12
Summary

Service provider–The organization providing the Carrier Ethernet service

Subscriber–The organization purchasing or using the Carrier Ethernet service

MEN–Metro Ethernet Network (synonymous with CEN)

CEN–Carrier Ethernet Network (synonymous with MEN)

CE–Customer edge (equipment)

UNI–User Network Interface (interface between CE and MEN/CEN; physical point of demarcation)

Service Frame–An Ethernet frame transmitted across the UNI toward the service provider or an Ethernet frame transmitted across the UNI toward the subscriber

CE-VLAN ID–Customer-edge VLAN ID

ENNI–External network-to-network interface (interface between two MEN/CENs)

EVC–Ethernet virtual connection (logical representation of a service connection between two or more UNIs)

E-Line–Point-to-point EVC

EPL–Ethernet private line (port-based E-Line service)

EVPL–Ethernet Virtual Private Line (VLAN-based E-Line service)

E-LAN–Multipoint-to-multipoint EVC

EP-LAN–Ethernet Private LAN (port-based E-LAN service)

EVP-LAN–Ethernet Virtual Private LAN (VLAN-based E-LAN service)

E-Tree–Rooted multipoint EVC (no traffic between leaves)

EP-Tree–Ethernet private tree (port-based E-Tree service)

EVP-Tree–Ethernet virtual private tree (VLAN-based E-Tree service)

Port-based service vs. VLAN-based service:

- Port-based: EPL, EP-LAN, EP-Tree
- VLAN-based: EVPL, EVP-LAN, EVP-Tree
- Key advantages of port-based services: (1) A subscriber can configure VLANs without any need to coordinate with the service provider, and (2) there is more service transparency compared to VLAN-based service.
- Key advantage of VLAN-based service: This service supports service multiplexing.

EVC Type:
- Point-to-point–EPL, EVPL
- Multipoint-to-multipoint–EP-LAN, EVP-LAN
- Rooted-multipoint–EP-Tree, EVP-Tree

Service Multiplexing–Association of multiple EVCs to a UNI

Bundling–The association of multiple CE-VLAN IDs to an EVC at a UNI

All-to-one bundling–The association of all service frames to one EVC at the UNI

2.13
Review

1. The interface between an Operator MEN (Operator CEN) and the subscriber is called a(an):

    a. ENNI

    b. OVC

    c. UNI

    d. EVC

    e. CE

2. Service multiplexing allows:

    a. Multiple services on one EVC

    b. Multiple EVCs on one UNI

    c. Multiple UNIs on one EVC

    d. Bundling

    e. Multiple CE-VLAN IDs to be mapped to an EVC

3. What is the MEF definition of a UNI?

    a. The logical representation of a service connection between two or more EVCs

    b. The physical demarcation point between the responsibility of the Service Provider and the responsibility of the Subscriber

    c. All functions required to connect the CE to the MEN (CEN)

    d. The interface between two Operator MEN/CENs

4. Equipment on the subscriber side of the UNI that connects to the MEN (CEN) is called the:

    a. Carrier edge

    b. Subscriber edge

    c. User network interface

    d. Consumer edge

e. Customer edge

f. Network edge

5. Which two Carrier Ethernet service combinations can be offered simultaneously at a given UNI?

a. EP-LAN and EVPL

b. EVP-Tree and EPL

c. EVPL and EVP-LAN

d. EPL and EP-Tree

6. A UNI is configured to support all-to-one bundling. Which three statements are true? (Choose three.)

a. Service multiplexing is not supported at the UNI.

b. Any number of EVCs can be bundled to this UNI.

c. Bundling is not supported at the UNI.

d. At this UNI, all CE-VLAN IDs map to the same EVC.

7. Which three service types are defined in MEF 6.1? (Choose three.)

a. Ethernet Virtual Connection (EVC)

b. Ethernet Line (E-Line)

c. Ethernet Tree (E-Tree)

d. Ethernet Access (E-Access)

e. Ethernet LAN (E-LAN)

f. TDM Line (T-Line)

8. An enterprise needs to connect three branch offices with its two payroll processing data centers. The two data centers require interconnectivity for data mirroring (protection). The service must ensure that the branch offices cannot see each other's sensitive payroll data. Which two of the following service arrangements supports this application? (Choose two.)

a. An EP-Tree service with the payroll data center UNIs designated as leaves and the branch office UNIs designated as roots

b. An EP-Tree service with payroll data center UNIs designated as roots and branch office UNIs designated as leaves

c. One EP-LAN service connecting all five sites

d. One EVPL service from each branch office to each payroll data center (six EVPLs in all), plus an EPL between the two payroll centers

9. A UNI is configured without bundling or all-to-one bundling and is also configured to support service multiplexing. How many EVCs can terminate at this UNI?

   a. 0

   b. 0 or 1

   c. Up to 64

   d. Up to 4094

10. Which kind of UNI-to-UNI connectivity is used for the Ethernet Private Line (EPL) service?

   a. Ethernet Private Circuit

   b. Point-to-point Ethernet Private Connection

   c. Point-to-point Ethernet Virtual Connection

   d. Point-to-point Ethernet Virtual Circuit

11. A customer with three locations requests a Carrier Ethernet service that provides the most transparent connectivity for all PDU types between sites. Which service is most suitable?

   a. A single Ethernet Private LAN (EP-LAN)

   b. A single Ethernet Virtual Private LAN (EVP-LAN)

   c. A single Ethernet Private Tree (EP-Tree)

   d. A single Ethernet Virtual Private Tree (EVP-Tree)

12. McBary's, a business enterprise with numerous independently-owned restaurant franchises, wants to connect each franchise to the corporate data center to obtain daily sales information. Which of the following service arrangements is most suitable?

   a. A single Ethernet Virtual Private LAN (EVP-LAN) service

   b. One Ethernet Virtual Private Line (EVPL) service between each franchise and the corporate data center

   c. A single Ethernet Private Tree (EP-Tree) service with the corporate data center UNI designated as the root and franchise UNIs designated as leaves

   d. One Ethernet Private Line (EPL) service between each franchise and the corporate data center

13. Which two statements about a VLAN-based (virtual private) service are true? (Choose two.)

- **a.** All UNIs must be configured to support service multiplexing.
- **b.** All UNIs must be configured to support bundling.
- **c.** All UNIs must be configured to support all-to-one bundling.
- **d.** All UNIs may be configured to support service multiplexing.
- **e.** All UNIs may be configured to support bundling.
- **f.** All UNIs may be configured to support all-to-one bundling.

14. A UNI is configured without bundling or all-to-one bundling and is also configured to support service multiplexing. What is the maximum number of CE-VLAN IDs that can be mapped to a given EVC at the UNI?

- **a.** 0
- **b.** 1
- **c.** 256
- **d.** 4094
- **e.** 4096

15. Which Carrier Ethernet services support bridging (frame forwarding based on MAC address learning)? (Choose all answers that are correct.)

- **a.** EPL
- **b.** EVPL
- **c.** EP-LAN
- **d.** EVP-LAN
- **e.** EP-Tree
- **f.** EVP-Tree

Related Links

Answers: MEF Services

3
How MEF Services are Defined (Part 1)

In this chapter:

Note: *This lesson relates to EVC-based services only (EPL, EVPL, EP-LAN, EVP-LAN, EP-Tree, and EVP-Tree services). OVC-based E-Access services (Access EPL and Access EVPL services) are defined in a later lesson.*

Related Links

E-Access Services

3.1
Categories of Carrier Ethernet Service Attributes

In this section:

MEF 6.1 associates every Carrier Ethernet service attribute to one of three categories:

- **UNI** service attributes
- **EVC per UNI** service attributes
- **EVC** service attributes

To understand why, consider the following figure.

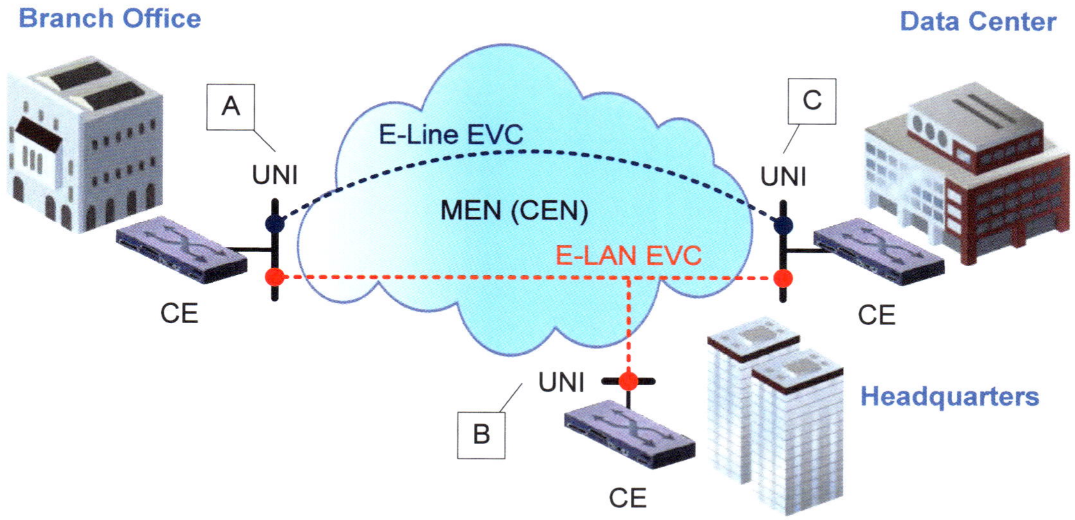

A **UNI service attribute** is an attribute that applies to a particular UNI (A, B, or C in the figure). To support the two services shown, UNIs A and C must be configured to support service multiplexing. *Service multiplexing* is one example of a UNI service attribute.

An **EVC service attribute** is an attribute that applies to an EVC as a whole. *CE-VLAN ID Preservation* is an example.

An **EVC per UNI service attribute** is an attribute that applies to a particular EVC at a particular UNI. You can think of it as an attribute that applies to an endpoint of an EVC (which is a UNI). In the figure, the E-LAN service has three UNIs: A, B, and C. An EVC per UNI service attribute, such as *CE-VLAN ID / EVC Map* (a listing of the CE-VLAN IDs that map to the EVC), is independently specified at each UNI.

Test your understanding: *Bundling = Yes/No is another example of a UNI service attribute. Suppose UNI A in the figure is configured with Bundling = No and UNI C is configured with Bundling = Yes. How many CE-VLAN IDs can the E-Line service between the two UNIs support?*

Answer: *Only 1 because UNI A supports only one CE-VLAN ID.*

Test your understanding: *You are configuring the E-LAN service shown in the previous figure. The Data Center houses two business critical servers on separate VLANs: one on VLAN 15, the other on VLAN 22. Headquarters is permitted access to both servers, but the branch office is only allowed to access the server on VLAN 22. Configure Bundling and CE-VLAN ID / EVC Map attributes to support this service.*

Answer: *UNIs B and C need to support two CE-VLAN IDs, so configure them with Bundling = Yes. UNI A only needs to support one CE-VLAN ID, so configure it with or without Bundling. Configure the CE-VLAN ID / EVC Map attribute for the EVC at UNIs B and C with two CE-VLAN IDs: VLAN 15 and VLAN 22. Configure the CE-VLAN ID / EVC Map attribute for the EVC at UNI A with VLAN 22 only.*

3.1.1
Bird's-Eye View of Service Attributes

MEF 10.2 defines 32 service attributes and their requirements for all six services: EPL, EVPL, EP-LAN, EVP-LAN, EP-Tree, and EVP-Tree. The following figure consolidates that information in a single image to give you a bird's-eye view of what it takes to define a MEF service. The rest of this lesson and the next lesson are dedicated to explaining all 32 service attributes in detail.

	Attribute	Requirement	
		Port-Based Service (EPL, EP-LAN, EP-Tree)	VLAN-Based Service (EVPL, EVP-LAN, EVP-Tree)
UNI Service Attributes	UNI Identifier	Arbitrary text string to identify the UNI.	
	Physical Medium	UNI Type 2 Physical Interface	
	Speed	10 Mbps, 100 Mbps, 10/100 Mbps Auto-negotiation, 10/100/1000 Mbps Auto-negotiation, 1 Gbps, or 10 Gbps.	
	Mode	Full Duplex	
	MAC Layer	IEEE 802.3-2005	
	UNI MTU Size	≥ 1522 bytes	
	Service Multiplexing	No	Should be supported at 1 or more UNIs
	Bundling	No	Yes/No. If Yes, CE-VLAN ID preservation must be Yes.
	All to One Bundling	Yes	No
	CE-VLAN ID for untagged and priority Service Frames	Not applicable	Must specify in range 1-4094
	Max. number of EVCs	1	≥1
	IBWPFL per UNI	Defer to next lesson	
	EBWPFL per UNI	Defer to next lesson	
	L2CP Processing	Peer, Discard, Pass to EVC, or Peer and Pass to EVC.	
EVC per UNI Service Attributes	UNI EVC ID	A string formed by the concatenation of the UNI ID and the EVC ID.	
	CE-VLAN ID / EVC Map	All service frames map to single EVC	MUST specify the mapping table of CE-VLAN IDs to the EVC at the UNI.
	IBWPFL per EVC	Defer to next lesson	
	IBWPFL per CoS Identifier	Defer to next lesson	
	EBWPFL per EVC	Defer to next lesson	
	EBWPFL per CoS Identifier	Defer to next lesson	
EVC Service Attributes	EVC Type	Point-to-Point (E-Line), Multipoint (E-LAN), or Rooted Multipoint (E-Tree)	
	EVC ID	An arbitrary string, unique across the MEN.	
	UNI List	UNI IDs + type (root/leaf), Type = root for all E-Line/ E-LAN UNIs	
	Maximum Number of UNIs	2 (E-Line) or ≥ 2 (E-LAN or E-Tree)	
	EVC MTU Size	(≥ 1522 bytes) AND (≤ Minimum of UNI MTU sizes)	
	CE-VLAN ID Preservation	Yes	Yes or No
	CE-VLAN CoS Preservation	Yes	Yes or No
	Unicast Service Frame Delivery	E-Line: Unconditionally. E-LAN / E-Tree: Conditionally or Unconditionally (with criteria specified)	
	Multicast Service Frame Delivery	E-Line: Unconditionally. E-LAN / E-Tree: Conditionally or Unconditionally (with criteria specified)	
	Broadcast Service Frame Delivery	E-Line: Unconditionally. E-LAN / E-Tree: Conditionally or Unconditionally (with criteria specified)	
	L2CP Processing	Tunnel or Discard	
	EVC Performance	Defer to next lesson	

3.1.2
Groups of Related Service Attributes

The following figure shows a bird's-eye view of service attributes color coded into groups that reflect their order of presentation in this study guide. Service attributes that are closely related or easier to understand as a group are grouped together.

Note: *These color-coded groups are defined in this study guide to aid understanding. You do not need to know these groups for the MEF-CECP exam. However, it is important to know the three categories of service attributes: UNI Service Attributes, EVC per UNI Service Attributes, and EVC Service Attributes.*

	Attribute	Requirement: Port-Based Service (EPL, EP-LAN, EP-Tree)	Requirement: VLAN-Based Service (EVPL, EVP-LAN, EVP-Tree)
UNI Service Attributes	UNI Identifier	Arbitrary text string to identify the UNI.	
	Physical Medium	UNI Type 2 Physical Interface	
	Speed	10 Mbps, 100 Mbps, 10/100 Mbps Auto-negotiation, 10/100/1000 Mbps Auto-negotiation, 1 Gbps, or 10 Gbps.	
	Mode	Full Duplex	
	MAC Layer	IEEE 802.3-2005	
	UNI MTU Size	≥ 1522 bytes	
	Service Multiplexing	No	Should be supported at 1 or more UNIs
	Bundling	No	Yes/No. If Yes, CE-VLAN ID preservation must be Yes.
	All to One Bundling	Yes	No
	CE-VLAN ID for untagged and priority Service Frames	Not applicable	Must specify in range 1-4094
	Max. number of EVCs	1	≥1
	IBWPFL per UNI	Defer to next lesson	
	EBWPFL per UNI	Defer to next lesson	
	L2CP Processing	Peer, Discard, Pass to EVC, or Peer and Pass to EVC.	
EVC per UNI Service Attributes	UNI EVC ID	A string formed by the concatenation of the UNI ID and the EVC ID.	
	CE-VLAN ID / EVC Map	All service frames map to single EVC	MUST specify the mapping table of CE-VLAN IDs to the EVC at the UNI.
	IBWPFL per EVC	Defer to next lesson	
	IBWPFL per CoS Identifier	Defer to next lesson	
	EBWPFL per EVC	Defer to next lesson	
	EBWPFL per CoS Identifier	Defer to next lesson	
EVC Service Attributes	EVC Type	Point-to-Point (E-Line), Multipoint (E-LAN), or Rooted Multipoint (E-Tree)	
	EVC ID	An arbitrary string, unique across the MEN.	
	UNI List	UNI IDs + type (root/leaf), Type = root for all E-Line/ E-LAN UNIs	
	Maximum Number of UNIs	2 (E-Line) or ≥ 2 (E-LAN or E-Tree)	
	EVC MTU Size	(≥ 1522 bytes) AND (≤ Minimum of UNI MTU sizes)	
	CE-VLAN ID Preservation	Yes	Yes or No
	CE-VLAN CoS Preservation	Yes	Yes or No
	Unicast Service Frame Delivery	E-Line: Unconditionally. E-LAN / E-Tree: Conditionally or Unconditionally (with criteria specified)	
	Multicast Service Frame Delivery	E-Line: Unconditionally. E-LAN / E-Tree: Conditionally or Unconditionally (with criteria specified)	
	Broadcast Service Frame Delivery	E-Line: Unconditionally. E-LAN / E-Tree: Conditionally or Unconditionally (with criteria specified)	
	L2CP Processing	Tunnel or Discard	
	EVC Performance	Defer to next lesson	

Legend

- Basic / Bookkeeping
- MTU Size
- Service Multiplexing, Bundling, and VLAN Tag Preservation
- Service Frame Mapping
- Service Frame Delivery
- L2CP Processing
- Bandwidth Profiles
- EVC Performance

3.2
The Carrier Ethernet Service Definition Framework

The following figure, from MEF 6.1, illustrates the framework for specifying an Ethernet service.

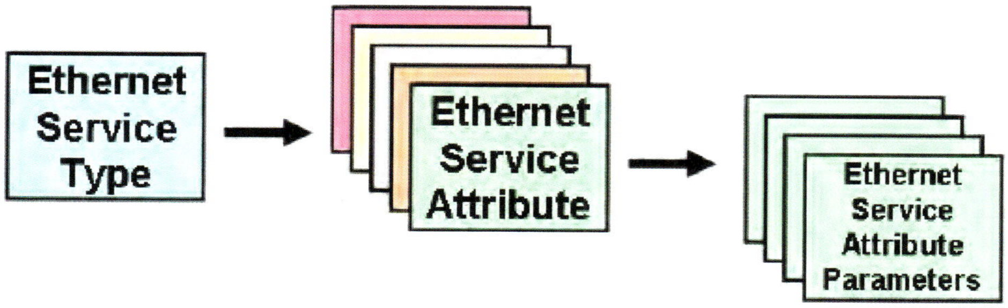

Reproduced with permission of the Metro Ethernet Forum

For each Ethernet service type (E-Line, E-LAN, or E-Tree as defined in the previous lesson), the MEF defines a set of *Ethernet service attributes*. Each service attribute represents a service characteristic that can be quite general. Because an Ethernet service attribute can be quite general, it in turn is defined by a set of *Ethernet service attribute parameters*.

Hint:

In the end, to specify a service, you simply choose an Ethernet service type and then specify values for a large set of parameters. Some Ethernet service attributes are defined by only one parameter that can be set to only a few possible values. Other Ethernet service attributes are specified by assigning values to a large set of well-defined parameters. In a few cases, the MEF intentionally avoids parametrizing the Ethernet service attribute, leaving it up to the service provider and subscriber.

Going forward, the focus of this study guide will be on Ethernet service attributes, and you will not see the term Ethernet service attribute parameter again. However, you should understand that an Ethernet service attribute is a general object that may require many parameters to define.

FUJITSU

3.3
Basic/Bookkeeping Service Attributes

	Attribute	Requirement					
		Port-Based Service			VLAN-Based Service		
		EPL	EP-LAN	EP-Tree	EVPL	EVP-LAN	EVP-Tree
UNI	UNI Identifier	Arbitrary text string to identify the UNI.					
UNI	Physical Medium	UNI Type 2 Physical Interface					
UNI	Speed	10 Mbps, 100 Mbps, 10/100 Mbps Auto-negotiation, 10/100/1000 Mbps Autonegotiation, 1 Gbps, or 10 Gbps.					
UNI	Mode	Full Duplex					
UNI	MAC Layer	IEEE 802.3-2005					
UNI	Max. number of EVCs	1			≥1		
EVC per UNI	UNI EVC ID	A string formed by the concatenation of the UNI ID and the EVC ID.					
EVC	EVC Type	Pt–Pt	Mpt–Mpt	Rooted Mpt	Pt–Pt	Mpt–Mpt	Rooted Mpt
EVC	EVC ID	An arbitrary string, unique across the MEN.					
EVC	UNI List	UNI IDs + type (root/leaf), Type = root for all E-Line/ E-LAN UNIs					
EVC	Maximum Number of UNIs	2	≥ 2		2	≥ 2	

These service attributes are grouped together in this study guide because they are mostly self-evident (do not need much explanation).

The **EVC Type** attribute declares the EVC Type: Point-to-Point (E-Line), Multipoint-to-Multipoint (E-LAN), or Rooted-Multipoint (E-Tree)

Four attributes declare identifiers for entities involved in the service:

- **UNI Identifier**–*Arbitrary text string to identify the UNI* (each UNI needs one)
- **UNI EVC ID**–*A string formed by the concatenation of the UNI ID and the EVC ID* (identifies an EVC endpoint; each EVC endpoint needs one)
- **EVC ID**–*An arbitrary string, unique across the MEN (CEN), for the EVC supporting the service instance* (identifies the EVC)
- **UNI List**–*A list of UNI Identifiers and type=root/leaf for each UNI associated with the EVC* (UNIs for type E-Line and E-LAN EVCs are type=root. UNIs for type E-Tree EVCs can be type root or leaf.)

Four attributes characterize UNI functionality:

- **Physical Medium**–*UNI Type 2 Physical Interface*. (Type 2 UNI functionality is described later in this study guide. For now, just observe that this attribute is currently fixed in the MEF specification; not configurable.)

- **Speed**–*10 Mb/s, 100 Mb/s, 10/100 Mb/s Auto-negotiation, 10/100/1000 Mb/s Auto-negotiation, 1 Gb/s, or 10 Gb/s* (specifies UNI rate capability)

- **Mode**–*Full Duplex* (this attribute is currently fixed, not configurable)

- **MAC Layer**–*IEEE 802.3-2005* (Ethernet standard)

Two attributes declare support limitations:

- **Maximum number of EVCs**–Declares how many EVCs the UNI can support. If the UNI is configured for *All-to-one Bundling*, it can support only one EVC. Otherwise, a number ≥1 is declared.

- **Maximum Number of UNIs**–Declares how many UNIs the EVC can support. For type E-Line services, the EVC supports two UNIs. Otherwise, a number ≥2 is declared.

FUJITSU

3.4
MTU Size Service Attributes

	Attribute	Requirement	
		Port-Based Service (EPL, EP-LAN, EP-Tree)	VLAN-Based Service (EVPL, EVP-LAN, EVP-Tree)
UNI	UNI MTU Size	≥ 1522 bytes	
EVC	EVC MTU Size	(≥ 1522 bytes) AND (≤ Minimum of UNI MTU sizes)	

There are two service attributes related to maximum transmission unit (MTU) size:

- **UNI MTU Size**—The maximum sized service frame allowed at the UNI
- **EVC MTU Size**—The maximum sized service frame allowed for an EVC

Requirements should be interpreted as follows:

- The *UNI MTU Size* (for any UNI) must be ≥ 1522 bytes.
- The *EVC MTU Size* (for a given EVC) must meet two requirements:
    1. EVC MTU Size ≤ minimum (*UNI MTU Size* across all UNIs associated to the EVC)
    2. EVC MTU Size ≥ 1522 bytes

If the size of a received frame exceeds the EVC MTU size, the frame may still be admitted to the service, but the service level specification (SLS), which governs delivery performance for the service, will not apply to it.[3]

Note: *In multi-network applications (the subject of another lesson), ENNIs and OVCs are required to support at least 1526 bytes (four more bytes than UNIs and EVCs).*

Related Links

The MEF Service Agreement Framework (SLA and SLS)
ENNI and OVC MTU Size

3 MEF 10.2 states: "When an ingress Service Frame has length greater than the EVC Maximum Transmission Unit Size, the SLS, if any, for this frame SHALL not apply to its delivery performance and the result of a Bandwidth Profile that applies to this Service Frame is not defined."

3.5
Service Multiplexing, Bundling, and VLAN Tag Preservation Service Attributes

	Attribute	Requirement	
		Port-Based Service (EPL, EP-LAN, EP-Tree)	VLAN-Based Service (EVPL, EVP-LAN, EVP-Tree)
UNI	Service Multiplexing	No	Should be supported at 1 or more UNIs
UNI	Bundling	No	Yes/No. If Yes, CE-VLAN ID preservation must be Yes.
UNI	All to One Bundling	Yes	No
EVC	CE-VLAN ID Preservation	Yes	Yes or No
EVC	CE-VLAN CoS Preservation	Yes	Yes or No

The concepts of *service multiplexing*, *bundling*, and *all to one bundling* were introduced in the previous lesson. MEF 10.2 captures these concepts in three UNI service attributes:

- **Service Multiplexing**—If Yes, the UNI can support multiple VLAN-based services and no port-based services.
- **Bundling**—If Yes, the UNI can support the association of multiple CE-VLAN IDs to an EVC.
- **All to One Bundling**—If Yes, the UNI can support one port-based service and no VLAN-based services.

Requirements for these UNI service attributes are interrelated with each other and are also interrelated with two EVC service attributes:

- **CE-VLAN ID Preservation**—If Yes, the EVC preserves the value of the VLAN ID (VID) in the CE-VLAN tag.
- **CE-VLAN CoS Preservation**—If Yes, the EVC preserves the value of the priority code point (PCP) bits in the CE-VLAN tag.

Recall that *CE-VLAN tag* refers to the IEEE 802.1Q customer tag in the Ethernet header of service frames.

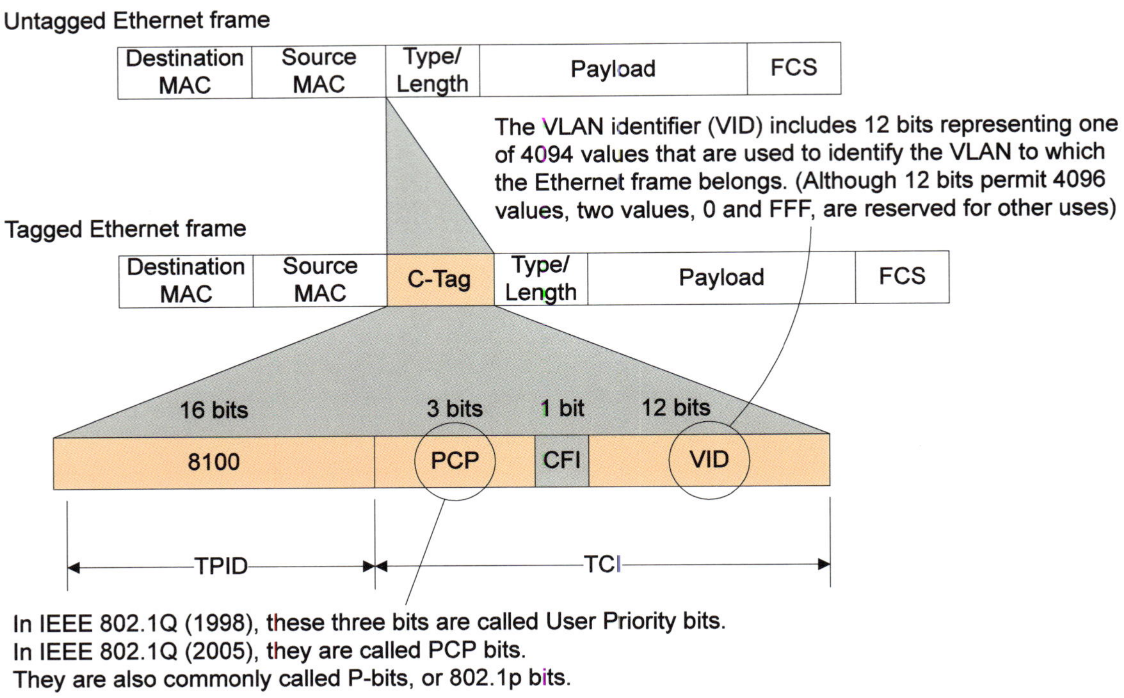

Figure 8: IEEE 802.1Q-2005 Ethernet Frame

About Ethernet Frame Terminology:

Prior to release of IEEE 802.1Q-2011, the frames depicted in the previous figure were commonly called *IEEE 802.1Q Ethernet frames* because they conform to IEEE 802.1Q-2005. However, with the release of IEEE 802.1Q-2011, the term *IEEE 802.1Q Ethernet frame*, now properly refers to an Ethernet frame that conforms to IEEE 802.1Q-2011, which is far more general.[4]

In this study guide, we avoid the term *IEEE 802.1Q Ethernet frame*. The following terms are used as required:

- **IEEE 802.1Q-2005 Ethernet frame** – An Ethernet frame that is either untagged or includes a single C-Tag (VLAN tag with TPID=8100), as shown in the previous figure.
- **IEEE 802.1Q Customer Tag** – A C-Tag (VLAN tag with TPID=8100), as shown in the previous figure.

4 IEEE 802.1Q-2011 incorporates many amendments to IEEE 802.1Q-2005, including IEEE 802.1ad (Provider Bridging), IEEE 802.1ah (Provider Backbone Bridging), and IEEE 802.1Qay (Provider Backbone Bridging with Traffic Engineering Extensions).

Preservation of CE-VLAN ID/CoS (VID/PCP bits) means the Carrier Ethernet service will not change the value. The Ethernet frame will exit the Carrier Ethernet service with the same value it arrived with. The key benefit of preservation is enhanced operational simplicity. If preservation is No, the service may be required to translate the value, adding complexity to the service implementation and requiring greater coordination between service provider and subscriber.

On the other hand, allowing translation (preservation = no) can be useful, for example, if a subscriber needs to interconnect VLANs on multiple campuses using IEEE 802.1Q bridges and each campus currently uses a different customer VLAN ID. Instead of renumbering the campus VLANs to a common VLAN ID (required if *CE-VLAN ID Preservation=Yes*), the service could be configured with *CE-VLAN ID Preservation=No* and then implemented to perform VLAN ID translations, as required.

MEF requirements for *CE-VLAN ID Preservation* and *CE-VLAN CoS Preservation* allow translation (preservation = No) under some conditions to accommodate special cases:

- The MEF requires *CE-VLAN ID Preservation = Yes* unless *Bundling = No*.
- The MEF requires *CE-VLAN CoS Preservation = Yes* unless *All to One Bundling = No*.

Test your understanding: *For each of the following MEF services, list CE-VLAN tag components that the service is required to preserve:*

1. *For a port-based service*
2. *For a VLAN-based service between two UNIs with both UNIs configured for Bundling=Yes*
3. *For a VLAN-based service between two UNIs and both UNIs configured for Bundling=No*

Answer:

1. *CE-VLAN ID (VID) and CoS (PCP bits)*
2. *CE-VLAN ID (VID)*
3. *Neither*

Test your understanding: *A subscriber wants to interconnect three VLANs that are on three separate campuses. Currently these VLANs use different VLAN IDs (101, 230, and 44), and the subscriber does not want to renumber them. Answer the following questions:*

1. *How should the UNIs be configured with respect to Bundling?*
2. *How should the UNIs be configured with respect to Service Multiplexing?*
3. *How should the EVC be configured with respect to CE-VLAN ID Preservation?*

Answer:

1. *Bundling must be No.*
2. *Service Multiplexing can be Yes or No.*
3. *CE-VLAN ID Preservation must be No.*

The following table, taken from MEF 10.2, shows valid combinations of *Service Multiplexing*, *Bundling*, and *All to One Bundling* UNI service attributes.

	Valid Combination 1	Valid Combination 2	Valid Combination 3	Valid Combination 4	Valid Combination 5
Service Multiplexing	☐	☑	☑	☐	☐
Bundling	☐	☐	☑	☑	☐
All to One Bundling	☐	☐	☐	☐	☑

Reproduced with permission of the Metro Ethernet Forum

Figure 9: Valid Combinations (Table 10 in MEF 10.2)

Observe that *All-to-One Bundling* precludes both *Bundling* and *Service Multiplexing*.

Test your understanding: *For each of the five valid combinations shown in the preceding figure, describe the support that the UNI provides for services.*

Answer:

1. *UNI supports one VLAN-based service with one CE-VLAN ID mapped to the service.*
2. *UNI supports multiple VLAN-based services with only one CE-VLAN ID mapped to each service.*
3. *UNI supports multiple VLAN-based services with one or many CE-VLAN IDs mapped to each service.*
4. *UNI supports one VLAN-based service with one or many CE-VLAN IDs mapped to it.*
5. *UNI supports one port-based service with all service frames mapped to the service.*

Test your understanding: *Which of the valid combinations shown in the previous figure support CE-VLAN ID translation?*

Answer: *Valid combinations 1 and 2 support CE-VLAN ID translation under the following conditions:*

- *The other UNIs participating in the service must also be configured with Bundling = No.*
- *The EVC must be configured with CE-VLAN ID Preservation = No.*

Related Links

Service Multiplexing
Bundling

3.6

Service Frame Mapping Service Attributes

Two service attributes are related to CE-VLANs: **CE-VLAN ID for untagged and priority tagged service frames** and **CE-VLAN ID / EVC Map**

Note: *A priority tagged service frame is an IEEE 802.1Q Ethernet frame tagged with customer VLAN identifier (VID) set to 0 (null).*

	Attribute	Requirement: Port-Based Service (EPL, EP-LAN, EP-Tree)	Requirement: VLAN-Based Service (EVPL, EVP-LAN, EVP-Tree)
UNI			
UNI	CE-VLAN ID for untagged and priority Service Frames	Not applicable	Must specify in range 1-4094
UNI			
EVC per UNI	CE-VLAN ID / EVC Map	All service frames map to single EVC	MUST specify the mapping table of CE-VLAN IDs to the EVC at the UNI.
EVC			

Neither attribute applies to port-based services (services on UNIs configured with all-to-one bundling) because all service frames in a port-based service (regardless of tagging or absence of tagging) are mapped to the only EVC that is supported at the UNI. However, VLAN-based services allow several EVCs to share a UNI, which leads to the following question: Which service frames should be mapped to which EVC? The *EVC per UNI* service attribute *CE-VLAN ID / EVC Map* resolves this question for tagged service frames as follows:

- At a given UNI, there is a *CE-VLAN ID / EVC Map* attribute for each EVC that is service multiplexed at the UNI.
- Each *CE-VLAN ID / EVC Map* attribute lists the CE-VLAN IDs that map to the EVC.
- Together, the collection of *CE-VLAN ID / EVC Map* attributes (one for each EVC sharing the UNI) defines how tagged service frames are mapped to EVCs at the UNI.

But what about untagged and priority tagged frames?

The *UNI* service attribute *CE-VLAN ID for untagged and priority tagged service frames* is used together with the *EVC per UNI* service attribute *CE-VLAN ID / EVC Map* to map untagged and priority service frames to an EVC.

The following figure illustrates the application of both attributes.

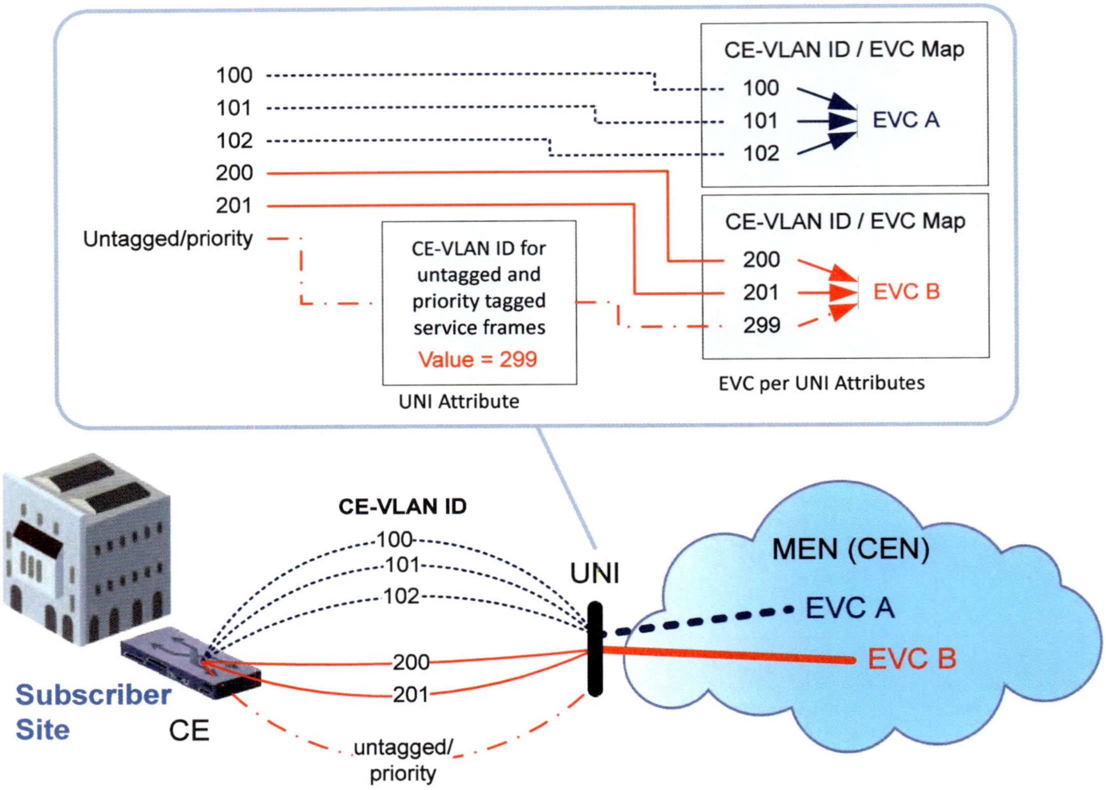

Figure 10: Example Application of VLAN-Related Service Attributes

In this example, untagged and priority tagged service frames are mapped to EVC B. The *UNI* service attribute *CE-VLAN ID for untagged and priority tagged service frames* is set to the value 299, and the *CE-VLAN ID / EVC Map* attribute for EVC B includes the value 299. The two attributes work together.

Note: *In effect, attribute **CE-VLAN ID for untagged and priority tagged service frames** can be imagined to push a pseudo CE-VLAN tag onto untagged service frames at MEN (CEN) ingress so that they can be assigned to an EVC through the **CE-VLAN ID / EVC Map** attribute. However, this pseudo tagging does not imply that a real C-Tag is actually pushed on the frames. The service provider has the option to push a real C-Tag onto these frames (as part of their service implementation), but if they do that, they must pop that C-tag on EVC egress.*

From the service definition perspective, the *CE-VLAN ID for untagged and priority tagged service frames* attribute is assigned only to allow untagged and priority service frames to be assigned to an EVC using the *CE-VLAN ID / EVC Map* attribute.

3.7 Service Frame Delivery Service Attributes

In this section:

3.7.1 Service Frames

The MEF defines a **service frame** as an Ethernet frame transmitted across the UNI toward the service provider or an Ethernet frame transmitted across the UNI toward the subscriber.

All service frames are either untagged or include an IEEE 802.1Q customer tag in the Ethernet header as shown in the following figure.

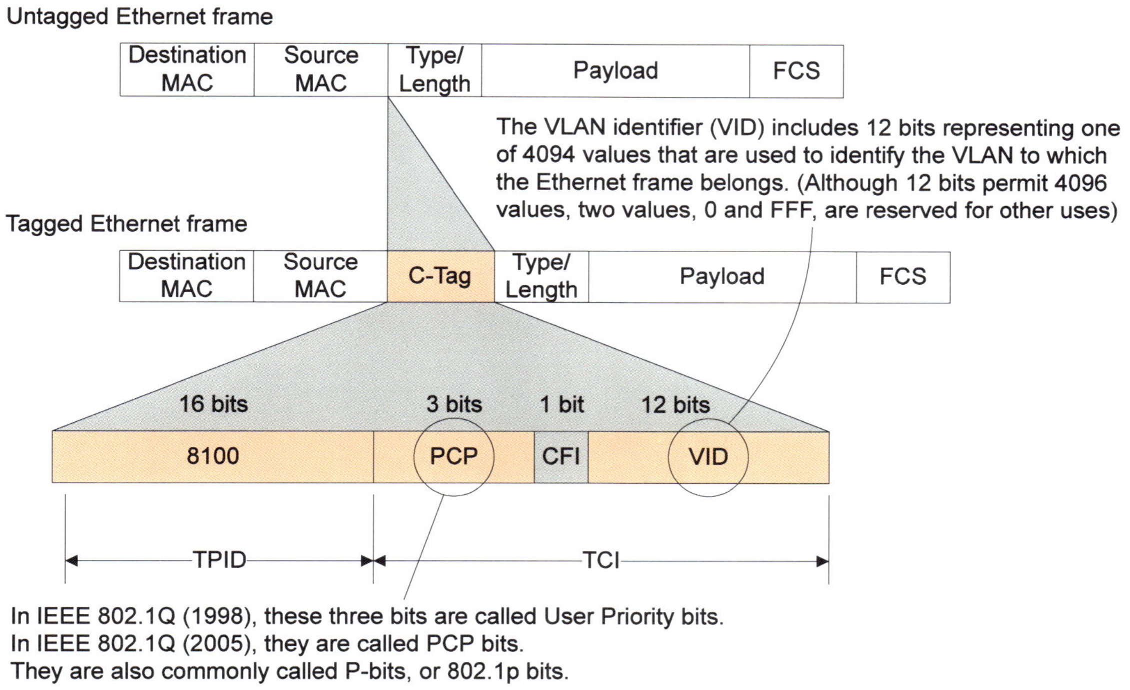

Figure 11: IEEE 802.1Q–2005 Ethernet Frame

About Ethernet Frame Terminology:

Prior to release of IEEE 802.1Q–2011, the frames depicted in the previous figure were commonly called *IEEE 802.1Q Ethernet frames* because they conform to IEEE 802.1Q–2005. However, with the release of IEEE 802.1Q–2011, the term *IEEE 802.1Q Ethernet frame* now properly refers to an Ethernet frame that conforms to IEEE 802.1Q–2011, which is far more general.[5]

In this study guide, we avoid the term *IEEE 802.1Q Ethernet frame*. The following terms are used as required:

- **IEEE 802.1Q–2005 Ethernet frame** – An Ethernet frame that is either untagged or includes a single C-Tag (VLAN tag with TPID=8100), as shown in the previous figure.
- **IEEE 802.1Q Customer Tag** – A C-Tag (VLAN tag with TPID=8100), as shown in the previous figure.

5 IEEE 802.1Q–2011 incorporates many amendments to IEEE 802.1Q–2005, including IEEE 802.1ad (Provider Bridging), IEEE 802.1ah (Provider Backbone Bridging), and IEEE 802.1Qay (Provider Backbone Bridging with Traffic Engineering Extensions).

3.7.2 Types of Service Frames

MEF 10.2 categorizes service frames, based on destination MAC address, as follows:

- **Unicast Service Frames**–Service frames with a unicast MAC address
- **Multicast Service Frames**–Service frames with a multicast MAC address (01-00-5E-00-00-00 through 01-00-5E-7F-FF-FF)
- **Broadcast Service Frames**–Service frames with a broadcast MAC address (FF-FF-FF-FF-FF-FF)
- **Layer 2 Control Protocol Service Frames**–Service frames with a L2CP MAC address (one of the MAC addresses shown in the following table)

Table 1: List of Standardized Layer 2 Control Protocols (Table 30 in MEF 6.1)

Destination MAC Addresses	Layer 2 Control Protocol
01-80-C2-00-00-00 through 01-80-C2-00-00-0F	Bridge Block of protocols
01-80-C2-00-00-20 through 01-80-C2-00-00-2F	GARP Block of protocols

Note: *Table 1 in MEF 10.2 provides a similar list of standard L2CP MAC addresses, but that table includes the "All Bridges Protocol" MAC address 01-80-C2-00-00-10, which is not included here because, as explained in MEF 6.1, " it has been depreciated in 802.1Q-2005" and "In the unlikely event that a customer may use this MAC DA, MEF services are expected to treat them as normal service frames."*

In addition, MEF 10.2 defines a **Data Service Frame** to be any service frame except an L2CP service frame.

3.7.3
Service Frame Disposition

Note: *The frame check sequence (FCS) is a set of checksum characters added to an Ethernet frame for error detection.*

MEF 10.2 defines four possible **dispositions** for an ingress service frame:

- **Discard**–All ingress service frames with an invalid FCS must be discarded. Certain L2CP service frames are also discarded.
- **Deliver Unconditionally**–The service frame is unconditionally admitted to the MEN/CEN (assuming correct FCS) and will be delivered across the MEN (CEN) to the proper egress UNI(s).
- **Deliver Conditionally**–The service frame is admitted (assuming correct FCS) and delivered across the MEN (CEN) to egress UNIs if certain conditions are met. The conditions can be anything that the service provider and subscriber agree to.
- **Tunnel**–This disposition is applicable to L2CP service frames only. Tunneling is the process by which an L2CP service frame is passed through the service provider network without being processed and is delivered unchanged at the proper egress UNI(s).

Important: *Regardless of the service frame's disposition, its delivery is still subject to bandwidth limitations as defined by bandwidth profiles and enforced by policing. Bandwidth profiles and policing are covered in the next lesson.*

Related Links

Bandwidth Profiles

3.7.4
Service Frame Transparency Requirements

Excluding changes to the CE-VLAN tag, MEF 10.2 requires that all fields of a service frame remain unchanged by the service. A service frame's content at egress must match its content at ingress.

If the CE-VLAN tag is modified (for example, if CE-VLAN ID and/or PCP bits are changed as is permitted if EVC service attributes *CE-VLAN ID Preservation* and *CE-VLAN CoS Preservation* are set to No), the frame check sequence (FCS) at egress must be recalculated.

3.7.5

Data Service Frame Delivery Service Attributes

Attribute	Requirement: Port-Based Service – EPL	EP-LAN	EP-Tree	Requirement: VLAN-Based Service – EVPL	EVP-LAN	EVP-Tree
EVC: Unicast Service Frame Delivery	Un-conditonal	Conditional* or Unconditional		Un-conditonal	Conditional* or Unconditional	
EVC: Multicast Service Frame Delivery	Un-conditonal	Conditional* or Unconditional		Un-conditonal	Conditional* or Unconditional	
EVC: Broadcast Service Frame Delivery	Un-conditonal	Conditional* or Unconditional		Un-conditonal	Conditional* or Unconditional	

Conditionally*–if delivery is conditional, the criteria for delivery must be specified

These attributes are used to set service frame disposition at MEN (CEN) ingress.

For E-Line type services (EPL or EVPL services), the MEF requires all data service frames (unicast, multicast, or broadcast) to be delivered unconditionally. There is no technical reason to preclude conditional delivery for such services. The MEF imposes this requirement to keep these services as simple and as transparent as possible.

For all other services (EP-LAN, EP-Tree, EVP-LAN, or EVP-Tree), the MEF allows two choices for each type of data service frame (unicast, multicast, or broadcast):

- **Deliver Unconditionally**–The service frame is unconditionally admitted to the MEN/CEN (assuming correct FCS) and will be delivered across the MEN (CEN) to the proper egress UNI(s).
- **Deliver Conditionally**–The service frame is admitted (assuming correct FCS) and delivered across to egress UNIs if certain conditions are met. The conditions can be anything that the service provider and subscriber agree to.

Important: *Regardless of the service frame's disposition, its delivery is still subject to bandwidth limitations as defined by bandwidth profiles and enforced by policing. Bandwidth profiles and policing are covered in the next lesson.*

Related Links

Bandwidth Profiles

3.8

L2CP Processing Service Attributes

In this section:

	Attribute	Requirement: Port-Based Service (EPL, EP-LAN, EP-Tree)	Requirement: VLAN-Based Service (EVPL, EVP-LAN, EVP-Tree)
UNI	L2CP Processing	Peer, Discard, Pass to EVC, or Peer and Pass to EVC.	
EVC	L2CP Processing	Tunnel or Discard	

There are two service attributes named *Layer 2 Control Protocol Processing*. One is a UNI service attribute, and the other is an EVC service attribute:

- **Layer 2 Control Protocols Processing** (UNI Attribute)—One of the following must be specified for each protocol: *Peer*, *Discard*, *Pass to EVC*, or *Peer and Pass to EVC*
- **Layer 2 Control Protocols Processing** (EVC Attribute)—One of the following must be specified for each protocol passed to the EVC: *Tunnel* or *Discard*

These attributes allow L2CP processing to be independently configured for each protocol and for each service.

3.8.1

How the Two L2CP Processing Attributes Work Together

The following figure illustrates how the two L2CP Processing service attributes work together to define L2CP processing for a protocol.

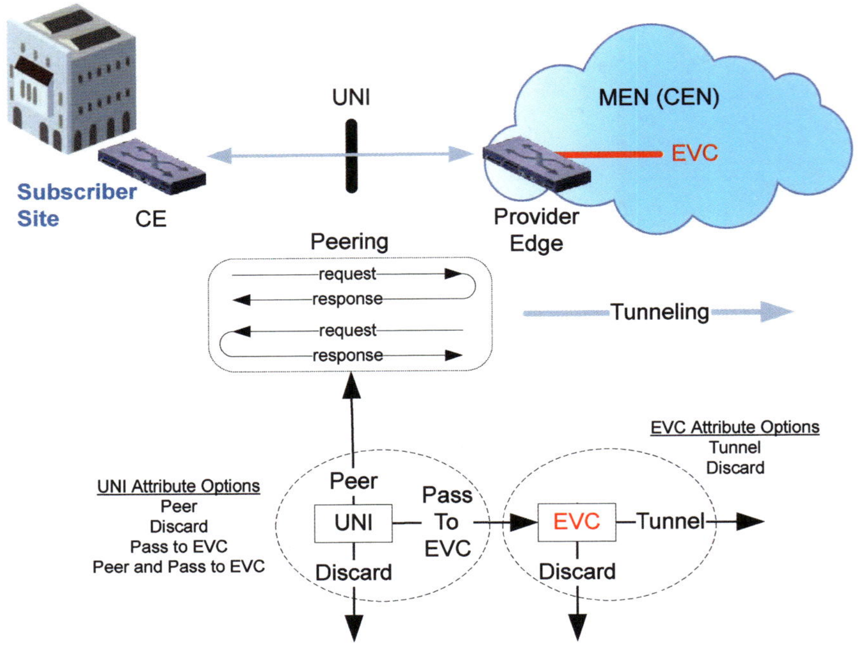

Figure 12: Layer 2 Control Protocol Processing

For each protocol, the two attributes can be configured to produce none, one, or both of the following outcomes:

- **Peering**–The MEN (CEN) will actively participate in the protocol.[6]
- **Tunneling**–The L2CP service frame received by the MEN (CEN) will be passed through the MEN (CEN) without being processed and delivered unchanged to the proper egress UNI(s).

As shown in the following table, values can be assigned to the two attributes in six ways. Each combination results in one of four outcomes: (1) *Discard*, (2) *Peer*, (3) *Tunnel*, or (4) *Peer and Tunnel*.

Attribute Value Combination	Attribute Values		Outcome
	UNI *L2CP Processing*	EVC *L2CP Processing*	
Combination 1	Discard	n/a	(1) Discard
Combination 2	Peer	n/a	(2) Peer
Combination 3	Pass to EVC	Discard	(1) Discard
Combination 4	Pass to EVC	Tunnel	(3) Tunnel
Combination 5	Peer and Pass to EVC	Discard	(2) Peer
Combination 6	Peer and Pass to EVC	Tunnel	(4) Peer and Tunnel

6 For example, LACP/LAMP, Link OAM, Port Authentication, and E-LMI might be peered by the MEN (CEN).

3.8.2
Per Protocol, Service-Specific L2CP Processing Requirements

Note: *The following requirements differ from L2CP processing requirements covered in the initial MEF-CECP exam. MEF-CECP 2.0 covers L2CP processing requirements as specified in MEF 6.1.1 (which has replaced Section 8 of MEF 6.1, the source of requirements for the initial MEF-CECP exam).*

MEF 6.1.1 specifies L2CP processing requirements for the following destination MAC address and protocols.

Table 2: L2CP MAC Addresses and Protocols

Destination MAC Address	Protocol Type	Ethertype/subtype
01-80-C2-00-00-00	STP/RSTP (IEEE 802.1D), MSTP (IEEE 802.1Q)	n/a
01-80-C2-00-00-01	PAUSE (IEEE 802.3)	0x8808
01-80-C2-00-00-02	LACP (IEEE 802.3)	0x8809/01
	LAMP (IEEE 802.3)	0x8809/02
	Link OAM (IEEE 802.3)	0x8809/03
	ESMC (ITU-T G.8264–2008)	0x8809/0A
01-80-C2-00-00-03	Port Authentication (IEEE 802.1X)	0x888E
01-80-C2-00-00-04	–	N/A
01-80-C2-00-00-05	–	N/A
01-80-C2-00-00-06	–	N/A
01-80-C2-00-00-07	E-LMI (MEF 16)	0x88EE
01-80-C2-00-00-08	–	N/A
01-80-C2-00-00-09	–	N/A
01-80-C2-00-00-0A	–	N/A
01-80-C2-00-00-0B	–	N/A
01-80-C2-00-00-0C	–	N/A
01-80-C2-00-00-0D	–	N/A
01-80-C2-00-00-0E	LLDP (IEEE 802.1AB)	0x88CC
	PTP Peer Delay (IEEE 1588–2008)	0x88F7
01-80-C2-00-00-0F	–	N/A
01-80-C2-00-00-20 through 01-80-C2-00-00-2F	GARP (IEEE 802.1Q), MRP (IEEE 802.1ak)	N/A

For all services, excluding EPL services, MEF 6.1.1 specifies the following L2CP processing requirements:

Table 3: L2CP Processing Requirements For EVPL, EP-LAN, EVP-LAN, EP-Tree and EVP-Tree Services

Destination MAC Address	Corresponding Protocols (Ethertype/subtype)	L2CP Action	
		EP-LAN, EP-Tree	EVPL, EVP-LAN, EVP-Tree
01-80-C2-00-00-00	STP/RSTP, MSTP	MUST Tunnel	Must Peer on all UNIs or Discard on all UNIs
01-80-C2-00-00-01	PAUSE (0x8808)	Must Discard on all UNIs	
01-80-C2-00-00-02	LACP (0x8809/01)	Must Peer or Discard per UNI	
	LAMP (0x8809/02)		
	Link OAM (0x8809/03)		
	ESMC (0x8809/0A)		
01-80-C2-00-00-03	Port Authentication (0x888E)	Must Peer or Discard per UNI	
01-80-C2-00-00-04	–	Must Not Tunnel	
01-80-C2-00-00-05	–		
01-80-C2-00-00-06	–		
01-80-C2-00-00-07	E-LMI (0x88EE)	Must Peer or Discard per UNI	
01-80-C2-00-00-08	–	Must Not Tunnel	
01-80-C2-00-00-09	–		
01-80-C2-00-00-0A	–		
01-80-C2-00-00-0B	–	Must Tunnel	Must Not Tunnel
01-80-C2-00-00-0C	–		
01-80-C2-00-00-0D	–		
01-80-C2-00-00-0E	LLDP (0x88CC)	Must Discard on all UNIs	
	PTP Peer Delay (0x88F7)	Must Peer on all UNIs or Discard on all UNIs	
01-80-C2-00-00-0F	–	Must Tunnel	Must Not Tunnel
01-80-C2-00-00-20 through 01-80-C2-00-00-2F	GARP/MRP	Must Tunnel on all UNIs	Should Peer on all UNIs or Tunnel on all UNIs[7]

7 For this case MEF 6.1.1 states that: "...one must consider MEF 10.2 requirements for processing of Service Frames. According to MEF 10.2, tunneling a L2CP frame must be done via the EVC that the frame maps to via the CE-VLAN ID/EVC Map. If an untagged frame ingresses at a UNI and if the CE-VLAN ID/EVC Map does not contain the CE-VLAN ID for untagged/priority tagged frames, MEF 10.2 mandates that the frame be discarded. If the CE-VLAN ID/EVC Map does contain the CE-VLAN ID for untagged/priority tagged frames, then MEF 10.2 mandates that the frame be tunneled on just that associated EVC."

For EPL service, MEF 6.1.1 provides two options for L2CP processing:

Table 4: L2CP Processing Requirements for EPL Service

Destination MAC Address	Corresponding Protocols (Ethertype/subtype)	L2CP Action	
		EPL (Option 1)	EPL (Option 2)
01-80-C2-00-00-00	STP/RSTP, MSTP	Must Tunnel on all UNIs	Must Tunnel on all UNIs
01-80-C2-00-00-01	PAUSE (0x8808)	Must Discard on all UNIs	Should Discard on all UNIs
01-80-C2-00-00-02	LACP (0x8809/01)	Must Peer or Discard per UNI	Should Tunnel on all UNIs
	LAMP (0x8809/02)		
	Link OAM (0x8809/03)		
	ESMC (0x8809/0A)		
01-80-C2-00-00-03	Port Authentication (0x888E)	Must Peer or Discard per UNI	Should Tunnel on all UNIs
01-80-C2-00-00-04	–	Must Not Tunnel	Not specified in MEF 6.1.1
01-80-C2-00-00-05	–		
01-80-C2-00-00-06	–		
01-80-C2-00-00-07	E-LMI (0x88EE)	Must Peer or Discard per UNI	Must Tunnel on all UNIs
01-80-C2-00-00-08	–	Must Not Tunnel	Not specified in MEF 6.1.1
01-80-C2-00-00-09	–		
01-80-C2-00-00-0A	–		
01-80-C2-00-00-0B	–	Must Tunnel	Not specified in MEF 6.1.1
01-80-C2-00-00-0C	–		
01-80-C2-00-00-0D	–		
01-80-C2-00-00-0E	LLDP (0x88CC)	Must Peer or Discard per UNI	Must Tunnel on all UNIs
	PTP Peer Delay (0x88F7)		
01-80-C2-00-00-0F	–	Must Tunnel	Not specified in MEF 6.1.1
01-80-C2-00-00-20 through 01-80-C2-00-00-2F	GARP/MRP	Must Tunnel on all UNIs	Must Tunnel on all UNIs

You do not have to memorize all of these requirements for the MEF-CECP exam, but you should at least understand those related to bridging.

L2CP Processing Requirements for Bridging

L2CP processing capability is especially important for subscribers who choose to deploy IEEE 802.1Q bridges (as opposed to routers) as CEs. The following figure shows three LANs that are to be interconnected through a MEN (CEN). The CEs are 802.1Q bridges.

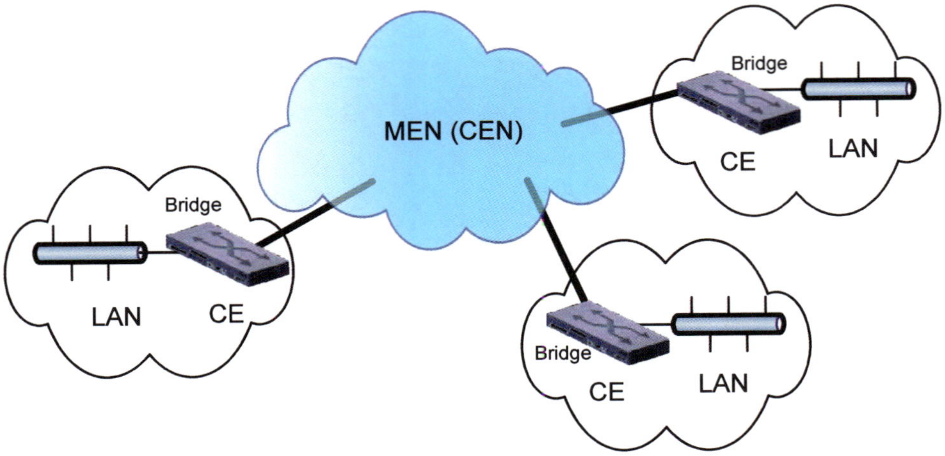

Figure 13: Example of LAN Interconnection with Bridge CEs (Not Routers)

In this example, the service must support the exchange of bridge protocol data units (BPDUs) between bridges so that spanning tree protocol (STP) works across the aggregation of connected LANs.

To the MEN (CEN), BPDUs are recognized as L2CP service frames with destination MAC address 01-80-C2-00-00-00. For transparent bridging, the service must tunnel them, that is, pass the BPDUs through the MEN (CEN) and deliver them unchanged to the proper (egress) UNI(s).

The following table summarizes L2CP processing requirements from MEF 6.1.1 for bridging protocols (L2CP service frames with destination MAC address 01-80-C2-00-00-00).

Table 5: L2CP Processing Requirements for BPDUs

Service		L2CP Requirement
Port-based	EPL	Must tunnel on all UNIs
	EP-LAN	
	EP-Tree[8]	
VLAN-based	EVPL	Must Peer on all UNIs or Discard on all UNIs
	EVP-LAN	
	EVP-Tree	

8 MEF 6.1.1 warns: "Since not all CEs in an EP-Tree service will see all BPDUs undesirable behavior may ensue. Service Providers should be careful to warn Subscribers about attaching bridges to such a service and expecting xSTP to work properly."

Key observations:

- BPDU tunneling is required for port-based services.
- BPDU tunneling is not allowed for VLAN-based services.

3.9
Review

1. The MEF refers to an Ethernet frame transmitted across a UNI towards the subscriber as a (an):
   - **a.** L2CP frame
   - **b.** Subscriber frame
   - **c.** Service frame
   - **d.** UNI frame
2. For an EVPL service, which two options are valid for Link OAM processing? (Choose two.)
   - **a.** Discard
   - **b.** Tunnel
   - **c.** Peer
   - **d.** Peer and Tunnel
   - **e.** Discard and Tunnel
3. An EVC is provisioned with a maximum transmission unit (MTU) size of 1526 bytes. A service frame containing 1530 bytes is transmitted to the UNI from the CE. Which statement is correct?
   - **a.** The frame should be dropped.
   - **b.** The frame should be marked yellow (drop-eligible) and forwarded.
   - **c.** The service level specification (SLS) does not apply to the frame.
   - **d.** The frame should be truncated to 1526 bytes and forwarded.
4. A broadcast service frame is transmitted from the subscriber to a UNI that is provisioned to support service multiplexing. Where will the service frame be delivered, assuming that it is not discarded?
   - **a.** To all UNIs of each EVC multiplexed at the UNI
   - **b.** To all UNIs in the EVC
   - **c.** To all EVCs multiplexed at the UNI
   - **d.** To all UNIs in the EVC, except the ingress UNI
   - **e.** To all UNIs in the EVC, except the egress UNI

5. Which two services are required to tunnel L2CP service frames containing Bridge Protocol Data Units (BPDUs) to support Spanning Tree Protocol (STP)? (Choose two.)

   a. EVP-LAN

   b. EP-Tree

   c. EVP-Tree

   d. EPL

   Note: *Tunneling is the process by which an L2CP service frame is passed through the service provider network without being processed and is delivered unchanged to the proper UNI(s).*

6. Which two of the following service attributes are *EVC per UNI* attributes? (Choose two.)

   a. Maximum number of EVCs

   b. Ingress Bandwidth Profile per CoS Identifier

   c. CE VLAN ID Preservation

   d. CE-VLAN ID / EVC Map

   e. Bundling

7. For an EPL service, which option is valid for L2CP processing of bridge protocol data units (BPDUs) used for spanning tree protocol (STP)?

   a. Discard

   b. Tunnel

   c. Peer

   d. Peer and Tunnel

8. A EVPL service has to support an application that uses IP packets of length up to 800 bytes. What is the smallest value of attribute *EVC MTU Size* that can be specified for this service?

   a. 800 bytes

   b. 818 bytes

   c. 822 bytes

   d. 1500 bytes

   e. 1518 bytes

   f. 1522 bytes

9. Which is a requirement of bundling?

   a. All CE-VLAN IDs are mapped to a single EVC.

   b. The UNI must be configured to support service multiplexing.

   c. The service must be an EVPL, EVP-LAN, or EVP-Tree service.

   d. More than one EVC must terminate at the UNI.

10. A UNI is configured without bundling or all-to-one bundling and is also configured to support service multiplexing. What is the maximum number of CE-VLAN IDs that can be mapped to a given EVC at the UNI?

    a. 4095

    b. 4094

    c. 64

    d. 1

11. A UNI supports service multiplexing. Which statement is not true?

    a. The UNI can support up to 4094 EVCs.

    b. Bundling is optional at the UNI.

    c. At the UNI, at least one CE-VLAN ID must be mapped to each EVC.

    d. At the UNI, all-to-one bundling is permitted.

12. Which statement is not true for a UNI configured with all-to-one bundling?

    a. Service multiplexing is No.

    b. Bundling is No.

    c. CE-VLAN ID preservation is Yes.

    d. CoS Preservation is Yes.

    e. The UNI can support up to 4094 EVCs.

13. An EPL service has to support jumbo frames with payload of up to 3000 bytes. What is the smallest value of attribute *EVC MTU Size* that can be specified for this service?

    a. 3000 bytes

    b. 3018 bytes

    c. 3022 bytes

d. 3026 bytes

e. 3030 bytes

14. The *CE-VLAN ID for Untagged and Priority Tagged Service Frames* service attribute applies to which three of the following services? (Choose three.)

a. EPL services

b. EVPL services

c. EP-LAN services

d. EVP-LAN services

e. EP-Tree services

f. EVP-Tree services

15. Which statement about Layer 2 Control Protocol (L2CP) handling is not true?

a. The UNI *L2CP Processing* attribute must specify *Peer*, *Discard*, *Pass to EVC*, or *Peer and Pass to EVC*.

b. For EVPL services, Link OAM L2CP frames must be peered or discarded.

c. The EVC *L2CP Processing* attribute must specify *Tunnel* or *Discard*.

d. For EVPL services, BPDUs for STP should be tunneled, or may be discarded.

16. The smallest value allowed for the *EVC MTU Size* attribute is:

a. 4096 bytes

b. 4094 bytes

c. 1526 bytes

d. 1522 bytes

Related Links

Answers: How MEF Services are Defined (Part 1)

4

How MEF Services are Defined (Part 2)

In this chapter:

Note: *This lesson relates to EVC-based services only (EPL, EVPL, EP-LAN, EVP-LAN, EP-Tree, and EVP-Tree services). OVC-based E-Access services (Access EPL and Access EVPL services) are defined in a later lesson.*

Note: *This lesson describes how the MEF defines classes of service in general (what is allowed, how details are specified). The MEF has also defined 12 standardized classes of service (referenced by* ***CoS Label*** *and* ***Performance Tier*** *values), which are the subject of a later lesson.*

Related Links

E-Access Services
MEF-Standardized Classes of Service

4.1 The MEF Service Agreement Framework (SLA and SLS)

The MEF framework for service agreements includes two components:

- **Service Level Agreement (SLA)**–The contract between the subscriber and service provider specifying the agreed to service-level commitments and related business agreements.
- **Service Level Specification (SLS)**–The part of the SLA that provides technical specifications for the service. Note that:
    - The SLS usually states service-level commitments in engineering terms.
    - The SLS usually includes the generic specification of one or more service levels offered by the service provider, not customized to the particular subscriber.
    - The SLS defines service performance attributes, such as frame delay, frame delay variation, frame loss ratio, and availability.

Note: ***Service Level Specification*** *is also the name of an OVC service attribute (to be described in a later lesson).*

Related Links

Service Level Specification (OVC Service Attribute)

4.2
Traffic Management and QoS

In this section:

Quality of service (QoS) refers to service performance goals specified by a set of parameters, such as frame loss, frame delay, and frame delay variation.

Traffic management is the set of mechanisms, tools, and policies used by the service provider to achieve QoS as defined in the SLS.

Class of service (CoS) is an indicator used to differentiate service frames with respect to QoS, allowing different traffic management treatment. Without class of service, all frames receive the same traffic management treatment.

Related Links

MEF-Standardized Classes of Service

4.2.1
The Multi CoS Framework

Recognizing that different types of traffic have different QoS requirements, the MEF defines a framework that allows service providers to offer more than one CoS. To illustrate this point, the following figure shows the implementation of a mobile backhaul service with only one CoS.

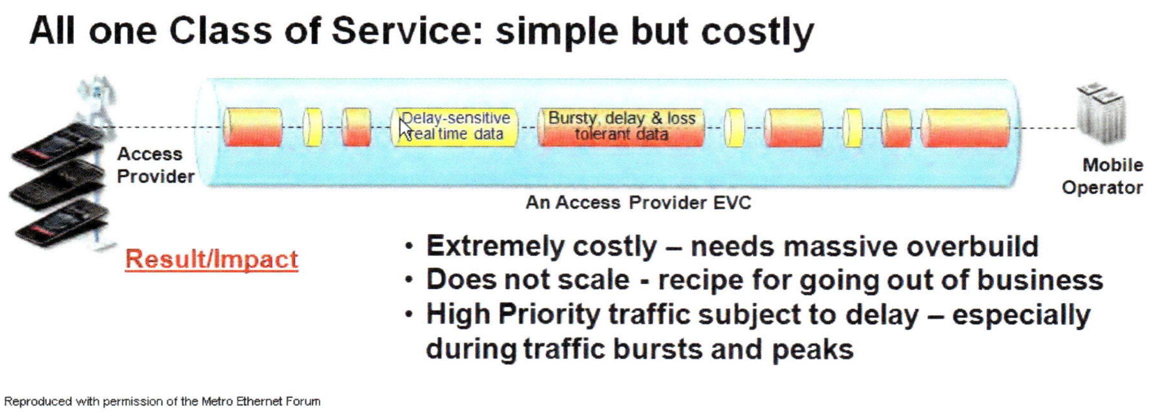

Figure 14: Example Mobile Backhaul Service with One Class of Service

With only one CoS, all traffic must be treated the same. Delay and loss-tolerant data, such a Internet traffic, must be treated the same as delay-sensitive real-time data, such as voice traffic. For the service provider, this is inefficient and costly.

The next figure shows the same application implemented in a network that supports multiple CoS. This network allows the mobile backhaul operator to use their network resources more efficiently. More revenue can be generated for the same cost by providing QoS only as required and charging for it.

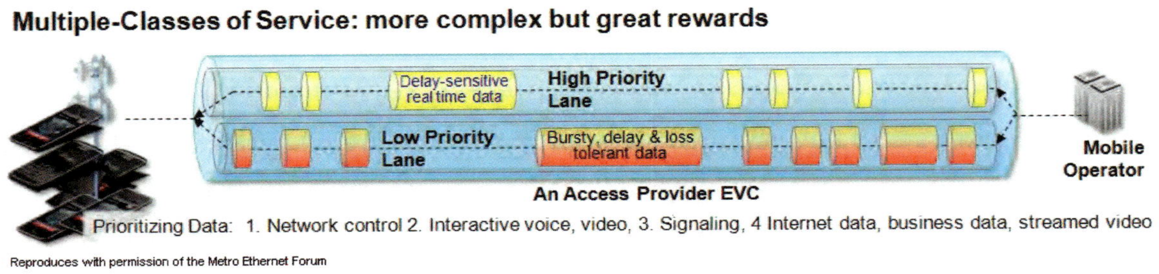

Figure 15: Example Mobile Backhaul Service with Multiple Classes of Service

In this framework:

- The service provider defines QoS for a small number of CoS to be offered in their network (normally a service provider supports eight or fewer CoS).
- The subscriber purchases a service with bandwidth quantified per-UNI, per-EVC, or per-CoS.
- The subscriber ensures that service frames are marked prior to MEN (CEN) ingress to reflect CoS (per prior agreement documented in the SLA).
- The service provider may police service frames at MEN (CEN) ingress (and possibly MEN/CEN egress) to monitor conformance to bandwidth agreements and to classify service frames with respect to drop-eligibility for traffic management purposes.
- The service provider delivers the service, ensuring QoS per CoS as specified in the SLA.

This framework serves both business parties (the subscriber and the service provider) by establishing a clear understanding about delivery commitments and QoS.

The rest of this lesson describes the components of this framework in detail.

4.2.2 CoS Identification at UNIs

CoS is an indicator that is used to differentiate service frames with respect to QoS.

A service provider might, for example, offer four CoS options named *gold*, *platinum*, *silver*, and *bronze*, with performance objectives for each option defined in the SLS.

With guidance from the service provider, the subscriber chooses the appropriate CoS options for their application and makes an agreement about how CoS will be identified at the UNIs.

MEF 10.2 defines four mechanisms for CoS identification at UNIs:

- CoS identification based on EVC
- CoS identification based on value of PCP bits in IEEE 802.1Q customer VLAN tag (CE-VLAN tag)
- CoS identification based on the value of the differentiated services code point (DSCP) in the IP header
- CoS identification based on Layer 2 Control Protocol

The SLA includes all details about how CoS is to be identified at the UNI.

Per MEF 23.1, if PCP or DSCP values are used for CoS identification, all PCP or DSCP values must be explicitly mapped to a CoS to ensure that customer frames are not inadvertently discarded. The following table provides examples of valid and invalid CoS maps using PCP values for CoS identification.

Table 6: Example of Valid and Invalid CoS Maps

CoS Maps	PCP Values			Notes
	Gold	Silver	Bronze	
Example 1	0	1	2, 3, 4, 5, 6, 7	Valid
Example 2	7, 5, 3, 1	6, 4, 2, 0	n/a	Valid
Example 3	7	6	5	Not valid because PCP values 0 through 4 are not mapped to a CoS.
Example 4	1, 2, 3, 4	4, 5, 6	7, 0	Not valid because PCP value 4 is mapped to two CoS.

To explicitly discard frames, a CoS can be equated to discard.

Note: *The MEF does not include CoS as a service attribute. However, three service attributes rely on it: CE-VLAN CoS Preservation, Ingress Bandwidth Profile per CoS, and Egress Bandwidth Profile per CoS.*

Note: *CoS propagation at ENNI using S-Tag PCP-bits is covered in another lesson.*

Related Links

Encoding Information in S-Tags

4.3
Bandwidth Profiles

In this section:

The MEF defines bandwidth profiles to characterize service bandwidth for business and rate enforcement purposes:

- To ensure that traffic, up to the contracted bandwidth, is allowed into the MEN (CEN) and is properly treated

- To protect the service provider from excess traffic in the MEN (CEN)

In a time division multiplexed (TDM) circuit, services are allocated specific time slots, and service bandwidth is easily defined.

In a packet-switched or frame-switched network, services share (or compete for) network resources that are **not** partitioned into time slots and are **not** dedicated to any particular service. Bandwidth sharing (based on statistical multiplexing) is adapted to accommodate the instantaneous traffic demands of multiple traffic streams. Without dedicated time slots, the concept of service bandwidth becomes more elusive.

MEF 10.2 characterizes service bandwidth using the **bandwidth profile**, which is defined by six parameters:

Table 7: Bandwidth Profile Parameters

Parameter		Units of Measure or Values
CIR	Committed Information Rate	Bits per second
CBS	Committed Burst Size	Bytes
EIR	Excess Information Rate	Bits per second
EBS	Excess Burst Size	Bytes
CF	Coupling Flag	N or Y
CM	Color Mode	color-blind or color-aware

These values are used in a **bandwidth profile algorithm** that processes service frames at a reference point and declares each service frame to be compliant or noncompliant with respect to the bandwidth profile. The level of compliance is expressed as one of three colors (green, yellow, or red), which are interpreted as follows:

- **Green**–The service frame is CIR/CBS-compliant, and the service provider is committed to completing its transport through the MEN (CEN).
- **Yellow**–The service frame is not CIR/CBS compliant but is EIR/EBS-compliant.

  Note: *The MEF does not associate any delivery commitment to yellow service frames, The service provider is free to characterize yellow service frame delivery commitment in the SLA, but is not required to do so.*

- **Red**–The service frame is not CIR/CBS-compliant or EIR/EBS-compliant. The service provider must drop the service frame.

It is important to understand that the bandwidth profile algorithm processes service frames at a specific reference point. In MEF service definitions, the reference point is always a UNI. In addition, because service frames travel in two directions at every UNI (ingress and egress), two bandwidth profiles can be applied to a traffic stream at a UNI:

- **Ingress bandwidth profile**–A bandwidth profile applied to service frames entering the MEN (CEN) from the CE
- **Egress bandwidth profile**–A bandwidth profile applied to service frames exiting the MEN (CEN) to the CE

The service provider may use bandwidth profiles for rate enforcement, or **policing**. In the following figure, functionality of the bandwidth profile operation is decomposed into two parts:

- **Compliance monitoring functionality**–The "Bandwidth Profiler" marks service frames red, yellow, or green as they pass by.
- **Policing functionality**–The policer stops and discards service frames marked red (non-compliant), preventing their delivery.

Reproduced with permission of the Metro Ethernet Forum

Figure 16: Monitoring and Policing Functionalities within the Bandwidth Profile

The MEF defines the bandwidth profile operation to include both functionalities (both are required; not optional).

4.3.1 The Bandwidth Profile Algorithm

MEF 10.2 defines the bandwidth profile algorithm in engineering terms using the following figure.

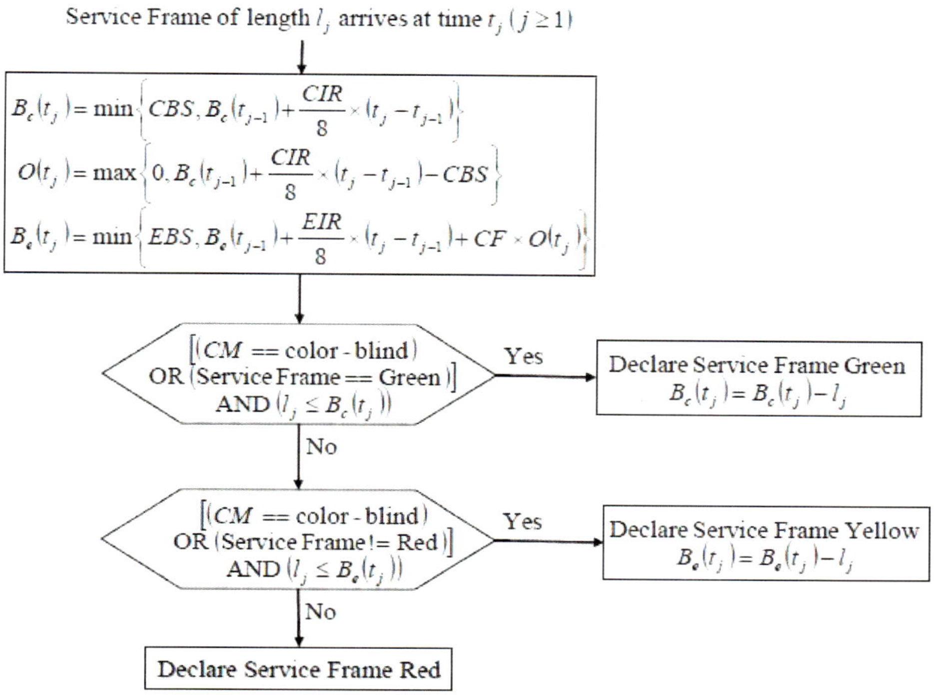

Figure 17: The Bandwidth Profile Algorithm (Figure 13 in MEF 10.2)

Do not be alarmed. This algorithm can be more easily understood using token buckets.

Token bucket and **leaky bucket** are two well-known algorithms that can be used to explain the bandwidth profile algorithm. In this lesson, we use the token bucket algorithm to explain the bandwidth profile algorithm.

First, we describe a simple token bucket algorithm, visualizing it as a process. Then we describe the bandwidth profile algorithm.

4.3.2 Visualizing the Token Bucket Algorithm

This visualization of the token bucket algorithm includes a token bucket and a traffic stream.

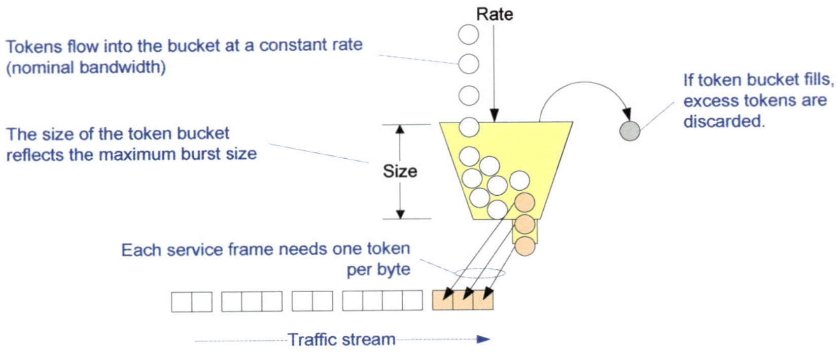

Figure 18: Visualization of the Token Bucket Algorithm

The token bucket is characterized by two parameters:

- **Rate**—Tokens flow into the token bucket at a constant rate, which is specified by a value called Rate.
- **Size**—The token bucket can hold only a limited number of tokens, which is specified by a value called Size.

Tokens can be removed from the bucket by the traffic stream (as will soon be explained). If the token bucket fills, overflowing tokens are discarded.

Service frames stream past the token bucket. As each service frame passes the token bucket, it is classified as **compliant** or **noncompliant** as follows:

1. If the token bucket contains one token for each byte in the current service frame, two actions are taken:
   a. The service frame is classified as compliant.
   b. Tokens, equal to the number of bytes in the current service frame, are drained from the token bucket.
2. Otherwise, the service frame is classified as noncompliant.

Test your understanding: *An 8-byte service frame arrives when the token bucket contains 13 tokens.* ***(a)*** *Is the service frame classified as compliant or noncompliant?* ***(b)*** *How many tokens are in the token bucket after the service frame is classified?*

Answers: ***(a)*** *Compliant.* ***(b)*** *Five (13 - 8 = 5).*

Test your understanding: *An 8-byte service frame arrives when the token bucket contains 7 tokens.* ***(a)*** *Is the service frame classified as compliant or noncompliant?* ***(b)*** *How many tokens are in the token bucket after the service frame is classified?*

Answers: ***(a)*** *Noncompliant.* ***(b)*** *Seven.*

Test your understanding: *A token bucket has Rate = 60 (tokens per second) and Size = 15 (tokens).* ***(a)*** *How many bytes might a compliant service frame contain?* ***(b)*** *How many bytes might a noncompliant service frame contain?* ***(c)*** *If the token bucket is empty, how long does it take to fill the token bucket, assuming no service frames arrive?*

Answers: ***(a)*** *Up to 15 bytes but no more than that because the token bucket can only hold up to 15 tokens (bytes).* ***(b)*** *Any number. If a service frame contains more than 15 bytes, it will definitely be classified as noncompliant. If it has 15 bytes or less, classification depends how many tokens are in the token bucket when the service frame arrives. If the number of tokens is less than the number of bytes, it will be classified as noncompliant.* ***(c)*** *0.25 seconds (60 tokens/second x 0.25 seconds = 15 tokens).*

4.3.3 Explaining the Bandwidth Profile Algorithm

The following table lists the bandwidth profile parameters.

Parameter		Units of Measure or Values
CIR	Committed Information Rate	Bits per second
CBS	Committed Burst Size	Bytes
EIR	Excess Information Rate	Bits per second
EBS	Excess Burst Size	Bytes
CF	Coupling Flag	N or Y
CM	Color Mode	color-blind or color-aware

Green and Yellow Token Buckets

The bandwidth profile algorithm is explained using two token buckets:

- A green token bucket with fill rate of CIR/8 and size CBS
- A yellow token bucket with fill rate of EIR/8 and size EBS

Note: *CIR and EIR are divided by 8 to determine fill rates because CIR and EIR are specified in bits per second and each token represents 1 byte (or 8 bits).*

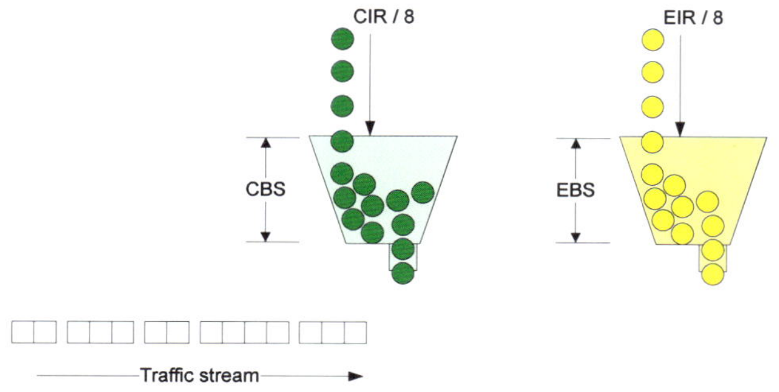

Incoming service frames stream past the two token buckets, which are used to determine compliance with respect to the bandwidth profile.

Coupling Flag

If the coupling flag (CF) is set to Y, tokens that overflow the green token bucket go into the yellow token bucket. Normally, CF is set to N, and such tokens are discarded.

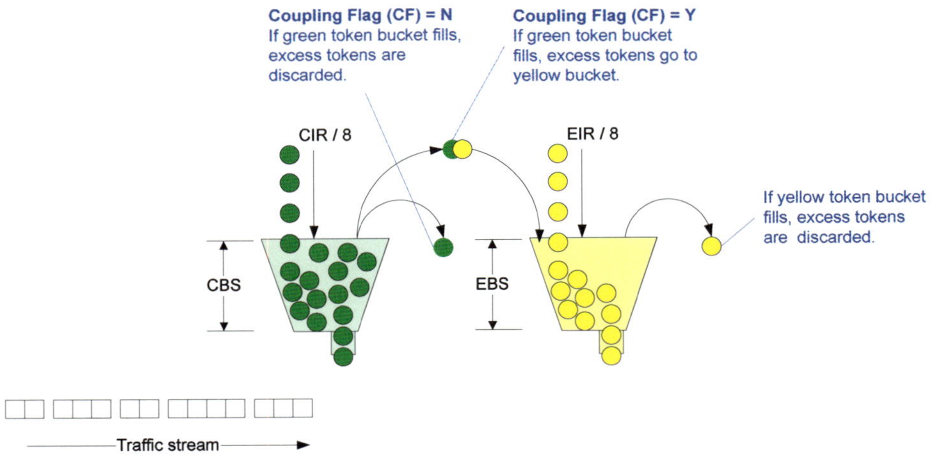

The coupling flag increases the rate of token flow into the yellow token bucket when the green token bucket is full. If CF=Y and the green token bucket is full, tokens flow into the yellow token bucket at a rate of EIR/8+CIR/8. Tokens that overflow the yellow token bucket are always discarded.

Color Mode

Color mode (CM) can be set to two values:

- **Color-blind**–The bandwidth profile algorithm disregards any prior coloring information (if present).
- **Color-aware**–The bandwidth profile algorithm uses prior coloring information (if present).

Here, *prior coloring information* refers to green/yellow drop-eligibility information that has already been encoded into service frames. If CM is set to color-aware, the SLA must specify the scheme used to encode drop-eligibility.

Note: *The subscriber can encode color (green/yellow drop-eligibility information) within the PCP bits of CE-VLAN tags or within DSCP bits of IP headers.*

Note: *CM=color-aware is mandatory at ENNI (as is explained in another lesson). At ENNI, coloring can be encoded in the DEI bit of the S-Tag or in the PCP bits of the S-Tag.*

Related Links

Color-Aware Bandwidth Profiles at ENNI

Color-Blind Bandwidth Profile Operation

If CM=color-blind, the bandwidth profile algorithm disregards any prior coloring information (if present) and works as follows:

1. If the green (CIR/CBS) token bucket contains one token for each byte in the current service frame, three actions are taken:

   - The service frame is classified as green (drop-ineligible).
   - The service frame is forwarded.
   - Tokens, equal to the number of bytes in the service frame, are drained from the green token bucket.

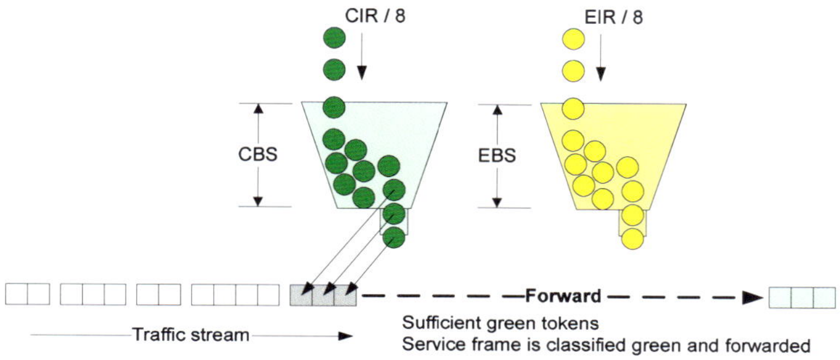

2. Otherwise, the green token bucket lacks enough tokens to process the current service frame, and processing continues as follows:

   a. If the yellow (EIR/EBS) token bucket contains one token for each byte in the current service frame, three actions are taken:

      - The service frame is classified as yellow (drop-eligible).
      - The service frame is forwarded.
      - Tokens, equal to the number of bytes in the service frame, are drained from the yellow token bucket.

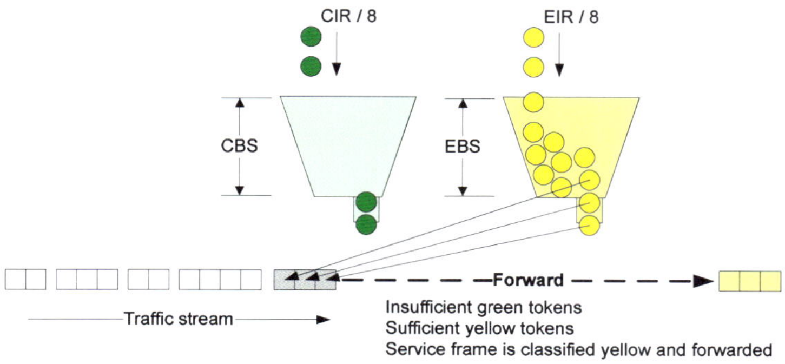

   b. Otherwise, both token buckets lack sufficient tokens to process the current service frame, and the current service frame is dropped (not forwarded).

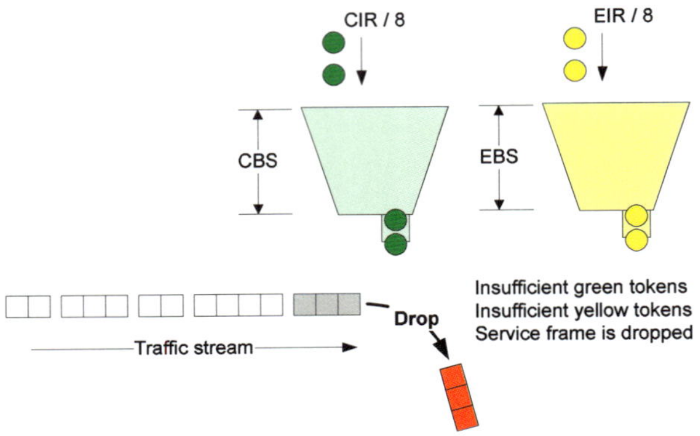

Color-Aware Bandwidth Profile Operation

If CM=color-aware, bandwidth profile operation is influenced by prior coloring that is already encoded in service frames.

If the current service frame has been classified as drop-eligible (yellow), the bandwidth profile algorithm processes the service frame as described in this section. Otherwise, the service frame is processed as described previously (Color-Blind Bandwidth Profile Operation).

If the current service frame is drop-eligible (yellow), the bandwidth profile algorithm works as follows:

1. The state of the green token bucket is disregarded.
2. If the yellow token bucket contains one token for each byte in the current service frame, three actions are taken:

- The service frame is classified as yellow (drop-eligible).
- The service frame is forwarded.
- Tokens, equal to the number of bytes in the service frame, are drained from the yellow token bucket.

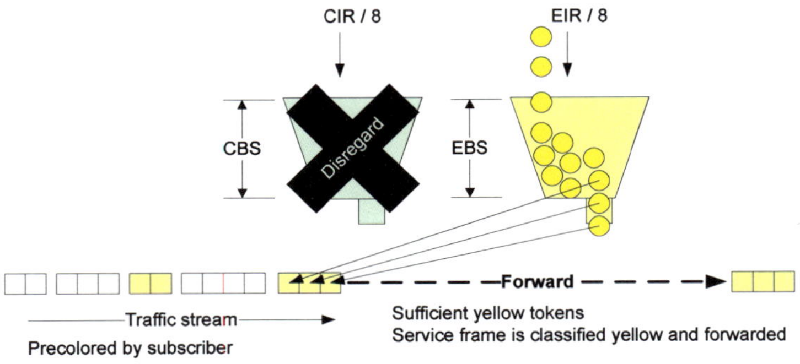

3. Otherwise, the yellow token bucket lacks enough tokens to process the current service frame, and the service frame is dropped (not forwarded).

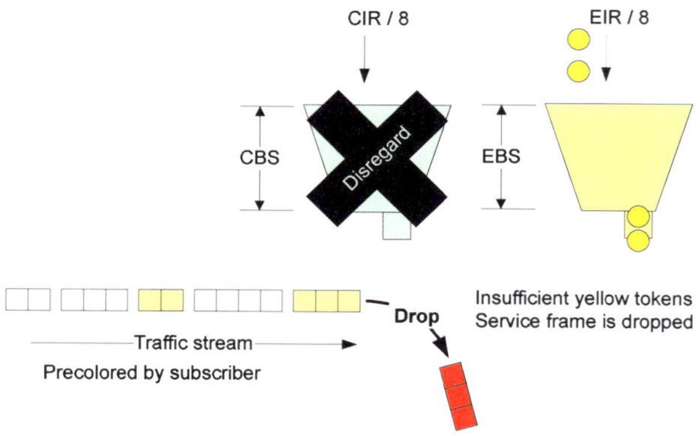

Related Links

Color-Aware Bandwidth Profiles at ENNI

4.3.4 Single-Rate versus Dual-Rate Bandwidth Profile Implementations

Some applications do not require both components (green and yellow token buckets) in the previously described **dual-rate** bandwidth profile algorithm. To accommodate single-rate applications, some components of the bandwidth profile are set to zero:

- In real-time applications (such as voice or video conferencing) or mission critical applications it is not appropriate to mark any service frames yellow (drop-eligible). However, bandwidth profiles are still needed for rate-enforcement purposes. For such applications, a single-rate bandwidth profile is defined using CIR and CBS components only (EIR and EBS are set to zero).

- In best-effort applications, including Internet traffic, all service frames may be marked yellow (drop-eligible). For such applications, a single-rate best-effort bandwidth profile is defined using EIR and EBS components only (CIR and CBS are set to zero).

4.4
Bandwidth Profile Service Attributes

In this section:

	Attribute	Requirement					
		Port-Based Service			VLAN-Based Service		
		EPL	EP-LAN	EP-Tree	EVPL	EVP-LAN	EVP-Tree
UNI	Ingress Bandwidth Profile per UNI	Must not specify	Optional (1)		Optional (1)	Optional (1)	
UNI	Egress Bandwidth Profile per UNI	Must not specify	Optional (3)		Optional (3)	Optional (3)	
EVC Per UNI	Ingress Bandwidth Profile per EVC	Optional (1)	Optional (1)		Optional (1)	Optional (1)	
EVC Per UNI	Ingress Bandwidth Profile per CoS Identifier	Optional (2)	Optional (2)		Optional (2)	Optional (2)	
EVC Per UNI	Egress Bandwidth Profile per EVC	Must not specify	Optional (3)		Must not specify	Optional (3)	
EVC Per UNI	Egress Bandwidth Profile per CoS Identifier	Must not specify	Optional (4)		Must not specify	Optional (4)	

(1) If supported, MUST specify <CIR, CBS, EIR, EBS, CM, CF>. MUST NOT be combined with any other type of ingress bandwidth profile

(2) If supported, MUST specify CoS ID and MUST specify <CIR, CBS, EIR, EBS, CM, CF> for each CoS. MUST NOT be combined with any other type of ingress bandwidth profile

(3) If supported, MUST specify <CIR, CBS, EIR, EBS, CM, CF>. MUST NOT be combined with any other type of egress bandwidth profile

(4) If supported, MUST specify CoS ID and MUST specify <CIR, CBS, EIR, EBS, CM, CF> for each CoS. MUST NOT be combined with any other type of egress bandwidth profile

Figure 19: Bandwidth Profile Service Attributes

MEF 10.2 defines six bandwidth profile service attributes: three for **ingress** and three for **egress**. For both ingress and egress, bandwidth profile attributes can be assigned **per UNI**, **per EVC**, or **per CoS Identifier**.

In each case, the bandwidth profile is defined by the six parameters listed in the following table.

Table 8: Bandwidth Profile Parameters

Parameter		Units of Measure or Values	Requirements
CIR	Committed Information Rate	Bits per second	CIR ≥ 0
CBS	Committed Burst Size	Bytes	If CIR > 0, then CBS ≥ maximum MTU over all EVCs to which the bandwidth profile applies
EIR	Excess Information Rate	Bits per second	EIR ≥ 0
EBS	Excess Burst Size	Bytes	If EIR > 0, then EBS ≥ maximum MTU over all EVCs to which the bandwidth profile applies
CF	Coupling Flag	N or Y	n/a
CM	Color Mode	color-blind or color-aware	n/a

Observations:

- CIR and EIR are specified in bits per second.
- CBS and EBS are specified in bytes (each byte is 8 bits).
- If CIR > 0, then CBS must be large enough to accommodate the largest applicable MTU.
- If EIR > 0, then EBS must be large enough to accommodate the largest applicable MTU.

Test your understanding: *Consider the following bandwidth profile:*

Parameter		*Value*
CIR	*Committed Information Rate*	*200 bits per second*
CBS	*Committed Burst Size*	*100 bytes*
EIR	*Excess Information Rate*	*800 bits per second*
EBS	*Excess Burst Size*	*300 bytes*
CF	*Coupling Flag*	*N*
CM	*Color Mode*	*color-blind*

1. *How many seconds are needed to fill the green token bucket if it is empty?*
2. *If a service frame contains 183 bytes, can it be marked green? Yellow? Red?*

Answer:

1. *The green token bucket holds 800 bits (CBS * 8 bits/byte). So, with a fill rate of 200 bits per second (CIR), it takes 4 seconds to fill.*

2. *The service frame cannot be marked green because the green bucket can only hold 100 bytes (CBS). It might be marked yellow because the yellow bucket can hold up to 300 bytes (EBS). However it might also be marked red because the yellow bucket might not contain 183 bytes when the service frame arrives.*

Test your understanding: *Explain why the MEF requires CBS to be ≥ maximum MTU if CIR>0.*

Answer: *Because, otherwise, a service frame of maximum MTU size would have no chance of being marked green. Also note that if CIR=0, the MEF places no restriction on CBS because it is irrelevant. CIR=0 is used if you want no service frames to be marked green (if you want all service frames to be drop-eligible), as is appropriate for a best-effort traffic stream.*

Test your understanding: *A subscriber transmits an 1800-byte service frame into a service that is configured with EVC MTU Size[9]=1522 bytes. The bandwidth profile at the ingress UNI is configured with CBS = 2000 bytes. When the frame arrives, the green token bucket contains 1900 tokens (bytes).*

1. *Can the frame be admitted to the service?*

2. *Can the frame be marked green?*

3. *Does the service level specification (SLS) apply to this frame?*

Answer:

1. *Yes*

2. *Yes*

3. *No (Refer to previous lesson about MTU Size Service Attributes.)*

Related Links

MTU Size Service Attributes

4.4.1

Ingress Bandwidth Profile Service Attributes

An ingress bandwidth profile is used to regulate the amount of ingress traffic at a particular UNI. An ingress bandwidth profile is defined for a particular UNI and applies to all or a subset of all ingress service frames at the UNI in question.

9 *EVC MTU Size is the maximum sized service frame allowed for an EVC.*

Ingress Bandwidth Profile per UNI Service Attribute

If the *per-UNI* ingress bandwidth profile service attribute is used, the bandwidth profile applies to all ingress service frames at the UNI. This bandwidth profile manages bandwidth indiscriminately for all EVCs at the UNI; some EVCs may get more bandwidth, while others may get less.

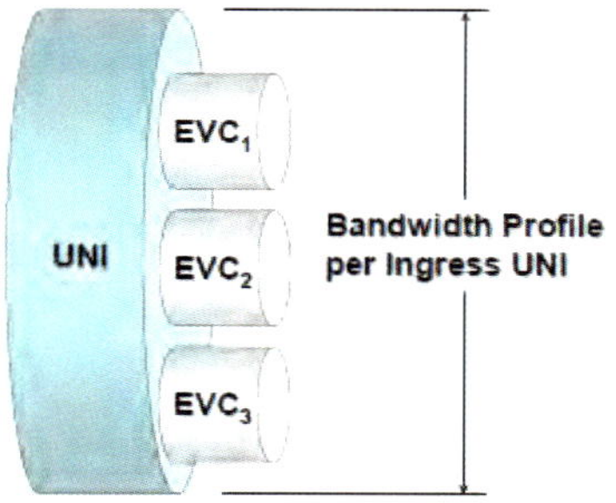

Reproduced with permission of the Metro Ethernet Forum

Figure 20: Ingress Bandwidth Profile per UNI (Figure 14 in MEF 10.2)

If a *per-UNI* ingress bandwidth profile is applied at a UNI, no other ingress bandwidth profiles can be applied at the same UNI.

Ingress Bandwidth Profile per EVC Service Attribute

The *per-EVC* ingress bandwidth profile service attribute applies an ingress bandwidth profile to a specific EVC at a UNI. The bandwidth profile applies to all ingress service frames mapped to the EVC. A different instance can be applied to each EVC at the UNI.

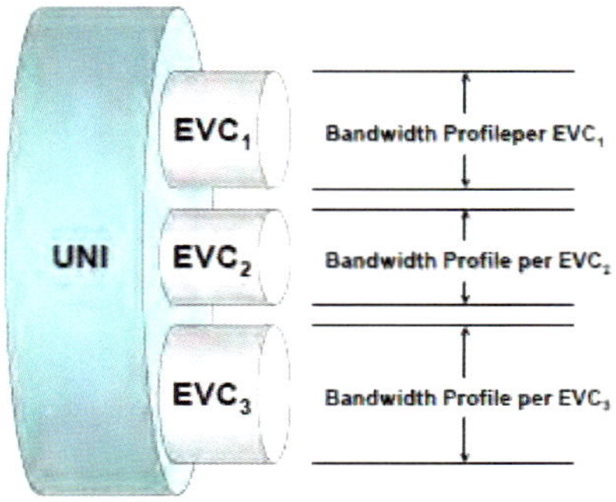

Reproduced with permission of the Metro Ethernet Forum

Figure 21: Ingress Bandwidth Profile per EVC (Figure 15 in MEF 10.2)

If a *per-EVC* ingress bandwidth profile is applied to an EVC at a UNI, no *per-CoS Identifier* ingress bandwidth profiles can be applied within that EVC at that UNI, and no *per-UNI* ingress bandwidth profile can be applied to that UNI.

Ingress Bandwidth Profile per CoS Identifier Service Attribute

The *per-CoS Identifier* ingress bandwidth profile service attribute applies an ingress bandwidth profile to a specific CoS identifier within a specific EVC at a UNI. The bandwidth profile applies to all ingress service frames that map to the CoS identifier and the EVC. A different instance can be applied to each CoS identifier of each EVC at the UNI.

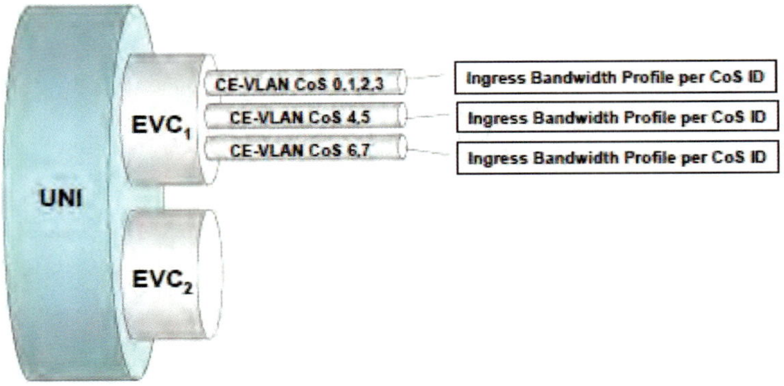

Reproduced with permission of the Metro Ethernet Forum

Figure 22: Ingress Bandwidth Profile per CoS Identifier (Figure 16 in MEF 10.2)

If a *per-CoS Identifier* ingress bandwidth profile is applied within an EVC at a UNI, no *per-EVC* ingress bandwidth profile can be applied to that EVC at that UNI, and no *per-UNI* ingress bandwidth profile can be applied to that UNI.

Simultaneous Application of Ingress Bandwidth Profiles

MEF 10.2 states that multiple ingress bandwidth profiles may exist simultaneously at a UNI subject to the following rules:

- If a *per-UNI* ingress bandwidth profile is applied at a UNI, no other ingress bandwidth profiles can be applied at that same UNI.
- If a *per-EVC* ingress bandwidth profile is applied to an EVC at a UNI, no *per-CoS Identifier* ingress bandwidth profiles can be applied within that EVC at that UNI.

These rules do not preclude the possibility of applying a *per-EVC* ingress bandwidth profile at one EVC and applying one or more *per-CoS Identifier* ingress bandwidth profiles within a different EVC at the same UNI.

4.4.2
Egress Bandwidth Profile Service Attributes

An egress bandwidth profile is used to regulate the amount of egress traffic at a particular UNI. An egress bandwidth profile is defined for a particular UNI and applies to all or a subset of all egress service frames at the UNI in question.

Egress Bandwidth Profile per UNI Service Attribute

If the *per-UNI* egress bandwidth profile service attribute is used, the bandwidth profile applies to all egress service frames at the UNI. This bandwidth profile manages bandwidth indiscriminately for all EVCs at the UNI; some EVCs may get more bandwidth, while others may get less.

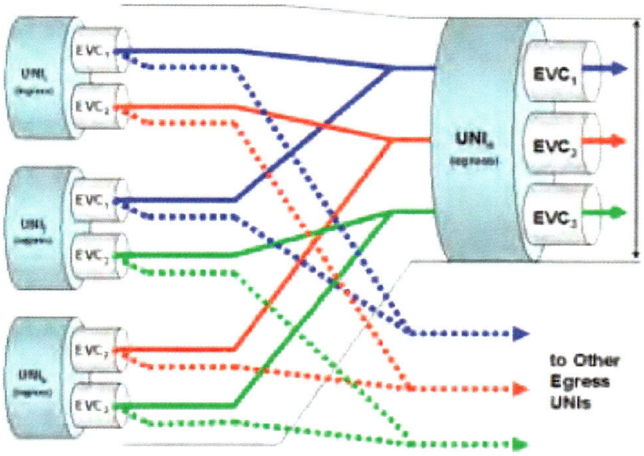

Reproduced with permission of the Metro Ethernet Forum

Figure 23: Egress Bandwidth Profile per UNI (Figure 17 in MEF 10.2)

If a *per-UNI* egress bandwidth profile is applied at a UNI, no other egress bandwidth profiles can be applied at the same UNI.

Egress Bandwidth Profile per EVC Service Attribute

The *per-EVC* egress bandwidth profile service attribute applies an egress bandwidth profile to a specific EVC at a UNI. The bandwidth profile applies to all egress service frames mapped to the EVC. A different instance can be applied to each EVC at the UNI.

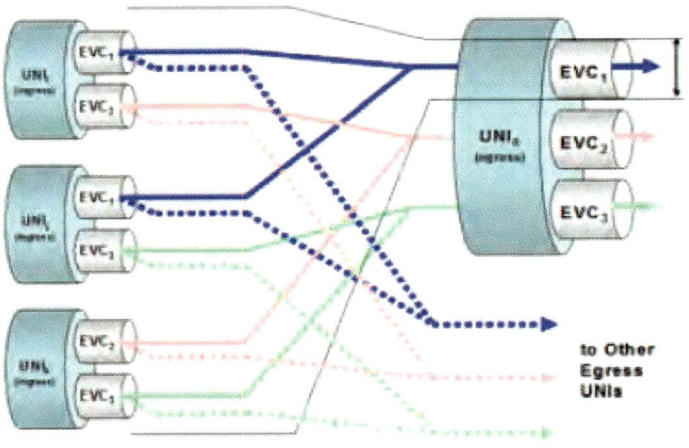

Reproduced with permission of the Metro Ethernet Forum

Figure 24: Egress Bandwidth Profile per EVC (Figure 18 in MEF 10.2)

If a *per-EVC* egress bandwidth profile is applied to an EVC at a UNI, no *per-CoS Identifier* egress bandwidth profiles can be applied within that EVC at that UNI, and no *per-UNI* egress bandwidth profile can be applied to that UNI.

Egress Bandwidth Profile per CoS Identifier Service Attribute

The *per-CoS Identifier* egress bandwidth profile service attribute applies an egress bandwidth profile to a specific CoS identifier within a specific EVC at a UNI. The bandwidth profile applies to all egress service frames that map to the CoS identifier and the EVC. A different instance can be applied to each CoS identifier of each EVC at the UNI.

If a *per-CoS identifier* egress bandwidth profile is applied within an EVC at a UNI, no *per-EVC* egress bandwidth profile can be applied to that EVC at that UNI, and no *per-UNI* egress bandwidth profile can be applied to that UNI.

Simultaneous Application of Egress Bandwidth Profiles

MEF 10.2 states that multiple egress bandwidth profiles may exist simultaneously at a UNI subject to the following rules:

- If a *per-UNI* egress bandwidth profile is applied at a UNI, no other egress bandwidth profiles can be applied at that same UNI.
- If a *per-EVC* egress bandwidth profile is applied to an EVC at a UNI, no *per-CoS Identifier* egress bandwidth profiles can be applied within that EVC at that UNI.

These rules do not preclude the possibility of applying a *per-EVC* egress bandwidth profile at one EVC and applying one or more *per-CoS Identifier* egress bandwidth profiles within a different EVC at the same UNI.

4.4.3 Egress Bandwidth Profiles for Point-to-Point Services

	Attribute	Requirement					
		Port-Based Service			VLAN-Based Service		
		EPL	EP-LAN	EP-Tree	EVPL	EVP-LAN	EVP-Tree
UNI	Ingress Bandwidth Profile per UNI	Must not specify	Optional (1)		Optional (1)	Optional (1)	
UNI	Egress Bandwidth Profile per UNI	Must not specify	Optional (3)		Optional (3)	Optional (3)	
EVC Per UNI	Ingress Bandwidth Profile per EVC	Optional (1)	Optional (1)		Optional (1)	Optional (1)	
EVC Per UNI	Ingress Bandwidth Profile per CoS Identifier	Optional (2)	Optional (2)		Optional (2)	Optional (2)	
EVC Per UNI	Egress Bandwidth Profile per EVC	Must not specify	Optional (3)		Must not specify	Optional (3)	
EVC Per UNI	Egress Bandwidth Profile per CoS Identifier	Must not specify	Optional (4)		Must not specify	Optional (4)	

(1) If supported, MUST specify <CIR, CBS, EIR, EBS, CM, CF>. MUST NOT be combined with any other type of ingress bandwidth profile

(2) If supported, MUST specify CoS ID and MUST specify <CIR, CBS, EIR, EBS, CM, CF> for each CoS. MUST NOT be combined with any other type of ingress bandwidth profile

(3) If supported, MUST specify <CIR, CBS, EIR, EBS, CM, CF>. MUST NOT be combined with any other type of egress bandwidth profile

(4) If supported, MUST specify CoS ID and MUST specify <CIR, CBS, EIR, EBS, CM, CF> for each CoS. MUST NOT be combined with any other type of egress bandwidth profile

Figure 25: Egress Bandwidth Profile Service Attributes for Point-to-Point Services

The MEF does not allow *per-EVC* or *per-CoS Identifier* egress bandwidth profiles for point-to-point EVCs (EPL or EVPL services). As explained in the following footnote taken from MEF 6.1, there is no use case for them:

For E-Line services, it is expected that an ingress bandwidth profile will be applied at the ingress UNI such that traffic on the EVC is already controlled; therefore, there is no need to apply an egress bandwidth profile per EVC or CoS at the egress UNI.

However, the MEF does allow the *per-UNI* egress bandwidth profile to be applied with EVPL services. An example use case for this application is shown in the following figure.

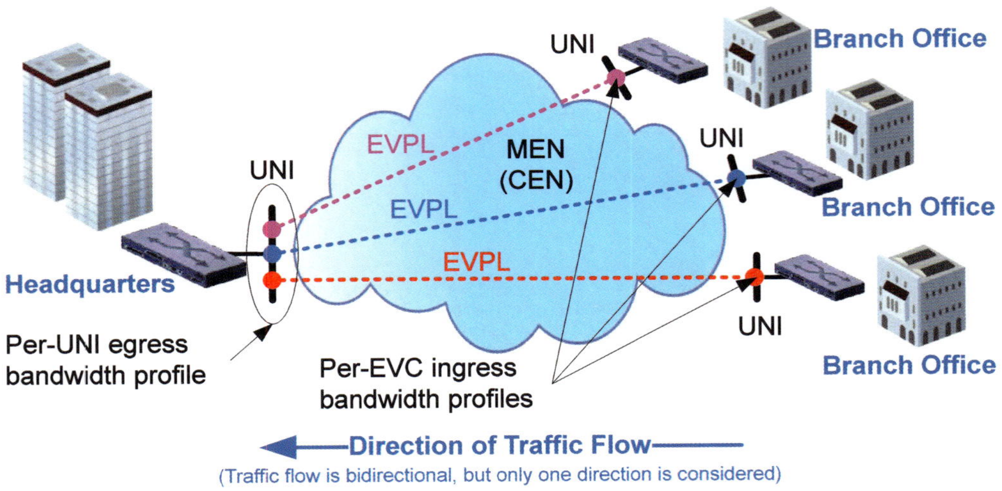

Figure 26: Example Use Case for per-UNI Egress Bandwidth Profile with EVPL Services

EVPLs from three branch offices are service multiplexed at headquarters to one UNI. Ingress bandwidth profiles applied at the branch offices limit egress traffic aggregating to headquarters. However, the worst-case aggregate total may be more than the headquarters UNI can support (or may cost more than the subscriber wishes to pay for).

Note: *For exam purposes, just remember that egress bandwidth profiles are not generally supported for point-to-point services.*

4.4.4 Egress Bandwidth Profiles in E-LAN and E-Tree Type Services

<table>
<tr><th rowspan="3"></th><th rowspan="3">Attribute</th><th colspan="6">Requirement</th></tr>
<tr><th colspan="3">Port-Based Service</th><th colspan="3">VLAN-Based Service</th></tr>
<tr><th>EPL</th><th>EP-LAN</th><th>EP-Tree</th><th>EVPL</th><th>EVP-LAN</th><th>EVP-Tree</th></tr>
<tr><td rowspan="2">UNI</td><td>Ingress Bandwidth Profile per UNI</td><td>Must not specify</td><td colspan="2">Optional (1)</td><td>Optional (1)</td><td colspan="2">Optional (1)</td></tr>
<tr><td>Egress Bandwidth Profile per UNI</td><td>Must not specify</td><td colspan="2">Optional (3)</td><td>Optional (3)</td><td colspan="2">Optional (3)</td></tr>
<tr><td rowspan="4">EVC Per UNI</td><td>Ingress Bandwidth Profile per EVC</td><td>Optional (1)</td><td colspan="2">Optional (1)</td><td>Optional (1)</td><td colspan="2">Optional (1)</td></tr>
<tr><td>Ingress Bandwidth Profile per CoS Identifier</td><td>Optional (2)</td><td colspan="2">Optional (2)</td><td>Optional (2)</td><td colspan="2">Optional (2)</td></tr>
<tr><td>Egress Bandwidth Profile per EVC</td><td>Must not specify</td><td colspan="2">Optional (3)</td><td>Must not specify</td><td colspan="2">Optional (3)</td></tr>
<tr><td>Egress Bandwidth Profile per CoS Identifier</td><td>Must not specify</td><td colspan="2">Optional (4)</td><td>Must not specify</td><td colspan="2">Optional (4)</td></tr>
</table>

(1) If supported, MUST specify <CIR, CBS, EIR, EBS, CM, CF>. MUST NOT be combined with any other type of ingress bandwidth profile

(2) If supported, MUST specify CoS ID and MUST specify <CIR, CBS, EIR, EBS, CM, CF> for each CoS. MUST NOT be combined with any other type of ingress bandwidth profile

(3) If supported, MUST specify <CIR, CBS, EIR, EBS, CM, CF>. MUST NOT be combined with any other type of egress bandwidth profile

(4) If supported, MUST specify CoS ID and MUST specify <CIR, CBS, EIR, EBS, CM, CF> for each CoS. MUST NOT be combined with any other type of egress bandwidth profile

Figure 27: Egress Bandwidth Profile Service Attributes for E-LAN and E-Tree Type Services

Egress bandwidth profiles are used in E-LAN and E-Tree type services to reign in delivery commitments for worst-case scenarios. The following figure illustrates an example EP-LAN service.

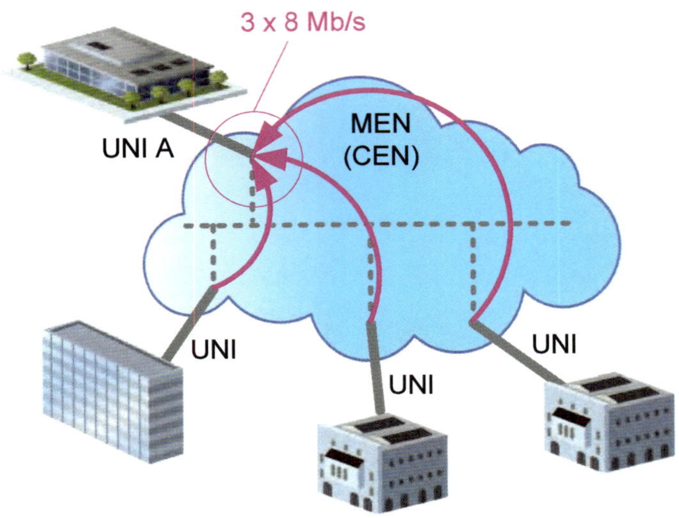

Figure 28: Example EP-LAN Service Illustrating Use of Egress Bandwidth Profiles

In this example:

- Each UNI supports 10 Mb/s.
- At each UNI, a per-UNI ingress bandwidth profile is applied with CIR=8 Mb/s.
- Any three UNIs can potentially simultaneously direct 8 Mb/s of CIR (green) traffic to the fourth UNI.

Without an egress bandwidth profile at each UNI, the service provider is formally committed to delivering traffic at the rate of 3×8=24 Mb/s through UNIs that can only support rates up to 10 Mb/s.

> For example, as shown in the figure, three UNIs might simultaneously forward 8 Mb/s of traffic to UNI A, resulting in 24 Mb/s of aggregated traffic at UNI A. However, UNI A can only support rates up to 10 Mb/s. Similarly, the same scenario applies to the other three UNIs.

One possible solution is to apply a per-UNI egress bandwidth profile (for example, with CIR=9 Mb/s) at each UNI. The application of such egress bandwidth profiles makes it clear that the service is not expected to deliver green service frames in all circumstances.

The Bottom Line: *When egress bandwidth profiles are used, service frames marked green at ingress can be dropped at egress due to egress traffic exceeding the agreed egress bandwidth profiles.*

4.5

About Line Rate and CIR

Important: *This topic is not included in the MEF-CECP exam, but practicing Carrier Ethernet professionals need to understand it.*

Note: *This material is adapted from Appendix A of MEF 33.*

The relationship between CIR and physical layer bandwidth (line rate) is commonly misunderstood. For example, a UNI with physical speed 100 Mb/s is commonly expected to support CIR values up to 100 Mb/s. However, the physical layer cannot support CIR values up to the line rate because the MEF bandwidth profile algorithm only counts service frame bits, not the bits used to separate frames in the physical layer (interframe gap and preamble bits).

As shown in the following figure, an Ethernet frame in the physical layer (Layer 1) includes 20 bytes (160 bits) which are not part of the MEF service frame and are therefore not counted by the MEF bandwidth profile algorithm.

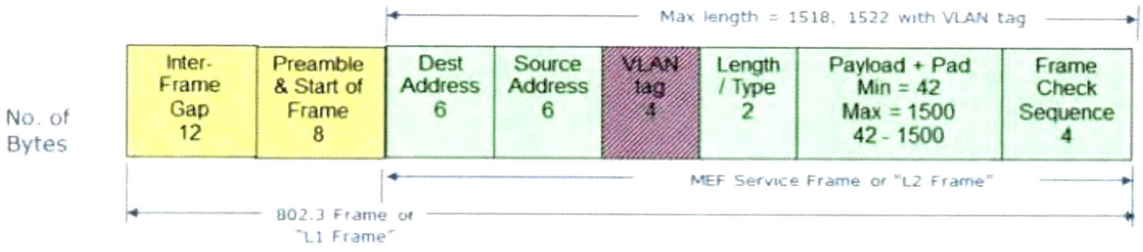

Reproduced with permission of the Metro Ethernet Forum

Figure 29: Length of an Ethernet Frame (Figure 6 in MEF 33)

These interframe bits consume physical layer bandwidth, but the MEF bandwidth profile algorithm does not account for them.

So, how close can CIR be set to the physical layer bit rate (line rate, UNI speed)?

The answer depends on the size of MEF service frames, as shown in the following graph for an example 100 Mb/s physical interface.

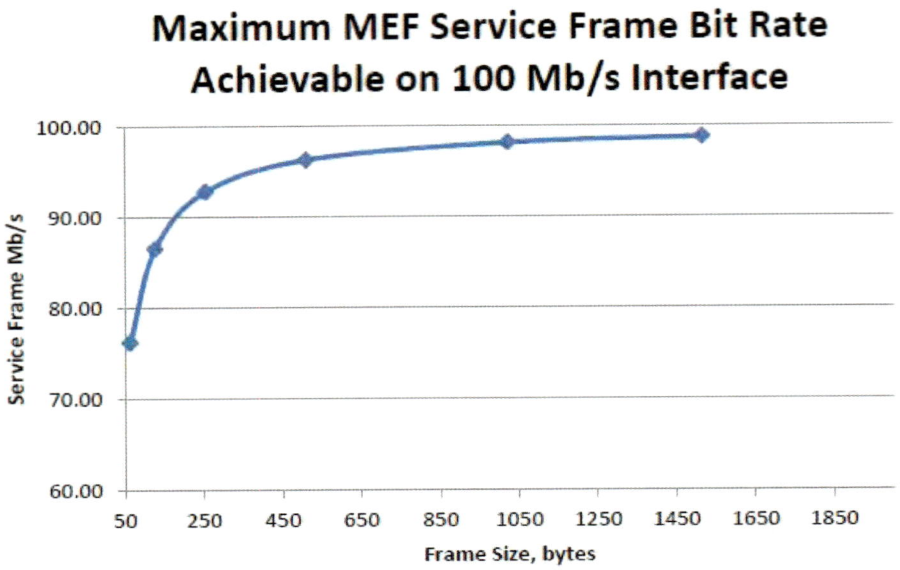

Reproduced with permission of the Metro Ethernet Forum

Figure 30: Maximum CIR versus Service Frame Size for 100 Mb/s Physical Interface (Figure 7 in MEF 33)

Observations:

- As frame size decreases, an increasing fraction of the line rate is consumed by interframe overhead.
- If the size of service frames (on average) is greater than 200 bytes, the physical layer can support CIR values up to about 90% of the line rate.
- If CIR < 76% of the line rate, the physical layer should support MEF services without issue, regardless of frame size.

Note: *This issue (line rate versus CIR) commonly surfaces during testing. For example, a subscriber might test a CIR=90 Mb/s EPL service implemented over 100 Mb/s UNIs using a test set configured to produce a 90 Mb/s stream of service frames. If the test is performed with 500-byte service frames, the service performs as expected (no dropped frames). However, if the test is performed with 50-byte service frames, frames are dropped, and the subscriber may incorrectly conclude that the service is not providing 90 Mb/s of CIR.*

Related Links

Layer 1 versus Layer 2 Ethernet Frames

4.6
EVC Performance Service Attributes

In this section:

	Attribute	Requirement	
		Port-Based Service (EPL, EP-LAN, EP-Tree)	VLAN-Based Service (EVPL, EVP-LAN, EVP-Tree)
EVC	EVC Performance	At least one CoS is REQUIRED. MUST specify CoS ID. MUST list values for each of the following attributes {Frame Delay, Inter Frame Delay Variation, Frame Loss Ratio, Availability, and Resiliency} for each CoS, where Not Specified (N/S) is an acceptable value.	

Figure 31: EVC Performance Service Attributes

EVC Performance service attributes specify quality of service (QoS), for each class of service (CoS) involved in the service.

For each CoS, five performance attributes are specified:

1. **Availability**–The percentage of time that a service is useable
2. **Resiliency**–The service's ability to continue to perform under fault conditions
3. **Frame Loss Ratio (FLR)** –The number of service frames lost as a percentage of those sent
4. **Frame Delay (FD)**–The time a service frame spends in transit between CEs
5. **Inter-Frame Delay Variation (IFDV)** [10]–The variation in the delays experienced by different service frames

Note: *Not Specified (N/S) is an acceptable value.*

10 MEF documents published before MEF 10.2 (including MEF 6.1, 14, 15, and 19) refer to **Inter-Frame Delay Variation (IFDV)** as Frame Delay Variation (FDV).

These descriptions are only high-level approximations to rigorous MEF technical definitions (to be described).

Note: *The MEF addresses the problem of measuring EVC performance using SOAM Performance Management (as described in a later lesson). The focus here is on defining EVC performance.*

Related Links

SOAM Performance Management

4.6.1

EVC Performance Attributes: Metrics, Objectives, and Parameters

The following figure illustrates how performance attributes are defined in MEF 10.2 and amendment 10.2.1 using performance metrics, performance objectives, and performance parameters.

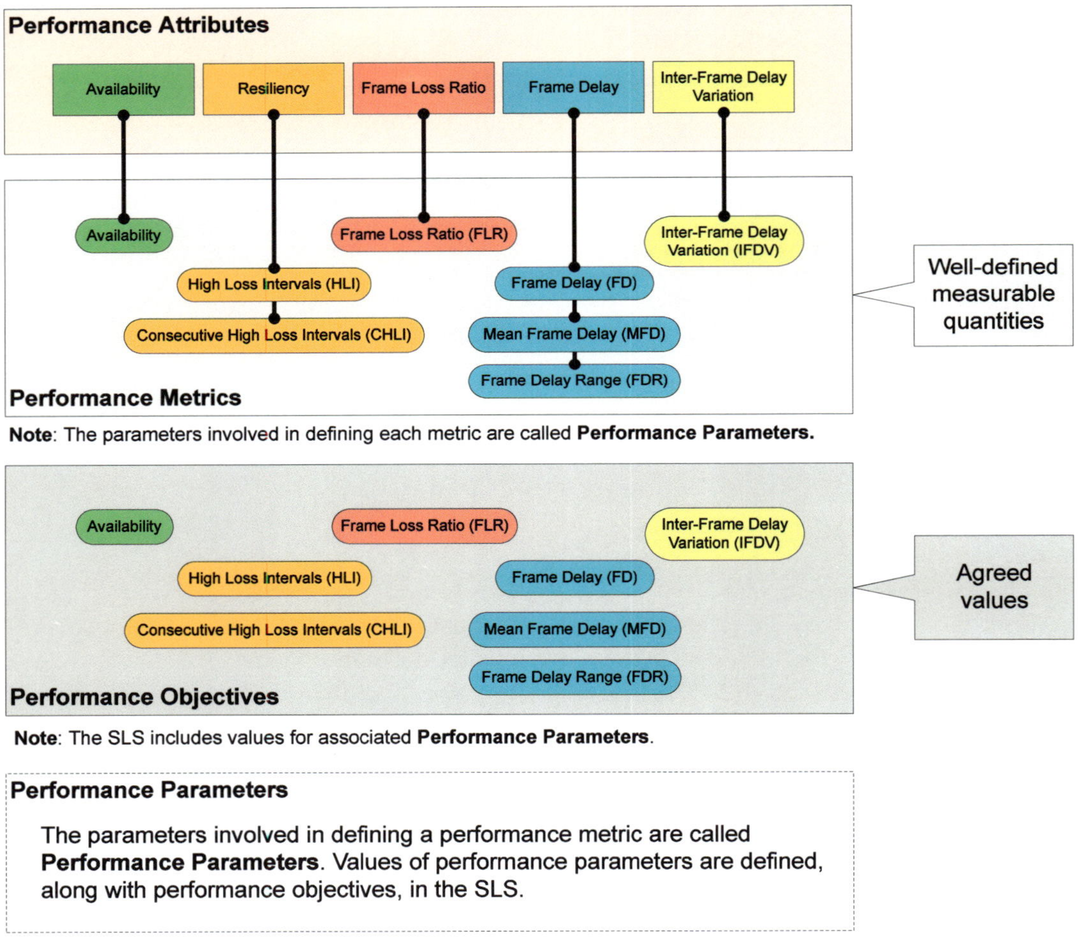

As shown, each performance attribute is associated with one or more performance metrics. For example, two performance metrics, HLI and CHLI, are associated with performance attribute Resiliency.

A **performance metric** is a well-defined measurable quantity.

A **performance objective** is the agreed value of the performance metric that is specified in the SLS.

A **performance parameter** is a quantity that is used in deriving and specifying a performance objective. The SLS specifies values for performance parameters and performance objectives.

Also notice that the same term is sometimes used for a performance attribute and a performance metric/objective. For example, Frame Delay is both a performance attribute and a performance metric/attribute.

4.6.2 Key Caveats

Four key caveats apply to all performance attributes:

- Performance attributes apply to **green service frames only**. They say nothing about delivery of yellow service frames.
- Performance attributes are measured over an agreed subset of **ordered UNI pairs**.
- When measurement is over multiple ordered UNI pairs, the **worst case** (lowest performing) **UNI pair** defines the measurement.
- **Maintenance intervals**[11] **are excluded** from performance measurements.

Additionally, all performance attributes, excluding Availability, are measured only when the service is judged to be "Available" (per criteria that will be explained).

4.6.3 Ordered UNI Pair

Every service frame travels one way between two UNIs: a source UNI and a destination UNI.

MEF 10.2 uses the term **ordered UNI pair** to refer to a unique combination (source UNI and destination UNI) in an EVC.

A point-to-point EVC between UNI A and UNI B has two ordered UNI pairs:

- (UNI A, UNI B)
- (UNI B, UNI A)

11 Maintenance intervals are periods of time when the service is not expected to perform well (or at all) per agreement between the service provider and subscriber. A maintenance window might be used, for example, to upgrade network equipment, to isolate a fault, or to change service features.

A multipoint-to-multipoint EVC with three UNIs has six ordered UNI pairs:

- (UNI A, UNI B), (UNI B, UNI A)
- (UNI A, UNI C), (UNI C, UNI A)
- (UNI B, UNI C), (UNI C, UNI B)

A multipoint-to-multipoint EVC with n UNIs has n*(n-1) ordered UNI pairs.

An ordered UNI pair can be visualized as a directed line segment (arrow) between two UNIs.

Test your understanding: *The following figure shows an EP-LAN service involving five UNIs and two CoS. The subscriber has nominated important paths for each CoS as shown.*

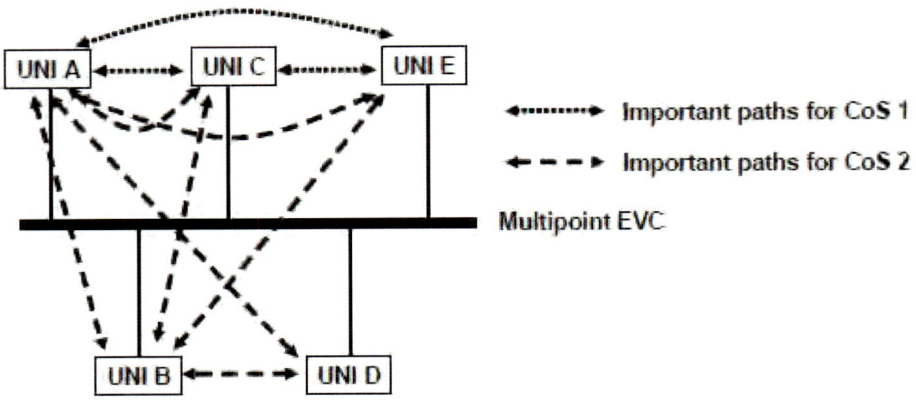

Reproduced with permission of the Metro Ethernet Forum

Figure 32: Multipoint EVC Example (Figure 30 in MEF 10.2)

1. *List the ordered UNI pairs that are important for CoS 1.*
2. *How many ordered UNI pairs are important for CoS 2?*
3. *What is the total number of ordered UNI pairs, considering all UNIs and all possible paths between them?*

Answer:

1. *(UNI A, UNI C), (UNI C, UNI A); (UNI A, UNI E), (UNI E, UNI A); (UNI C, UNI E), (UNI E, UNI C)*
2. *Fourteen. 7 bidirectional paths x 2 ordered UNI pairs per path = 14 ordered UNI pairs*
3. *Twenty. Using the formula n*(n-1) with n=5, you get 5*(5-1)=20.*

4.6.4
Availability Performance

Availability performance is based on the concept that a service at any instant of time is in one of two states:

- Available (working satisfactorily)
- Unavailable (not working satisfactorily)

With this concept, availability performance is easily defined:

Availability Performance—The percentage of time within a specified time interval (T) during which the service is available

However, this definition side steps the difficult part of defining availability: determining which state (available or unavailable) the service is in at each instant of time.

Important: *Defining availability state is critical to defining all service performance attributes (not just Availability performance). FD, IFDV, FLR, and Resiliency performance are each measured during "available" periods only.*

Availability Specification

The MEF has defined a rigorous, but flexible, framework for defining availability based on frame loss ratio performance. Per MEF 10.2.1, availability performance is defined in an SLS for a particular CoS by specifying values for five parameters and one performance objective:

Parameter	Description	Example Value
T	The time interval over which the SLS applies	30 days
S	Non-empty subset of ordered UNI pairs	All
Δt	A time interval much smaller than T	2 minutes
C	Unavailability frame loss ratio threshold	0.1
n	Number of consecutive small time intervals, Δt, for assessing availability	10
Objective	**Description**	**Example Value**
Availability	Availability performance objective expressed as a percentage	99.9%

Availability performance is measured over a time interval *T* (for example, 30 days) that is divided into much smaller time intervals Δt (for example, 2 minutes), as specified in the SLS.

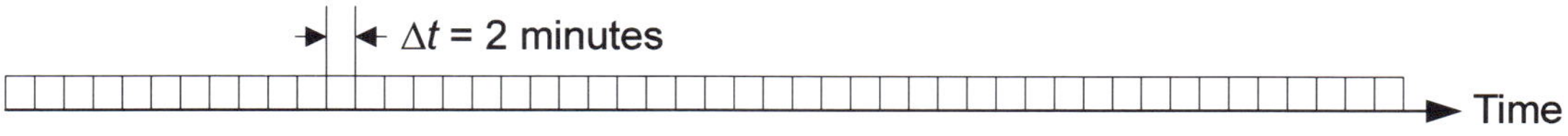

How Availability is Determined

For each Δt, the one-way service between an ordered pair of UNIs is declared to be either *unavailable* or *available*, based on frame loss ratio measurements and the following process.

Note: *Refer to MEF 10.2.1 for a formal precise description of the process. This informal description of the process is meant to be easier to understand.*

1. Each Δt is assigned a color, white or grey, as follows:

   If frame loss ratio (*flr*) during time interval Δt exceeds threshold *C*, Δt is assigned color grey. Otherwise, Δt is assigned color white.

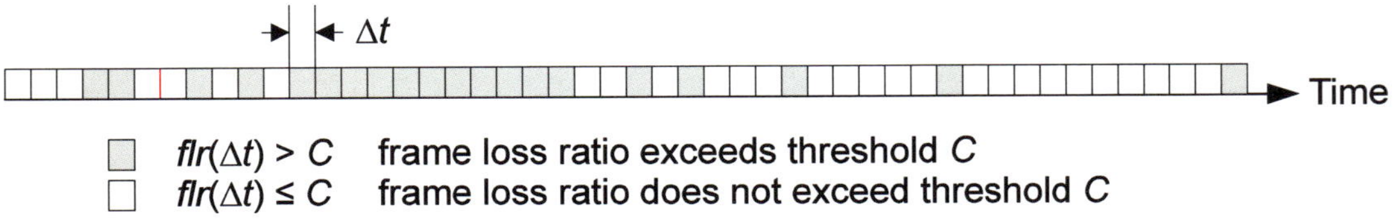

- **flr(Δt)**– The number of service frames lost as a fraction of the number of service frames sent over time interval Δt

Note: *The color, white or grey,* <u>*does not*</u> *represent availability during interval Δt. Availability is assigned in the next step.*

2. The availability state of each time interval Δt is determined by a process (algorithm) that involves two things:

   - A sliding window of *n* time intervals Δt that is imagined to advance through time (from left to right)
   - A binary window state, white or grey, that is initialized to value *white* to start the process

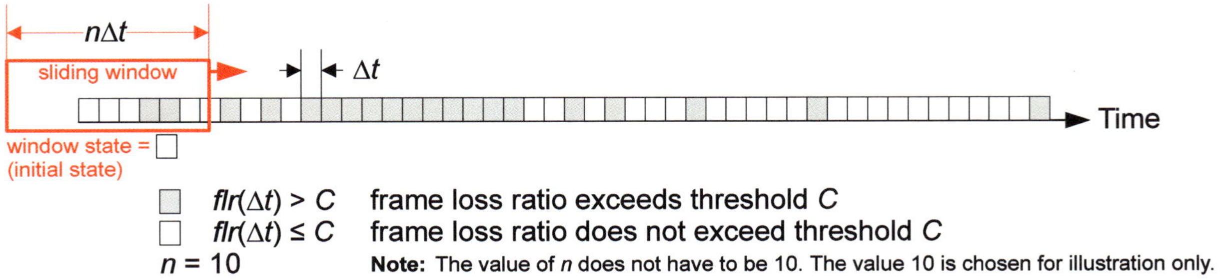

Each time the left side of the sliding window reaches a new time interval Δt, availability for that time interval is determined by a two-step process:

a. If the color (white or grey) of all Δt's in the sliding window is opposite to the window state, then change the window state to match the color of all Δt's within the window.

b. Set availability value for the left-most Δt to:

- 1 (available), if the window state is white
- 0 (not available), if the window state is grey

Assignment of availability, for this example, progresses as follows.

Left side of the sliding window reaches the first time interval.

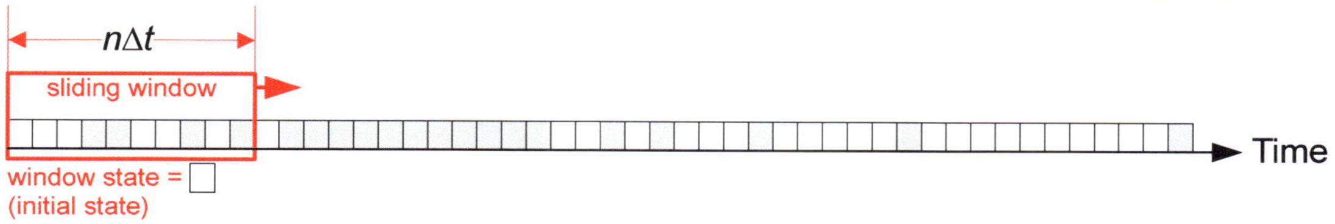

Window state remains white because not all Δt's in the window are grey. State of first Δt is set to 1 (available) because the window state is white. The window increments forward.

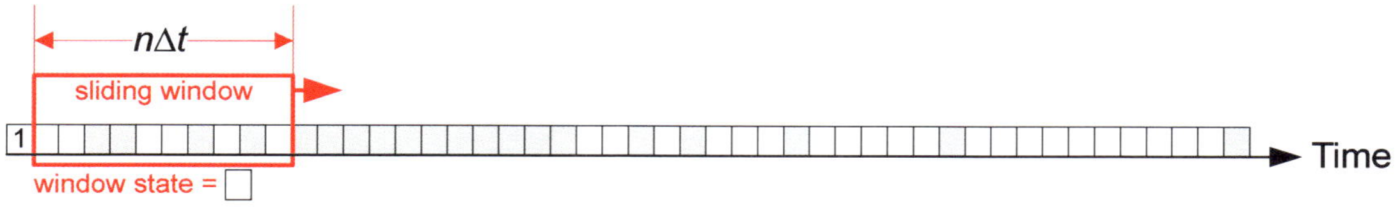

Again the window state remains white (because not all Δt's in the window are grey). The state of the left-most Δt is set to 1 (available). The sliding window increments forward.

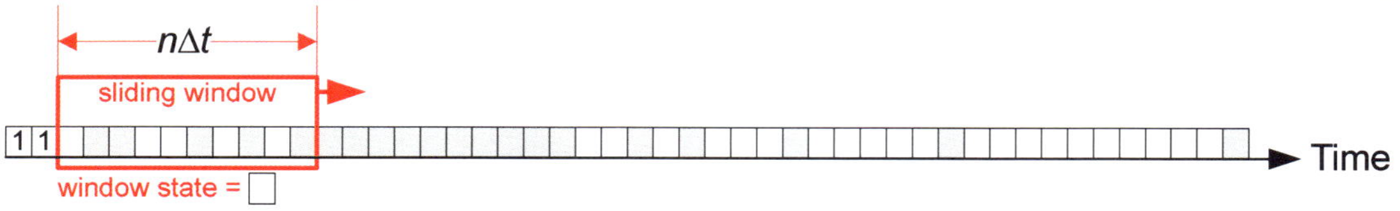

Eventually the window reaches a position where all Δt's in the window are grey.

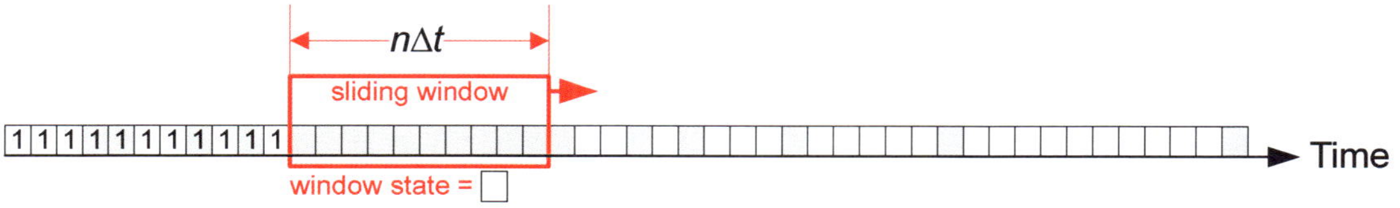

The window state is now changed from white to grey. The state of the left-most **Δ***t* is set to 0 (unavailable) because the window state is grey. The window increments forward.

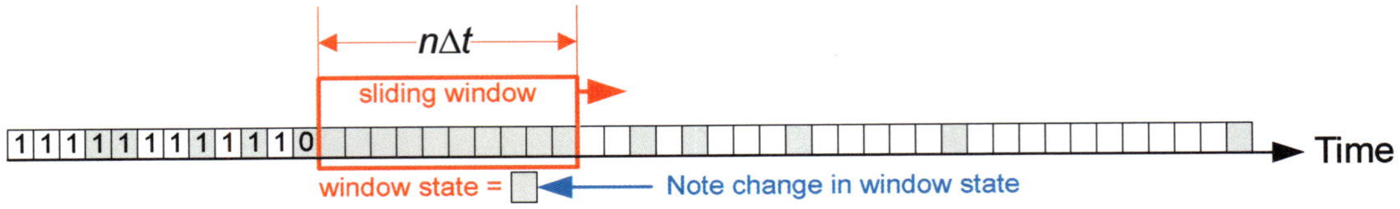

Now many subsequent **Δ***t*'s are set to 0 (unavailable) as the window advances because the window state is grey and the **Δ***t*'s within the window are not all white (opposite color to the window state).

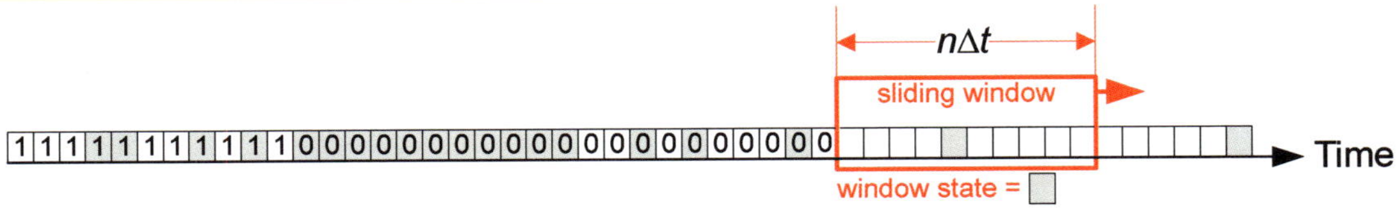

Eventually, the sliding window reaches a position where all **Δ***t*'s in the window are white.

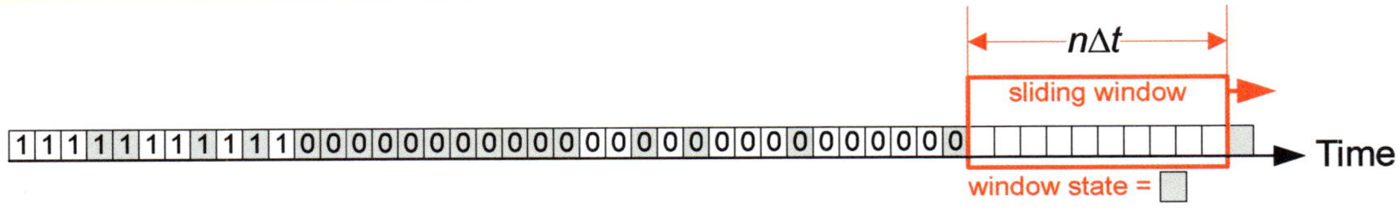

The window state is now changed from grey to white. The state of the left-most **Δ***t* is set to 1 (available) because the window state is white. The window increments forward.

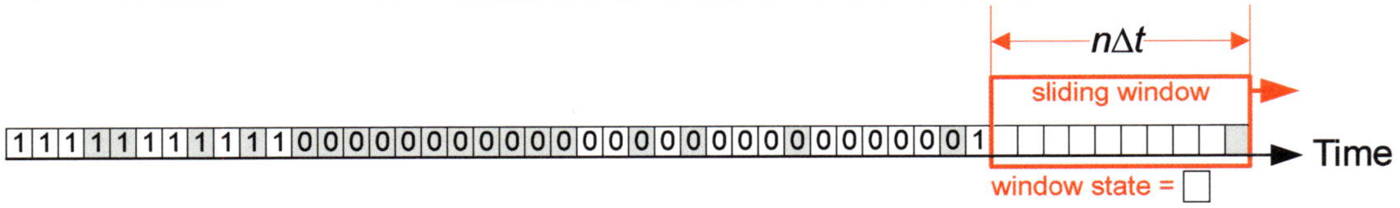

The process eventually completes, resulting in the assignment of availability to each **Δ***t*.

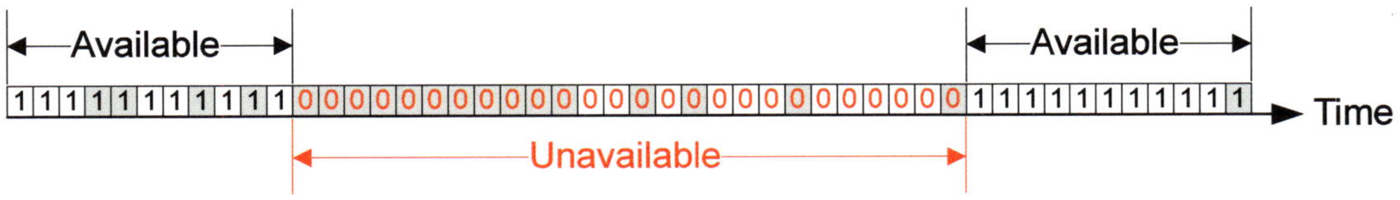

Summary

Availability performance is measured over a time interval *T* that is divided into small time intervals **Δ***t* that are much smaller than *T*.

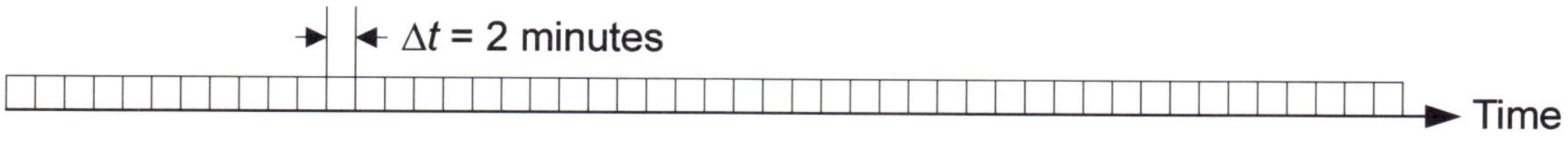

Parameter	Description	Example Value
T	The time interval over which the SLS applies	30 days
Δ*t*	A time interval much smaller than T	2 minutes

For each **Δ***t*, the service is declared to be unavailable or available, based on frame loss ratio measurements and using an algorithm specified in MEF 10.2.1, which depends on three parameters specified in the SLS.

Parameter	Description	Example Value
S	Non-empty subset of ordered UNI pairs	All
C	Unavailability frame loss ratio threshold	0.1
n	Number of consecutive small time intervals, **Δ***t*, for assessing availability	10

The algorithm involves a sliding window and is not easily explained, but the result is just the assignment of an availability state to each **Δ***t*.

Once availability has been assigned to each **Δ***t*, availability performance is computed as the percentage of available **Δ***t*'s, compared to all **Δ***t*'s over the interval T.

The availability performance objective specified in the SLS declares the level of availability performance that the service is expected to deliver.

Objective	Description	Example Value
Availability	Availability performance objective expressed as a percentage	99.9%

Note: *As with all performance attributes, four key caveats apply:*

- *Performance attributes apply to* ***green service frames only****. They say nothing about delivery of yellow service frames.*
- *Performance attributes are measured over an agreed subset of* ***ordered UNI pairs****.*
- *When measurement is over multiple ordered UNI pairs, the* ***worst-case*** *(lowest performing)* ***UNI pair*** *defines the measurement.*
- ***Maintenance intervals are excluded*** *from performance measurements.*

4.6.5

Resiliency Performance

Resiliency performance is defined using the same framework that was used to define Availability performance.

Per MEF 10.2.1, resiliency performance is defined in an SLS for a particular CoS by specifying values for five parameters and two performance objectives:

Parameter	Description	Example Value
T	The time interval over which the SLS applies	30 days
S	Non-empty subset of ordered UNI pairs	All
Δt	A time interval much smaller than T	2 minutes
C	Unavailability frame loss ratio threshold	0.1
p	The number of consecutive small time intervals for assessing CHLI, where $p < n$	3
Objective	**Description**	**Example Value**
HLI	HLI performance objective expressed as an integer	100
CHLI	CHLI performance objective expressed as an integer	20

The first four parameters (*T*, *S*, Δt, and *C*) are the same ones used to define Availability performance. They are defined in the same way and are required to have the same values, per CoS, within an SLS.

Parameter *n* is not used to define Resiliency. Instead, parameter *p* is used (which serves a similar role).

Conceptually, Resiliency performance characterizes the service's ability to continue to perform under fault conditions. Formally, it is defined by two numbers:

- **Resiliency Performance** – The number of HLIs (High Loss Intervals) and the number of CHLIs (Consecutive High Loss Intervals) in *T*

MEF 10.2.1 does not provide concise definitions for HLI or CHLI. Instead, it provides equations and algorithms to define how HLIs and CHLIs are counted. However, the following definitions are consistent with those equations and algorithms:

- **HLI** – High Loss Interval. A small time interval Δt over which service has been declared *Available* even though its frame loss ratio $flr(\Delta t)$ exceeds C (the unavailability frame loss ratio threshold).
- **CHLI** – Consecutive High Loss Interval. A sequence of p or more HLIs.

Note: *As stated in MEF 10.2.1, Resiliency attributes HLI and CHLI are similar to the definitions of Severely Errored Seconds (SES) and Consecutive SES in ITU-T standard Y.1563.*

For better clarity, let's unpack these definitions. Recall that Availability classifies service during each small time interval, Δt, as *Available* or *Unavailable*.

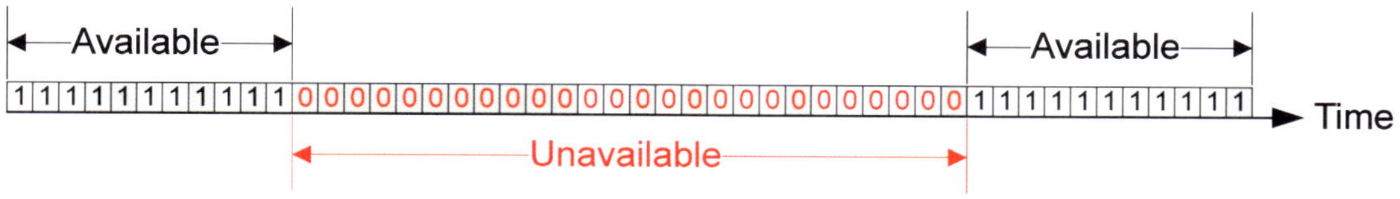

Also recall that prior to classification of Δt's as *Available*/*Unavailable*, each Δt is assigned a color, white or grey.

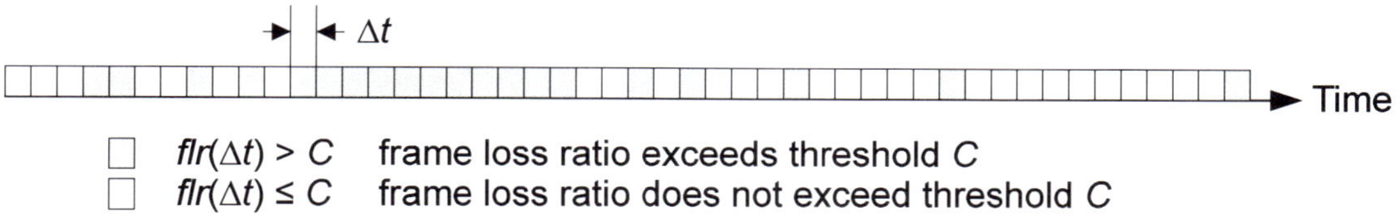

A High Loss Interval (HLI) is a Δt that is both *grey* and *Available*. This example contains five HLIs.

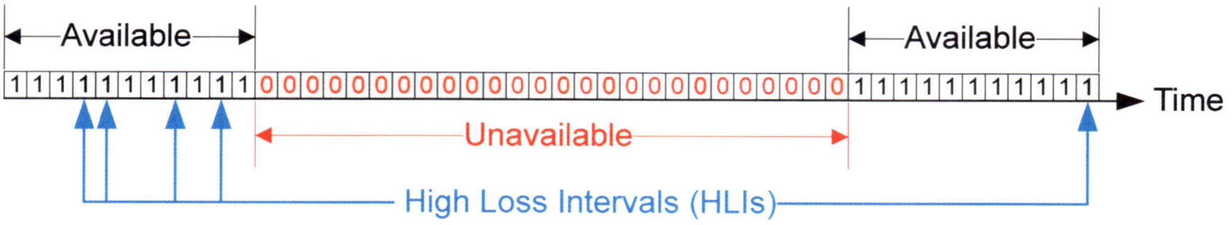

A Consecutive High Loss Interval (CHLI) is a sequence of p or more HLIs. If p=2, this example contains one CHLI.

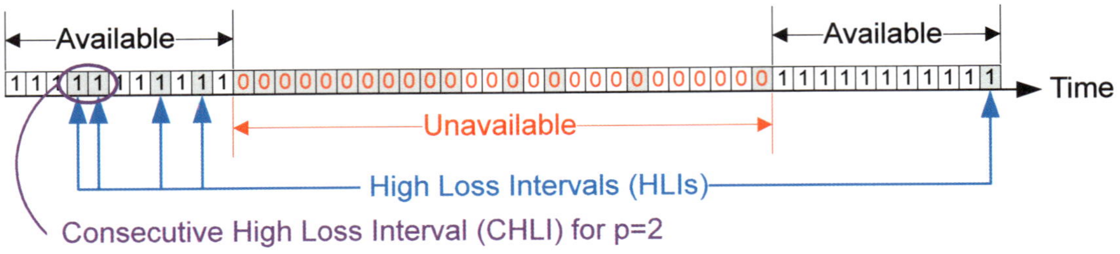

For *p*>2, this example contains no CHLIs. The following example contains two CHLIs for *p*=2 or *p*=3.

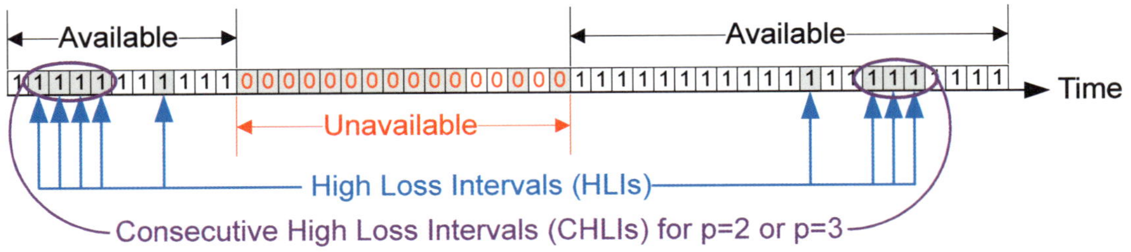

For *p*=4, this example contains one CHLI. For *p*>4, it contains no CHLIs.

Hierarchy of Service Time Classification

The following figure from MEF 10.2.1 illustrates how service time is classified in relation to Availability and Resiliency.

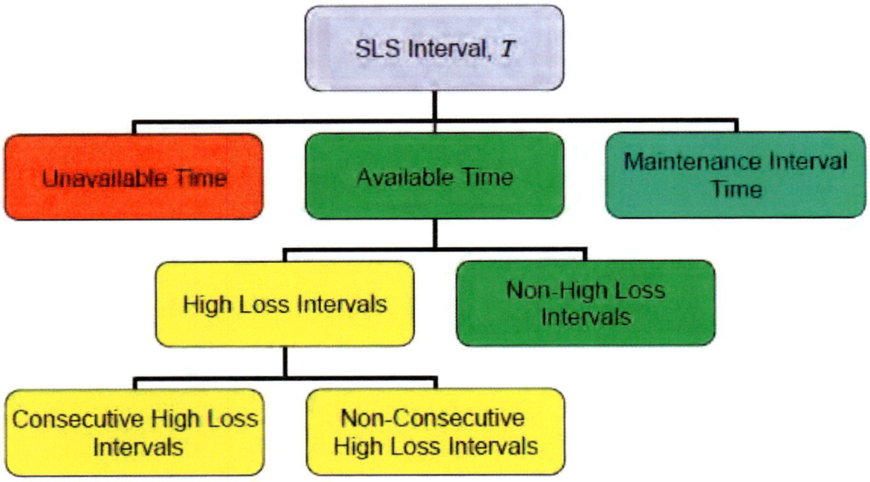

Figure D – Hierarchy of Time Showing the Resiliency Attributes

Time within the big time interval T (over which Availability and Resiliency are defined) consists of three types of time: *Unavailable* time, *Available* time, and *Maintenance Interval* time.

Available time is further classified into two types: *HLI* time and *Non-HLI* time.

HLI time is further classified into two types: *CHLI* time and *Non-CHLI* time.

Summary

Resiliency performance is measured per CoS using **five parameters** specified in the SLS: four parameters (T, S , Δt, and C) used for Availability performance measurement, plus a new integer parameter p (where $p < n$).

Resiliency performance is defined as the number of HLIs and the number of CHLIs in T, where:

- **HLI** – High Loss Interval. A small time interval Δt over which service has been declared *Available* even though its frame loss ratio $flr(\Delta t)$ exceeds C (the unavailability frame loss ratio threshold).
- **CHLI** – Consecutive High Loss Interval. A sequence of p or more HLIs.

The SLS, per CoS, specifies two resiliency performance objectives that declare the level of resiliency performance that the service is expected to deliver.

Objective	Description	Example Value
HLI	HLI performance objective expressed as an integer	100
CHLI	CHLI performance objective expressed as an integer	20

Note: *As with all performance attributes, four key caveats apply:*

- *Performance attributes apply to **green service frames only**. They say nothing about delivery of yellow service frames.*
- *Performance attributes are measured over an agreed subset of **ordered UNI pairs**.*
- *When measurement is over multiple ordered UNI pairs, the **worst-case** (lowest performing) **UNI pair** defines the measurement.*
- ***Maintenance intervals are excluded** from performance measurements.*

4.6.6
Frame Loss Ratio Performance

Per MEF 10.2, **Frame Loss Ratio (FLR) performance** is defined in an SLS for a particular CoS by specifying values for two parameters and one performance objective:

Parameter	Description	Example Value
T	The time interval over which the SLS applies	30 days
S	Non-empty subset of ordered UNI pairs	All
Objective	**Description**	**Example Value**
FLR	Frame loss ratio performance objective expressed as a percentage	0.1% (1 frame per 1000)

Note: *As with all performance attributes, four key caveats apply:*

- *Performance attributes apply to* ***green service frames only****. They say nothing about delivery of yellow service frames.*
- *Performance attributes are measured over an agreed subset of* ***ordered UNI pairs****.*
- *When measurement is over multiple ordered UNI pairs, the* ***worst-case*** *(lowest performing)* ***UNI pair*** *defines the measurement.*
- ***Maintenance intervals are excluded*** *from performance measurements.*

4.6.7

Frame Delay Performance

In general terms, Frame Delay (FD) performance quantifies the time a green service frame spends in transit between CEs.

MEF 10.2 uses the following figure to explain how one-way frame delay is measured for a single service frame.

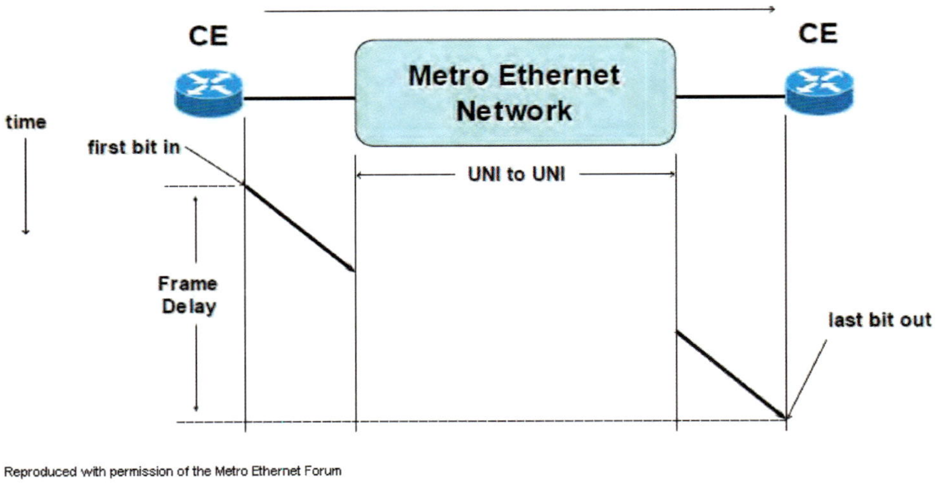

Figure 33: One-Way Frame Delay for Service Frame (Figure 5 in MEF 10.2)

One-way frame delay is a measurement of elapsed time:

- The timer starts when the first bit of the service frame leaves the source CE.
- The timer stops when the last bit of the service frame arrives at the destination CE.

Per MEF 10.2, **Frame Delay (FD) performance** is defined in an SLS for a particular CoS by specifying values for 4 parameters and 3 performance objectives:

Parameter	Description	Example Value
T	The time interval over which the SLS applies	30 days
S	Non-empty subset of ordered UNI pairs	All
P_d	Percentile defining nominal frame delay performance	99.5
P_r	Percentile defining frame delay range performance	99.99
Objective	**Description**	**Example Value**
MFD	Mean frame delay performance objective	200 ms
FD	Frame delay performance objective	230 ms
FDR	Frame delay range performance objective	40 ms

Note: *As with all performance attributes, four key caveats apply:*

- *Performance attributes apply to **green service frames only**. They say nothing about delivery of yellow service frames.*
- *Performance attributes are measured over an agreed subset of **ordered UNI pairs**.*
- *When measurement is over multiple ordered UNI pairs, the **worst case** (lowest performing) **UNI pair** defines the measurement.*
- ***Maintenance intervals are excluded** from performance measurements*

4.6.8 Inter-Frame Delay Variation Performance

Inter-Frame Delay Variation (IFDV) for any two green service frames is the difference between their one-way frame delays.

Note: *MEF documents published before MEF 10.2 (including MEF 6.1, 14, 15, and 19) refer to **Inter-Frame Delay Variation** as "Frame Delay Variation" (FDV).*

Note: *Please do not call it jitter. Jitter is a well-defined term in electronics, not in network communications.*

Per MEF 10.2, **Inter-Frame Delay Variation (IFDV) performance** is defined in an SLS for a particular CoS by specifying values for four parameters and one performance objective:

Parameter	Description	Example Value
T	The time interval over which the SLS applies	30 days
S	Non-empty subset of ordered UNI pairs	All
P	The performance percentile for which IFDV is defined	99.9
Δt	The separation between frame pairs for which IFDV is defined	1 sec
Objective	**Description**	**Example Value**
IFDV	IFDV performance objective	32 ms

Note: *As with all performance attributes, four key caveats apply:*

- *Performance attributes apply to **green service frames only**. They say nothing about delivery of yellow service frames.*
- *Performance attributes are measured over an agreed subset of **ordered UNI pairs**.*
- *When measurement is over multiple ordered UNI pairs, the **worst-case** (lowest performing) **UNI pair** defines the measurement.*
- ***Maintenance intervals are excluded** from performance measurements.*

4.6.9 Example of EVC Performance Attribute Specification

The following table (based on Table 38 in MEF 6.1) shows an example of how the EVC performance service attribute might be specified for a service involving three CoS.

EVC Performance Attribute	Parameter/Objective	Class of Service Offering		
		Krypton	Argon	Neon
CoS ID (Priority Code Point value)	N/A	PCP=5	PCP=4	PCP=0
Frame Delay (FD)	Subset of ordered UNI pairs (S)	All	All	All
	FD Objective	X ms	Y ms (Y>X)	Z ms (Z>Y)
	MFD Objective	N/S	N/S	N/S
	FDR Objective	N/S	N/S	N/S
	Percentile (P_d)	95%	95%	95%
	Percentile (P_r)	N/A	N/A	N/A
	Time Interval (T)	1 hour	1 hour	1 hour

EVC Performance Attribute	Parameter/Objective	Class of Service Offering		
		Krypton	Argon	Neon
Inter-Frame Delay Variation (IFDV)	Subset of ordered UNI pairs (S)	All	All	All
	IFDV Objective	Q ms	N/S	N/S
	Percentile (P)	95%	N/S	N/S
	Time Interval (T)	1 hour	N/S	N/S
	Pair Interval (Δt)	100 ms	N/S	N/S
Frame Loss Ratio (FLR)	Subset of ordered UNI pairs (S)	All	All	All
	FLR Objective	A %	B% (B>A)	C% (C>B)
	Time Interval (T)	1 hour	1 hour	1 hour
Availability	Subset of ordered UNI pairs (S)	All	All	All
	Availability Objective	α%	β% ($\beta<\alpha$)	γ% ($\gamma<\beta$)
	Time Interval (T)	1 month	1 month	1 month
	Number of consecutive small time intervals (n)	10	10	10
	Small time interval (Δt)	2 minutes	2 minutes	2 minutes
	Unavailability frame loss ratio threshold (C)	0.1	0.3	0.75
Resiliency	Subset of ordered UNI pairs (S)	All	All	N/S
	HLI Objective	50	100	N/S
	CHLI Objective	10	20	N/S
	Time Interval (T)	1 month	1 month	N/S
	Small time interval (Δt)	2 minutes	2 minutes	N/S
	Unavailability frame loss ratio threshold (C)	0.1	0.3	N/S
	The number of consecutive small time intervals for assessing CHLI, where $p < n$	3	3	N/S

4.7
Review

1. At UNIs, which three of the following are MEF-permissible CoS identifiers? (Choose three.)

    a. Ethernet DEI bit

    b. Differentiated services code point (DSCP) in the IP header

    c. PCP bits in IEEE 802.1Q customer VLAN tag (CE-VLAN tag)

    d. Ethernet CFI bit

    e. EVC identifier

2. Performance monitoring estimates frame delivery performance for:

    a. All service frames admitted to the EVC

    b. Yellow service frames

    c. All service frames, except L2CP service frames

    d. Green service frames, per CoS

    e. Green and Yellow service frames

3. Which three performance objectives are defined using a sliding window? (Choose three.)

    a. Availability

    b. Frame Delay (FD)

    c. Mean Frame Delay (MFD)

    d. Inter-Frame Delay Variation (IFDV)

    e. High Loss Intervals (HLI)

    f. Consecutive High Loss Intervals (CHLI)

4. What is the maximum number of ordered UNI pairs in an EP-Tree service that has four leaf UNIs and two root UNIs?

    a. 6

    b. 8

    c. 9

d. 10

e. 18

5. Which one of the following statements is false?

a. The SLS is considered part of the SLA.

b. The SLS contains technical details in engineering terms.

c. The SLA includes all business agreements related to the service.

d. The SLA is considered part of the SLS.

6. The one-way Frame Delay performance objective for an EP-LAN service is specified as 80 ms. Which statement best describes the meaning of this performance objective?

a. During an agreed time period, green service frames on average should spend no more than 80 ms in transit between CEs.

b. For each ordered UNI pair in the pre-agreed subset of UNI pairs, the average time a green service frame spends in transit between CEs should not exceed 80 ms.

c. No service frame should spend more than 80 ms in transit between CEs.

d. For each ordered UNI pair in the set of all ordered UNI pairs, the average time a green service frame spends in transit between CEs should not exceed 80 ms.

e. No green service frame should spend more than 80 ms in transit between CEs.

7. A Carrier Ethernet service for a bank needs to deliver critical data that cannot be dropped. Bandwidth profiles should be set with:

a. CIR>0, CBS=0, EIR=0, EBS=0

b. CIR>0, CBS>0, EIR=0, EBS=0

c. CIR>0, CBS>0, EIR>0, EBS>0

d. CIR=0, CBS>0, EIR=0, EBS=0

e. CIR=0, CBS=0, EIR>0, EBS>0

8. FLR (Frame Loss Ratio) performance is determined by:

a. [1 + (green service frames delivered)/(green service frames sent)] ×100

b. [1 + (total service frames delivered)/(total service frames sent)] ×100

c. [1 – (total service frames delivered)/(total service frames sent)] ×100

d. [1 – (green service frames delivered)/(green service frames sent)] /100

e. [1 – (green service frames delivered)/(green service frames sent)] ×100

9. A Service Level Agreement (SLA) is which one of the following?

a. A contract between the subscriber and service provider specifying a business agreement

b. An agreement stating service-level commitments in engineering terms

c. A technical specification for the service

d. A specification of service levels offered by the service provider

e. An agreement about service performance attributes, including frame delay, frame delay variation, frame loss ratio, and availability

10. Given a continuous stream of service frames with rate greater than EIR + CIR bits per second, approximately how many bytes of service frames will the bandwidth profile algorithm declare to be Yellow over a time interval of T seconds.

a. EIR * T bytes

b. EIR/8 * T bytes

c. CIR/8 * T bytes

d. (CIR+EIR) * T bytes

e. (EIR-CIR)/8 * T bytes

11. A Subscriber requires an EVC that supports three Classes of Service: *Silver*, *Gold*, and *Diamond*. Which one of the following schemes is acceptable for identifying CoS at UNIs using PCP bit values (0-7)?

a. Silver (5); Gold (6); Diamond (7)

b. Silver (7, 6); Gold (5, 4); Diamond (3, 2, 1, 0)

c. Silver (6); Gold (4); Diamond (2)

d. Silver (0,1, 2); Gold (5); Diamond (6)

e. Silver (4); Gold (5); Diamond (4)

12. Which one of the following statements is <u>not</u> true?

a. A performance attribute can include more than one performance objective.

b. Performance is judged by comparing the measured value of the performance metric to the performance objective.

c. A performance attribute can include more than one performance metric.

d. A performance metric is defined by a set of performance objectives (possibly null) and one performance parameter.

13. When a performance attribute is measured over an agreed subset of UNI pairs (more than one UNI pair), how is the final value of the performance attribute calculated?

    a. By taking the best case (highest performing) UNI pair measurement

    b. By averaging the UNI pair measurements

    c. By taking the worst-case (lowest performing) UNI pair measurement

    d. By determining the median value within the set of UNI pair measurements

14. Which set of parameter settings ensures that the bandwidth profile algorithm will never declare a service frame green?

    a. CIR>0, EIR=0

    b. EBS=0, CBS>0

    c. CBS=0, EBS>0

    d. EIR>0, EBS>0

    e. EBS>0, CBS>0

15. Four EVCs are service multiplexed to a UNI that is configured with the following per-UNI ingress bandwidth profile.

CIR (Committed Information Rate) = 200 Mb/s
CBS (Committed Burst Size) = 20,000 bytes
EIR (Excess Information Rate) = 800 Mb/s
EBS (Excess Burst Size) = 25,000 bytes
CF (Coupling Flag) = N
CM (Color Mode) = Color-Blind

Which statement is accurate?

    a. The rate of yellow frames for each EVC is limited to 200 Mb/s.

    b. The total rate of green frames for all EVCs is limited to 200 Mb/s.

    c. The size of green frames is limited to 25,000 bytes.

    d. The rate of green frames for each EVC is limited to 50 Mb/s.

    e. The size of yellow frames is limited to 20,000 bytes.

16. Availability for a particular one-way service between two UNIs is defined with a sliding window of n=13 small time intervals **Δ***t*. Which condition ensures that the first **Δ***t* in the sliding window is declared to be **available**? (No **Δ***t* is part of a maintenance interval.)

- **a.** The frame loss ratio (*flr*) for at least 7 **Δ***t*'s in the sliding window is greater than threshold *C*.
- **b.** The frame loss ratio (*flr*) for at least 7 **Δ***t*'s in the sliding window is less than threshold *C*.
- **c.** The frame loss ratio (*flr*) for all **Δ***t*'s in the sliding window is less than threshold *C*.
- **d.** The frame loss ratio (*flr*) for all **Δ***t*'s in the sliding window is greater than threshold *C*.

17. What happens to an Ethernet service frame that is that is marked Green by the bandwidth profile algorithm?

- **a.** It is discarded.
- **b.** It is placed in a higher class of service.
- **c.** It becomes drop-eligible.
- **d.** It is forwarded.
- **e.** It is buffered.

18. If a service frame is marked Yellow at MEN (CEN) ingress, which one of the following statements is true?

- **a.** The subscriber is exceeding CIR/CBS, but not EIR/EBS.
- **b.** The subscriber is exceeding CIR/CBS and EIR/EBS.
- **c.** The subscriber is exceeding EIR/EBS, but not CIR/CBS.
- **d.** The subscriber is not exceeding EIR/EBS or CIR/CBS.

19. Which two performance metrics are defined as a percentage?

- **a.** *Availability* and *Frame Loss Ratio (FLR)*
- **b.** *High Loss Interval (HLI)* and *Consecutive High Loss Interval (CHLI)*
- **c.** *Inter-Frame Delay Variation (IFDV)* and *Mean Frame Delay (MFD)*
- **d.** *Frame Delay (FD)* and *Mean Frame Delay (MFD)*

20. Which four services allow specification of per-CoS Egress Bandwidth Profiles at UNIs? (Choose four.)

- **a.** EPL
- **b.** EP-LAN

c. EP-Tree

d. EVPL

e. EVP-LAN

f. EVP-Tree

21. What is the maximum number of ordered UNI pairs in a EP-LAN service that includes four UNIs?

a. 4

b. 6

c. 8

d. 12

e. 16

Related Links

Answers: How MEF Services are Defined (Part 2)

5

UNI Requirements

In this chapter:

The user-to-network interface (UNI) serves both as a boundary between the subscriber and the service provider (point of demarcation) and as a point of connection (or link) between the subscriber and the service provider.

So far, in describing the MEF service definitions, we have focused on the UNI as a point of demarcation:

- The UNI locates where service begins and ends.
- The UNI locates where most service attributes are defined (*UNI service attributes* and *EVC per UNI service attributes*).

Note: *The UNI demarcation point is not abstract or conceptual. The MEF allows the UNI demarcation point to be located anywhere between the CE and the MEN (CEN), but its location must be precisely specified in the SLA. Typically, the UNI demarcation point is a port on the CE or a port on the MEN (CEN) edge device that connects to the CE.*

Now we focus on the UNI as a point of connection (or link) between the subscriber and the service provider.

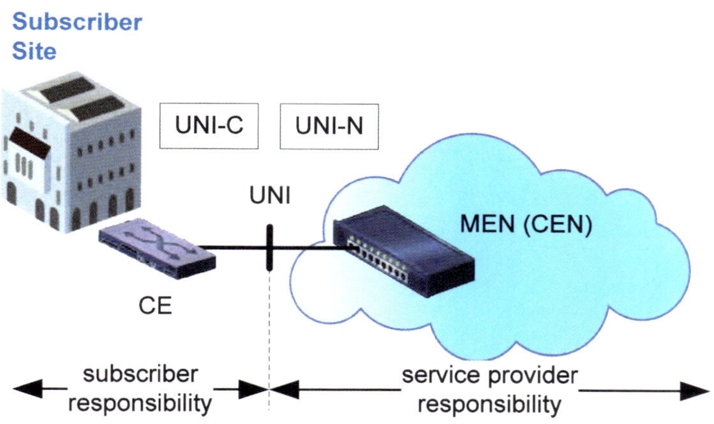

On either side of the UNI demarcation point, the MEF defines two functional elements:

- **UNI-C (UNI Client)**–Represents all of the subscriber-side functions required to connect the CE to the MEN (CEN). Individual functions in a UNI-C are entirely in the subscriber domain but may in some cases be managed by the service provider.
- **UNI-N (UNI Network)**–Represents all of the MEN (CEN)-side functions required to connect the MEN (CEN) to the CE. The individual functions in a UNI-N are entirely in the service provider/network operator domain.

Together the UNI-C and UNI-N provide all linking functionality, including IEEE 802.3 Ethernet PHY/MAC interface support and additional data, management, and control plane functions defined by the MEF.

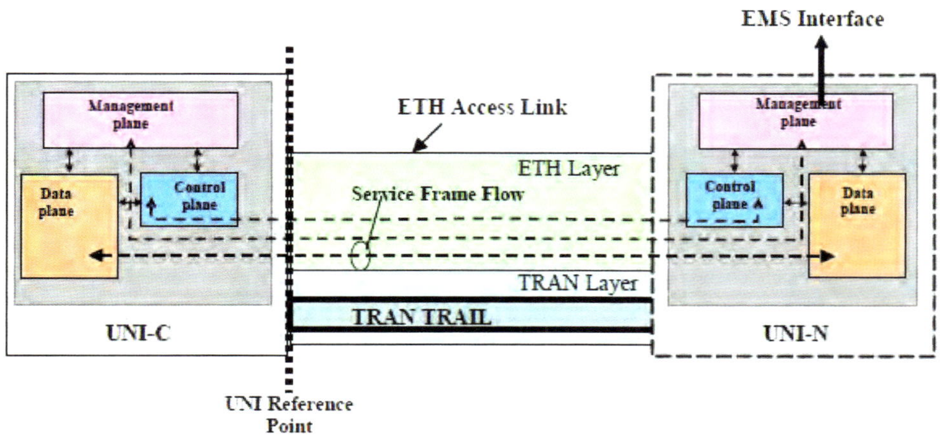

Reproduced with permission of the Metro Ethernet Forum

Figure 34: UNI Functionality

MEF Specifications and UNI Functionality

The MEF carefully defines UNI functionality to promote interoperability between equipment suppliers. The UNI-C (CE equipment) and UNI-N (MEN/CEN edge equipment) may come from different manufacturers, but must still work together.

As shown in the following figure, quite a few MEF specifications are devoted to UNI functionality.

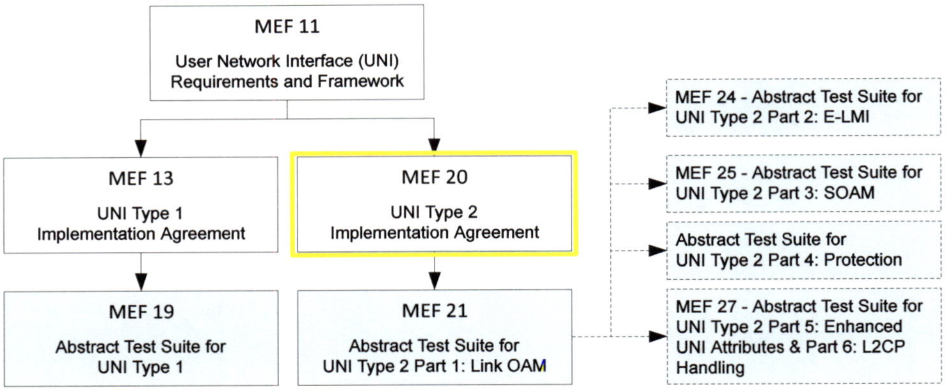

Of these, MEF 20 is the most important for the MEF-CECP 2.0 exam.

Type 1 and Type 2 UNIs

Currently, the MEF defines two types of UNIs:

- UNI Type 1–Provides basic functionality and is only manually configurable
- UNI Type 2–Adds the following functionalities: SOAM, Enhanced UNI attributes, L2CP handling, E-LMI (optional), LOAM (optional), and Protection (optional)

UNI Type 2 is backward compatible with UNI Type 1.

If a Type 2 UNI includes all of the *optional* functionalities, it is classified as Type 2.2. Otherwise, it is classified at Type 2.1.

Note: *Type 1 UNI requirements (specified in MEF 13) basically say that the UNI must provide the functionality needed to support MEF services as defined in MEF 6.1 and MEF 10.2. Type 1 UNI functionality was delivered by enterprise-class Ethernet equipment before carrier-class Ethernet equipment was available. Today, all carrier-class Ethernet equipment delivers Type 2 UNI functionality. Here,* ***enterprise-class equipment*** *refers to Ethernet switches that are intended to be used on an enterprise LAN.* ***Carrier-class equipment*** *refers to equipment designed to support Ethernet as a metro or wide-area service.*

Related Links

UNI Functionality

5.1 Type 2 UNI Requirements

In this section:

The following list summarizes requirements for Type 2 UNIs.

SOAM, Service operations, administration, and management (OAM) – allows the subscriber and the service provider to monitor and diagnose end-to-end connectivity and performance

Enhanced UNI Attributes – adds support for the newest UNI attributes, including bandwidth profile per egress UNI and MTU size, as defined in MEF 10.2 and MEF 6.1

L2CP Handling – governs the passing, processing, or filtering of Layer 2 control protocols to the EVC

E-LMI (optional), Ethernet Local Management Interface – allows UNI-C to retrieve EVC status and service attributes from UNI-N as specified in MEF 16

LOAM (optional), Link OAM – allows the subscriber and the service provider to monitor and diagnose link-level connectivity issues between UNI-C and UNI-N per clause 57 of IEEE 802.3

Link Protection (optional) – supports link aggregation per clause 43 of IEEE 802.3 to protect the UNI against port failure

Note: *Type 2 UNIs support all Type 1 UNI requirements and are backward compatible with Type 1 UNIs.*

5.1.1 SOAM Requirement for Type 2 UNIs

Note: *This requirement is mandatory for UNI Type 2.1 and 2.2.*

A Type 2 UNI is required to support Service operations, administration, and management (OAM) as defined by the MEF based on ITU-T Y.1731 and IEEE 802.1Q[12] standards. SOAM includes mechanisms for end-to-end connectivity fault detection and performance management.

SOAM functionality is covered later in this study guide in a larger context that includes more than just UNI functionality. For now, just note that SOAM support is an important requirement for Type 2 UNIs.

Related Links

Service OAM Overview

12 IEEE 802.1ag-2007 referenced in various MEF specifications has been incorporated into and superseded by IEEE 802.1Q-2011.

5.1.2 Enhanced UNI Attributes Requirement for Type 2 UNIs

Note: *This requirement is mandatory for UNI Type 2.1 and 2.2.*

Type 2 UNIs must support the newest UNI attributes, including *bandwidth profile per egress UNI* and *MTU size*, as defined in MEF 10.2 and MEF 6.1. Basically, this requirement adds functionality needed to support enhancements to the MEF service definitions (MEF 6.1 and MEF 10.2) that occurred after Type 1 UNI requirements were defined.

Related Links

MTU Size Service Attributes
Bandwidth Profile Service Attributes

5.1.3 L2CP Handling Requirement for Type 2 UNIs

Note: *This requirement is mandatory for UNI Type 2.1 and 2.2.*

Type 2 UNIs must support the L2CP processing as defined in MEF 10.2 and MEF 6.1.

Recall that the *UNI L2CP Processing Attribute* was described in a previous lesson. Type 2 UNIs must include functionality to support all four values of this attribute: *Peer*, *Discard*, *Pass to EVC*, or *Peer and Pass to EVC*.

Related Links

L2CP Processing Service Attributes

5.1.4 LOAM Requirement for Type 2.2 UNIs

Note: *This requirement is optional for UNI Type 2.1 and is mandatory for UNI Type 2.2.*

Type 2.2 UNIs are required to support Link operations, administration, and management (OAM) as defined in MEF 20, based on clause 57 of IEEE 802.3. LOAM allows the subscriber and the service provider to monitor and diagnose link-level connectivity issues between UNI-C and UNI-N.

LOAM functionality is covered later in this study guide within a larger context that includes all types of links (not just UNI-C to UNI-N links). For now, just note that LOAM support is an important requirement for Type 2.2 UNIs.

Related Links

Link OAM

5.1.5
E-LMI Requirement for Type 2.2 UNIs

Note: *This requirement is optional for UNI Type 2.1 and is mandatory for UNI Type 2.2.*

The Ethernet Local Management Interface (E-LMI) protocol allows UNI-C to acquire configuration information from UNI-N so that it can automatically configure itself.

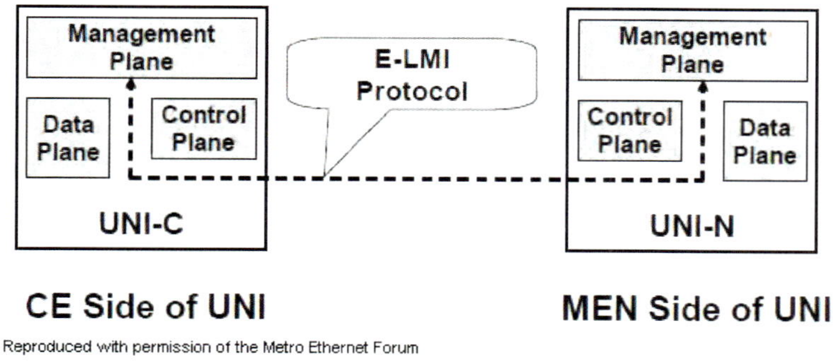

Figure 35: E-LMI Termination and Functional Components (Figure 2 in MEF 16)

The E-LMI protocol also notifies the UNI-C of changes to EVC status, including:

- Notification of the addition of an EVC
- Notification of the deletion of an EVC
- Notification of a change in EVC state (Active, Not Active, or Partially Active)
- Communication of UNI and EVC attribute values

Technical requirements for E-LMI are specified in MEF 16.

Note: *The E-LMI protocol is based on ITU-T Q.933, X.36 and other relevant recommendations as well as Frame Relay Local Management Interface (FR-LMI) Implementation Agreement document defined by the Frame Relay Forum and related ITU-T recommendations.*

5.1.6
Link Protection Requirement for Type 2.2 UNIs

Note: *This requirement is optional for UNI Type 2.1 and is mandatory for UNI Type 2.2.*

To be Type 2.2, UNI equipment (UNI-C or UNI-N) must be capable of protecting against port failure, using link aggregation as currently defined in IEEE 802.1AX (or as previously defined in IEEE 802.3ad or clause 43 of IEEE 802.3).

Note: *MEF 20 cites clause 43 of IEEE 802.3 for link aggregation requirements.*

Link aggregation associates two or more links (port-cable-port connections) between two neighboring NEs to form a link aggregation group (LAG) that functions like a single link.

There are two types of LAGs:

- 1:1 LAG—Association of two links, one active and one standby, for 1:1 protection.

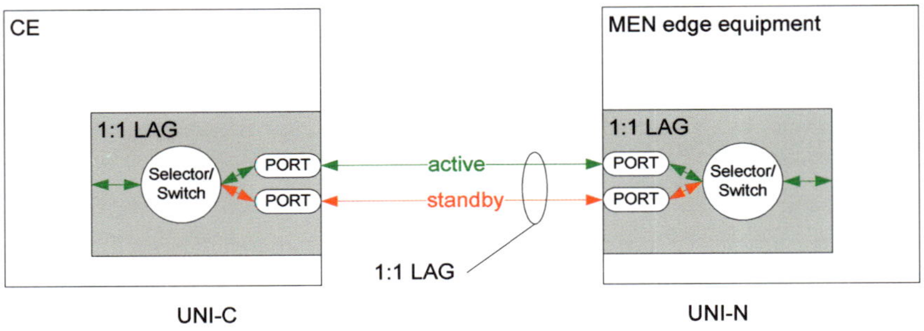

Figure 36: 1:1 LAG

- 0:N LAG—Association of N links to form a pseudo link with greater bandwidth. This mechanism is primarily intended to support link bandwidth scalability but also provides limited protection (if one link fails, traffic is distributed over the remaining links).

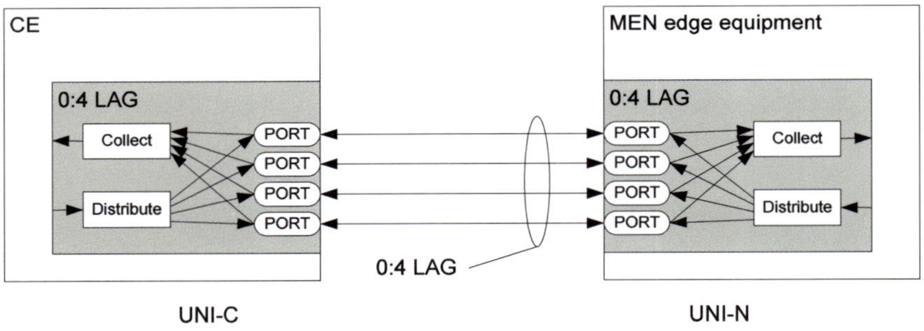

Figure 37: 0:4 LAG

Link bandwidth scalability refers to the problem of supporting links in a graduation of sizes (1, 2, 3, ..., 10 Gb/s) using ports that are available in only a few sizes (1 Gb/s, 10 Gb/s). For example, 0:3 LAG of three 1 Gb/s ports provides a link with a bandwidth of 3 Gb/s.

0:N LAG includes hashing algorithms to distribute traffic evenly between ports and to avoid reordering packets within an information stream.

5.2
Summary

This lesson described MEF requirements for UNI functionality. Key points include:

- The MEF carefully defines UNI functionality to promote interoperability between equipment suppliers.
- MEF 20 defines Type 2 UNI requirements and subdivides those requirements to define Type 2.1 and Type 2.2 UNIs.
- The following features are mandatory for all Type 2 UNIs:
  - Backward compatibility with Type 1 UNIs
  - Service OAM
  - Enhanced UNI Attributes
  - L2CP Handling
- The following features are optional for Type 2.1 UNIs but mandatory for Type 2.2 UNIs:
  - Link OAM
  - E-LMI
  - LAG protection
- SOAM allows the subscriber and service provider to monitor end-to-end service connectivity and performance.
- The L2CP handling requirement refers to support for L2CP processing as defined in MEF 10.2 and MEF 6.1.
- E-LMI enables autoconfiguration of UNI-C (CE) to support MEF services.
- LOAM allows the subscriber and service provider to monitor and diagnose UNI connectivity issues.
- Type 2.2 UNIs are required to support link aggregation for the purpose of port failure protection.

5.3
Review

1. A MEF Type 2.2 UNI is required to support which type of protection?

    a. Layer 2 Control Protocol (L2CP)

    b. Link OAM per clause 57 of IEEE 802.3

    c. G.8031 Ethernet Linear Protection Switching

    d. Link aggregation per clause 43 of IEEE 802.3

    e. G.8032 Ethernet Ring Protection Switching

2. Which of the following provides port protection?

    a. L2CP

    b. Link aggregation

    c. E-LMI

    d. Link OAM

    e. Type 1 UNI

3. Which two of the following statements about the Ethernet Local Management Interface (E-LMI) are true? (Choose two.)

    a. E-LMI protects against port failure using link aggregation.

    b. E-LMI is required for Type 2.1 UNIs.

    c. E-LMI notifies the UNI-C of changes to EVC status.

    d. E-LMI allows the subscriber and the service provider to monitor and diagnose link-level connectivity issues between UNI-C and UNI-N.

    e. E-LMI allows UNI-C to acquire configuration information from UNI-N so that it can automatically configure itself.

4. Which three of the following requirements are mandatory for Type 2.2 UNIs, but optional for Type 2.1 UNIs? (Choose three.)

    a. Link aggregation support

    b. Service OAM support

c. Link OAM support

d. Support of multiplexed UNI for services such as EVPL

e. Backward compatibility with UNI Type 1

f. E-LMI support

5. Which one of the following statements about Type 2 UNIs is not true?

a. Type 2 UNIs support Service OAM.

b. MEF 20 describes two types of Type 2 UNIs.

c. Type 2 UNI requirements apply only to the UNI-N (they are not applicable to the UNI-C).

d. Type 2 UNIs are backward compatible with Type 1 UNIs.

e. Type 2 UNIs support L2CP handling as defined in MEF 6.1 and MEF 10.2.

6. UNI Type 2.2 required protection mechanisms protect against which three types of failure? (Choose three.)

a. Port failure on UNI-N

b. Loss of timing

c. Loss of "green" service frames due to bandwidth oversubscription

d. E-LMI denial of service (DOS) attack

e. Port failure on UNI-C

f. Disconnection or failure of a cable connecting the CE and MEN (CEN)

Related Links

Answers: UNI Requirements

6

Extending MEF Services over Multiple Operator MENs

In this chapter:

Related Links

General Reference Model

6.1
Introduction

So far we have focused on defining MEF services in a simple business context involving two stakeholders:

- **Subscriber**–The organization purchasing the Carrier Ethernet service
- **Service Provider**–The organization providing the Carrier Ethernet service

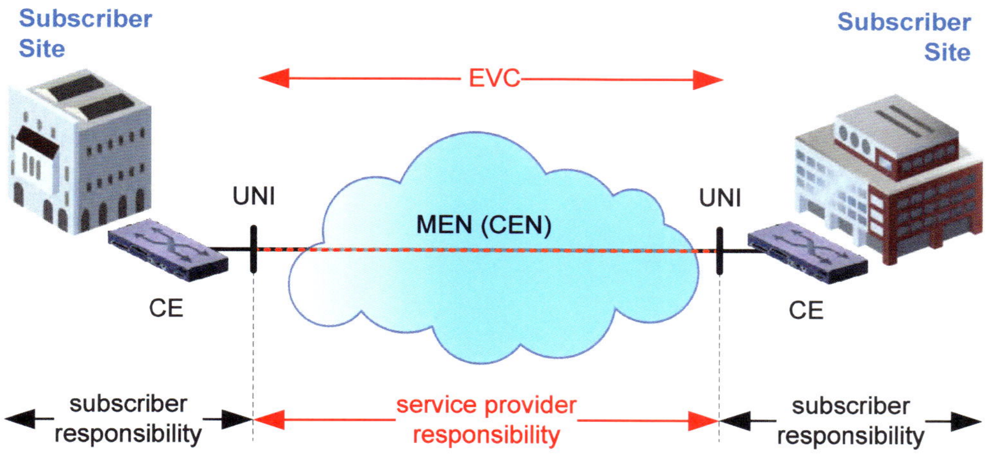

Figure 38: Basic Reference Model

However, in some cases, the service provider network may not be able to reach (or connect to) all of the subscriber sites, so collaboration is required. For this reason, MEF 26.1 extends the basic reference model to include additional participants called operators.

- **Operator**–The administrative organization responsible for a particular network in a multi-network Carrier Ethernet service implementation

The following figure shows a service implemented across three independent networks.

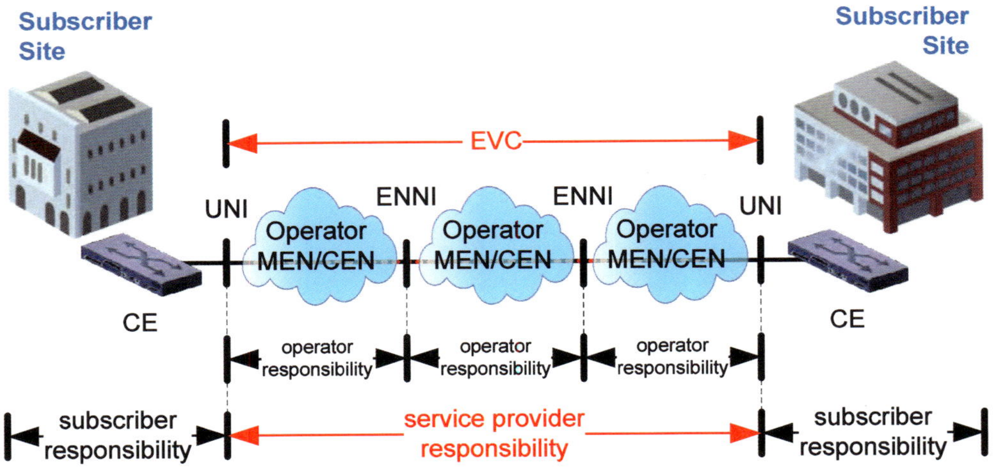

Figure 39: General Reference Model

The introduction of operators enhances the service implementation framework, but it does not change the Carrier Ethernet service from the subscriber's perspective:

- There is still only one Carrier Ethernet service provider responsible for the service as a whole.
- The Carrier Ethernet service is still implemented with exactly one EVC connecting two or more UNIs.
- The basic reference model is sufficient to define the service from the subscriber's perspective.
- The general reference model adds details of implementation that are only important to the service provider and subcontracting operators.

Although nothing changes from the subscriber's perspective, the service provider now has a new relationship with one or more operators. If an operator is internal to the service provider's business organization, the relationship can be informal. Otherwise, a formal business relationship is generally required. For this purpose the operator's contribution to the overall service is elevated to the status of an Ethernet service:

- **Operator Service**—An Ethernet service deployed by an operator and sold to a service provider

This lesson describes how the MEF defines operator services in general. The next lesson describes two standardized operator services (the **Access EPL** and the **Access EVPL**) that the MEF has defined for Ethernet access applications.

Related Links

MEF Model for Service Across an Operator MEN/CEN
E-Access Services

6.2
Terminology

In this section:

6.2.1
Service Provider and Operators

In MEF 26.1, an independent network that participates in providing a service is called an **Operator MEN** or an **Operator CEN** (not a MEN or a CEN). In this study guide, we follow the same convention:

- **MEN (CEN)**–A metro Ethernet network comprising a single administrative domain
- **Operator MEN (Operator CEN)**– An independent network that participates in providing the service

Other MEF specifications (and reference presentations) are sometimes less careful with terminology. For example, the following figure shows two service providers participating in a service.

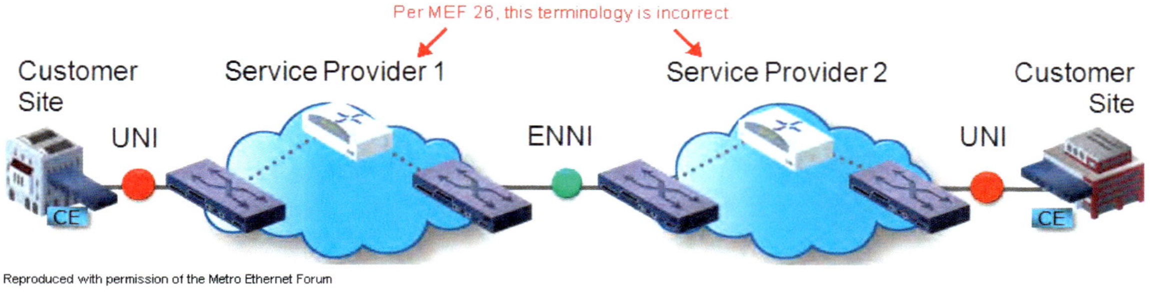

Figure 40: Incorrect Terminology

Terminology in the preceding figure is incorrect because a MEF service can have only one service provider (responsible to the subscriber for the end-to-end service as a whole). Knowing this, it is clear that the "service providers" shown in the preceding figure should be interpreted as operators.

Note: *Terminology in the preceding figure is arguably correct from a generic perspective (any organization that provides services can be called a service provider). However, the MEF defines service provider more narrowly as the organization responsible to the subscriber for the end-to-end service as a whole.*

The following figure shows an example of correct terminology. The collaborating networks are associated to operators (not service providers). The network labeled *Operator 1* also belongs to the service provider (responsible for the end-to-end service), but the other network does not.

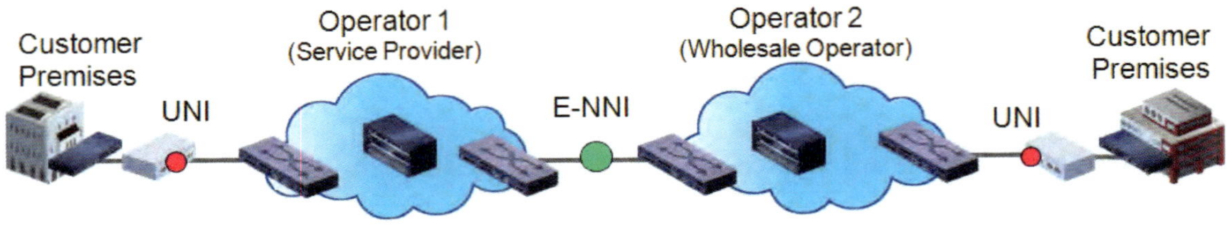

Figure 41: Correct Terminology

Normally, the service provider owns and operates one or more of the Operator MENs (Operator CENs) involved in the service. However, this is not a requirement. The service provider might not operate any of the Operator MENs (Operator CENs), but only coordinate operators and take responsibility for the overall service.

6.2.2 ENNI and ENNI-N

The MEF defines the ENNI (external network-to-network interface) as a reference point representing the boundary between two Operator MENs (Operator CENs).

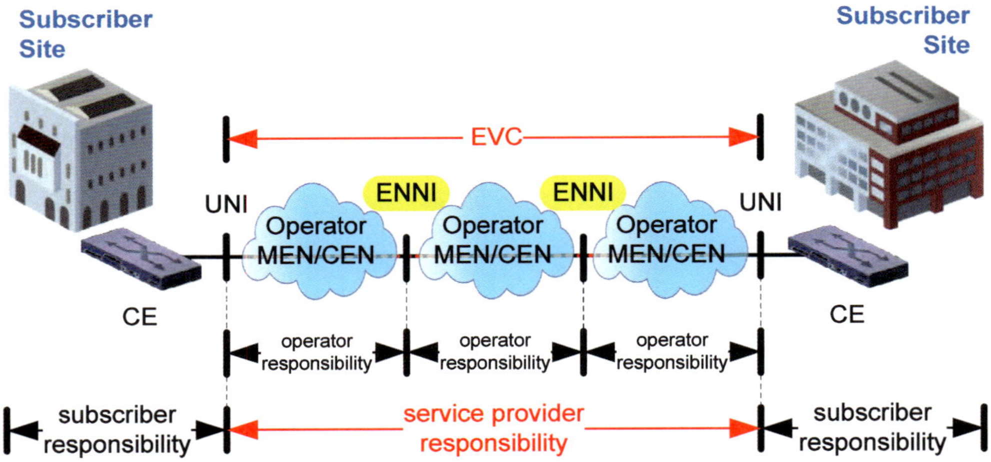

Figure 42: ENNI

Similar to the UNI, the ENNI serves two roles:

- As a boundary between two Operator MEN/CENs (point of demarcation)
- As a point of connection (or link) between two Operator MEN/CENs

The ENNI demarcation point is located at a specific place between two Operator MENs (Operator CENs). Typically, it is a port on one of the Operator MEN (Operator CEN) edge devices.

On each side of the ENNI demarcation point, the MEF defines a functional component, called an ENNI-N (one in each Operator MEN/CEN). Together the two ENNI-Ns support a IEEE 802.3 Ethernet PHY/MAC interface, plus additional functionality, described later in this lesson.

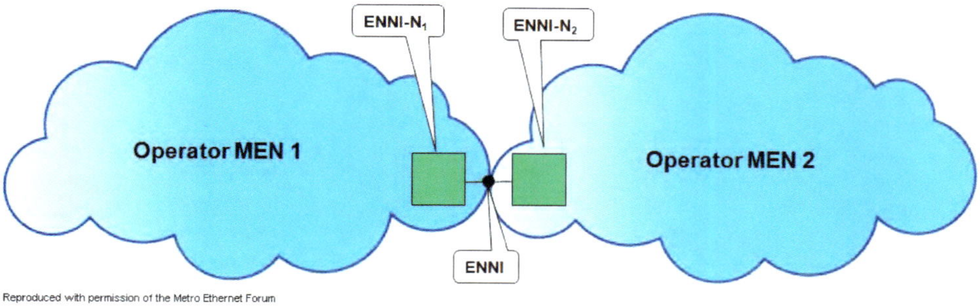

Figure 43: ENNI-N

6.2.3 OVC

An **OVC (Operator Virtual Connection)** is the association of two or more external interfaces (UNI or ENNI) within an Operator MEN, where at least one of the external interfaces is an ENNI. An OVC must include at least one ENNI because otherwise it would be an EVC.

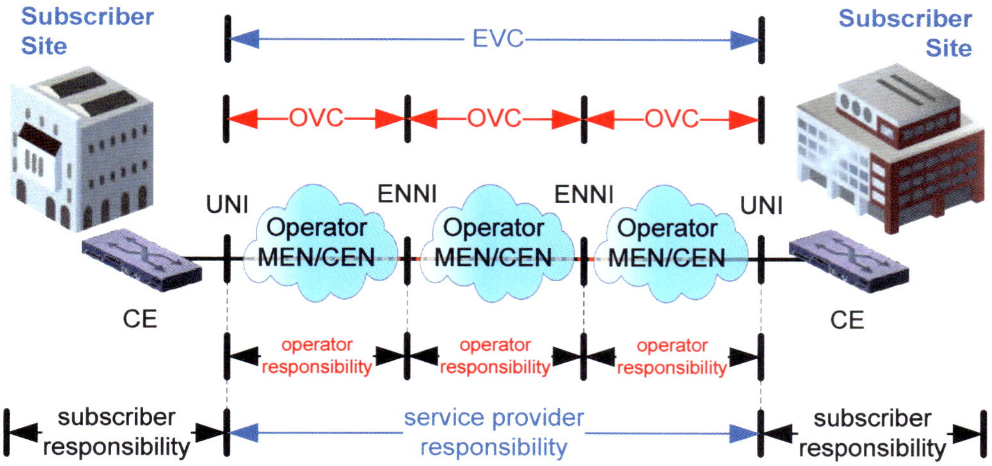

Figure 44: EVC Composed of Point-to-Point OVCs

OVCs can be type point-to-point, multipoint-to-multipoint, or rooted-multipoint. The following figure includes one multipoint-to-multipoint OVC and two point-to-point OVCs.

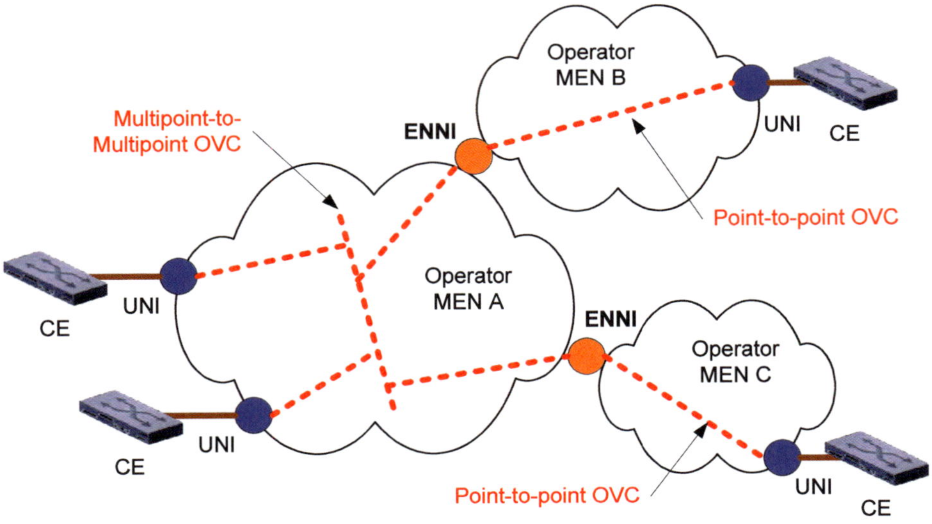

Figure 45: EVC with Multipoint-to-Multipoint OVC

Note: *The initial MEF-CECP exam did not include rooted-multipoint OVCs. They were introduced in MEF 26.1.*

An EVC is formed by the concatenation of OVCs. The EVC for the previous example looks as follows:

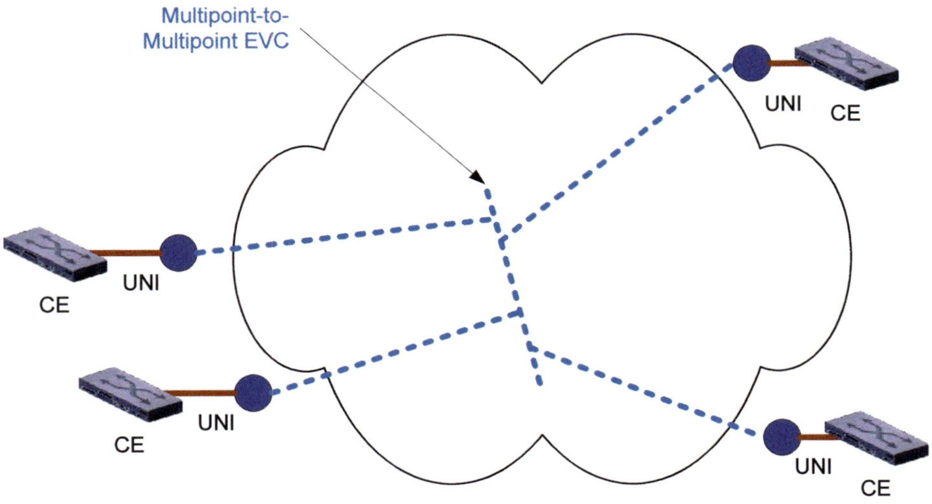

Figure 46: Multipoint-to-Multipoint EVC

6.2.4
OVC End Point

An **OVC End Point** associates an OVC with a specific external interface (UNI or ENNI).

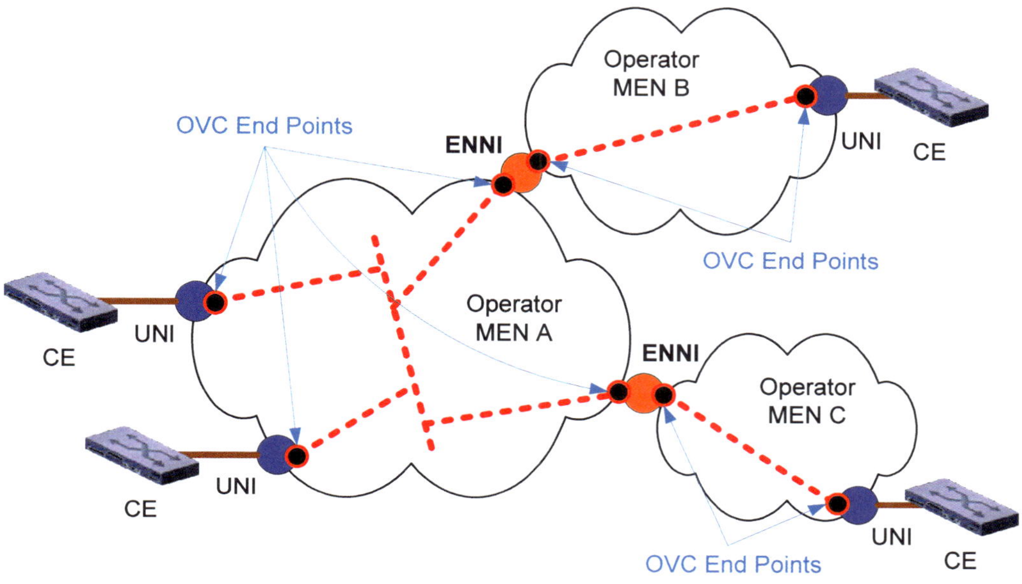

Figure 47: EVC End Points

The OVC end point is not the UNI or ENNI. UNIs and ENNIs are not OVC end points.

6.2.5
EI (External Interface)

Several MEF specifications, including MEF 26.1, use the term EI. It is defined as follows:

- **External Interface (EI)** – Either a UNI or an ENNI

6.2.6
Ingress and Egress Frames

- **Ingress ENNI frame**–An ENNI frame received by an Operator MEN/CEN is called an ingress ENNI frame (from the perspective of the receiving Operator MEN/CEN).

- **Egress ENNI frame**–An ENNI frame transmitted by an Operator MEN/CEN is called an egress ENNI frame (from the perspective of the transmitting Operator MEN/CEN).

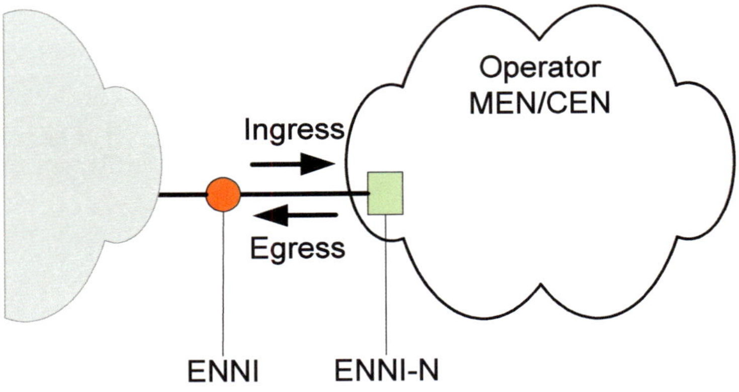

Figure 48: Ingress and Egress ENNI Frames

At an OVC end point:

- An **Ingress frame** is a frame **received by the OVC** through the interface (UNI or ENNI)
- An **Egress frame** is a frame **sent from the OVC** through the interface (UNI or ENNI)

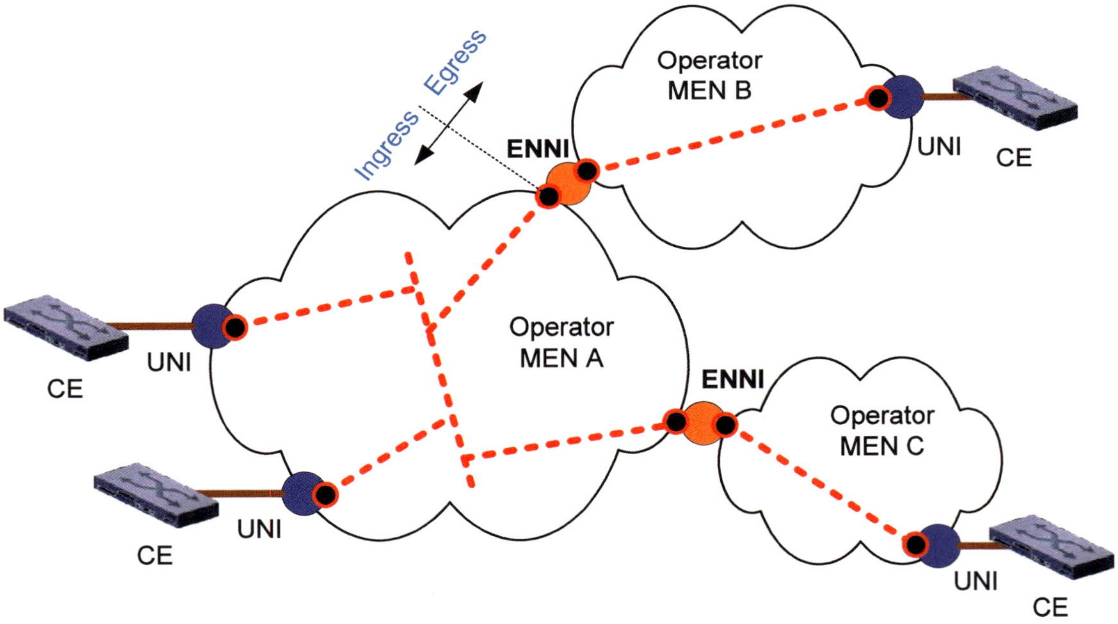

6.2.7
OVC End Point Role

Note: *The initial MEF-CECP exam did not include OVC end point roles. They were introduced in MEF 26.1 to support rooted-multipoint OVCs.*

Per MEF 26.1, every OVC end point is assigned an OVC end point role (Root, Leaf, and Trunk), but only the rooted-multipoint OVC supports more than one OVC end point role:

OVC Type	Supported OVC End Point Roles
Point-to-point	Root
Multipoint-to-multipoint	Root
Rooted-multipoint	Root, Leaf, and Trunk

Note: *For point-to-point and multipoint-to-multipoint OVCs, OVC end point role assignment is trivial. It is done for consistency only.*

Rooted-multipoint OVCs support three OVC end point roles (Root, Leaf, and Trunk) for the same reason that rooted-multipoint EVCs support two UNI designations (Root and Leaf). In both cases, the purpose is to identify E-Tree-service-related frame forwarding constraints.

The following example illustrates the use of OVC end point roles Root and Leaf, which are easy to explain compared to OVC end point role Trunk.

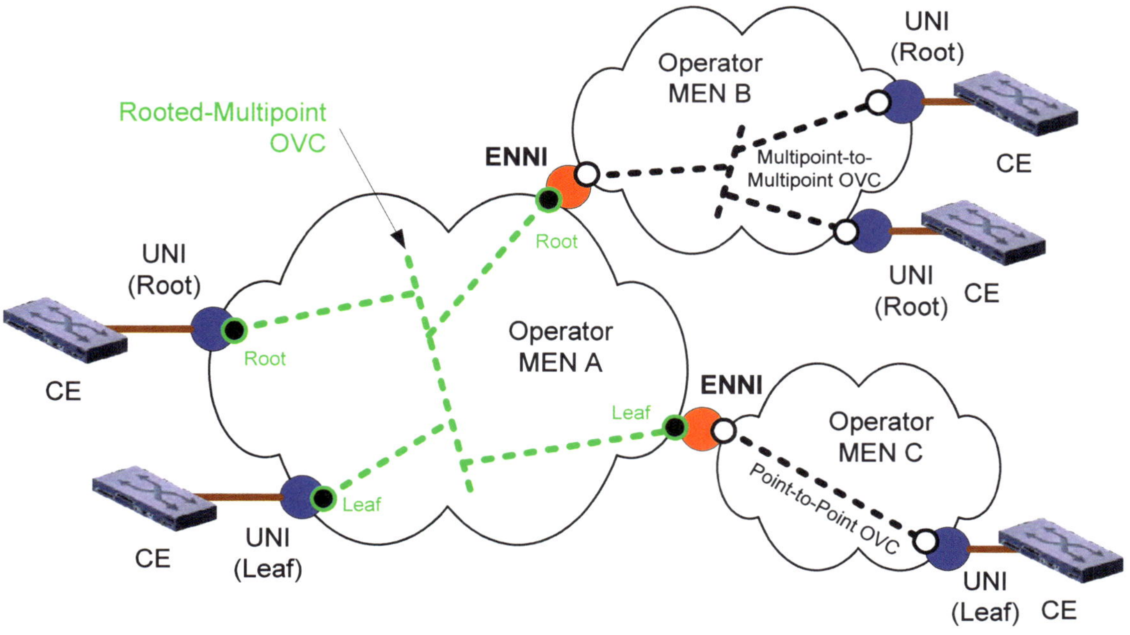

In this example:

- OVC end point role **Root** – Indicates that all frames ingressing to the OVC end point originated at a Root UNI and can be forwarded to any OVC end point (Root, Leaf, or Trunk)
- OVC end point role **Leaf**–Indicates that all frames ingressing to the OVC end point originated at a Leaf UNI and must not be forwarded to a Leaf OVC end point.

OVC end point role Trunk is used when ingress frames have mixed origin (some originating at a Root UNI and others originating at a Leaf UNI), as in the following example:

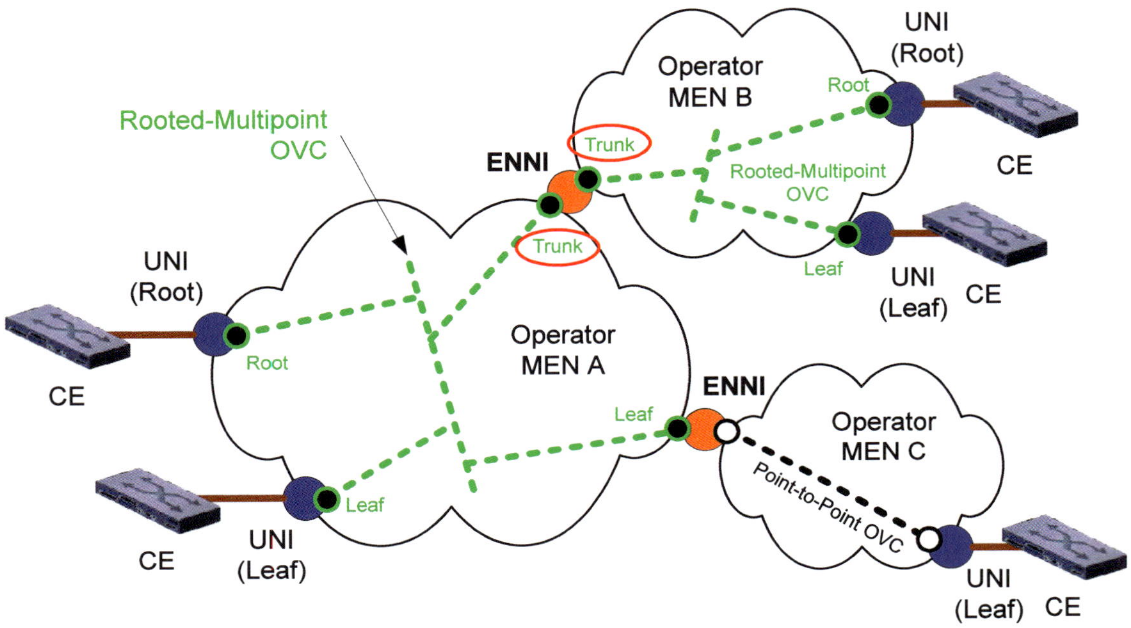

OVC end point roles are explained in more detail later in this lesson.

Related Links

Rooted-Multipoint OVCs
OVC Service Attributes

6.2.8 Hairpin Switch

The following figure illustrates a hairpin switch:

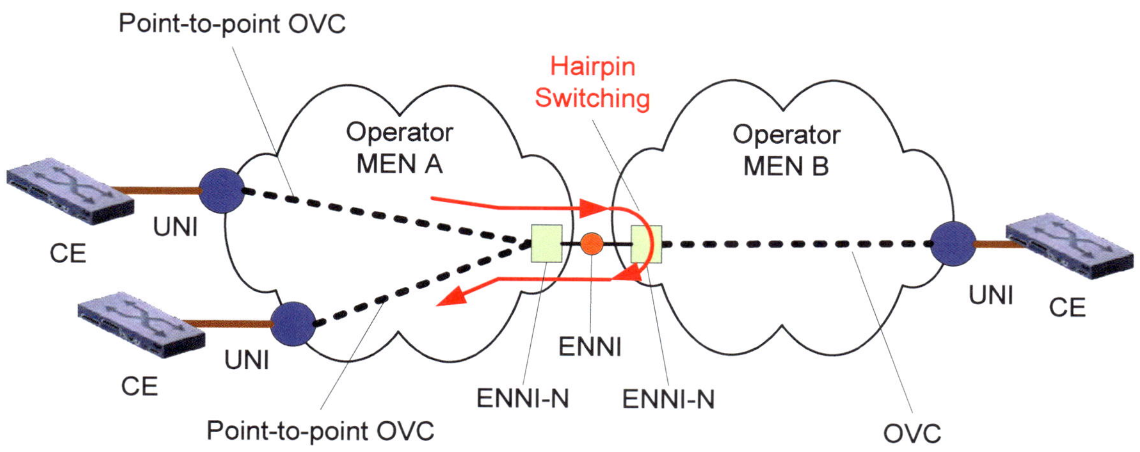

Figure 49: Hairpin Switching

ENNI-N in Operator MEN B receives an ingress ENNI frame from Operator MEN A and then returns the frame with changes (defined later in this lesson) that cause the frame to be associated to a different OVC upon return to Operator MEN A.

To understand why hairpin switching is needed, consider the following example application.

A subscriber requires the following EVP-LAN service.

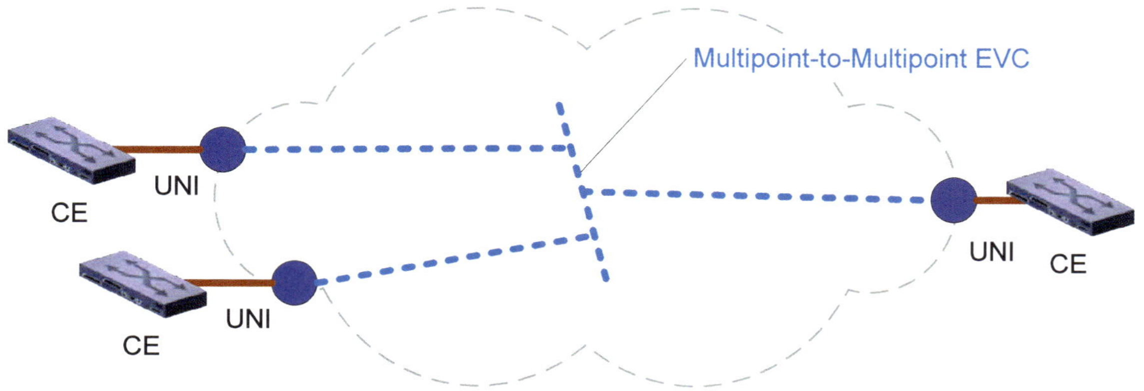

Figure 50: Example EVP-LAN Service

However, the service is implemented with three OVCs in two Operator MENs as follows.

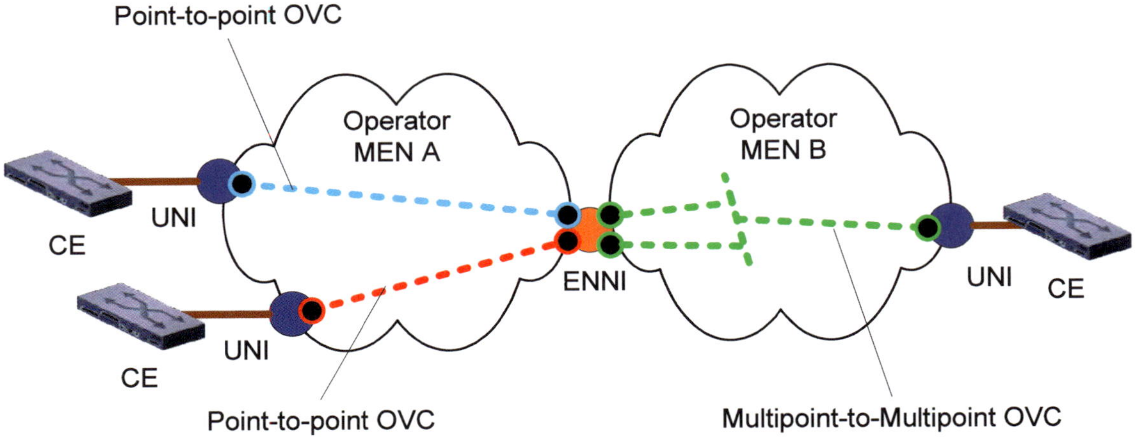

Figure 51: Implementation with Three OVCs

Because the two UNIs in Operator MEN A are not connected within Operator MEN A, hairpin switching (within Operator MEN B) is needed to support the exchange of service frames.

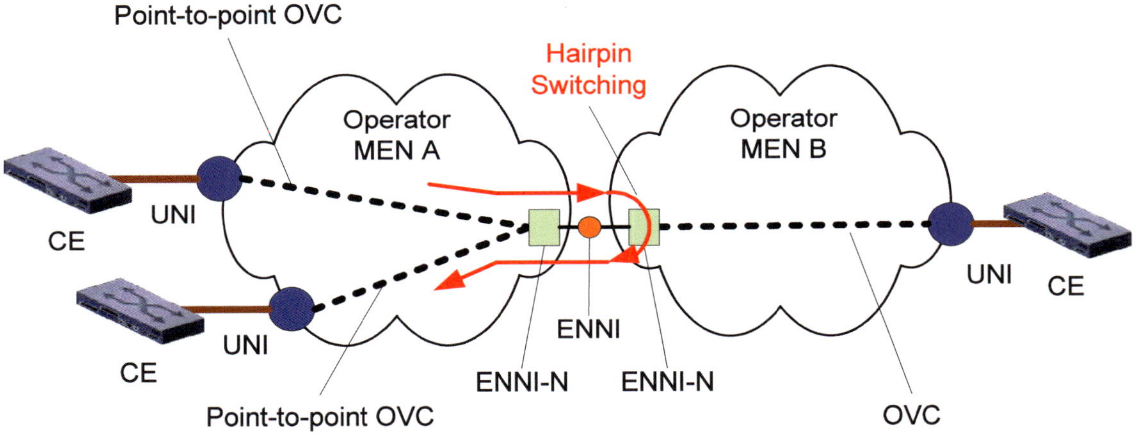

Figure 52: Hairpin Switching

Additional details are presented later in this lesson.

Related Links

Hairpin Switching

6.3
Service Handoff at ENNI

In this section:

Service handoff at ENNIs is key to defining services across Operator MEN/CENs. ENNIs are defined to be high bandwidth (1 Gb/s or 10 Gb/s) and to support multiple services.

The following figure shows two EPL services implemented across two Operator MENs. The ENNI must support both services (and, potentially, many others).

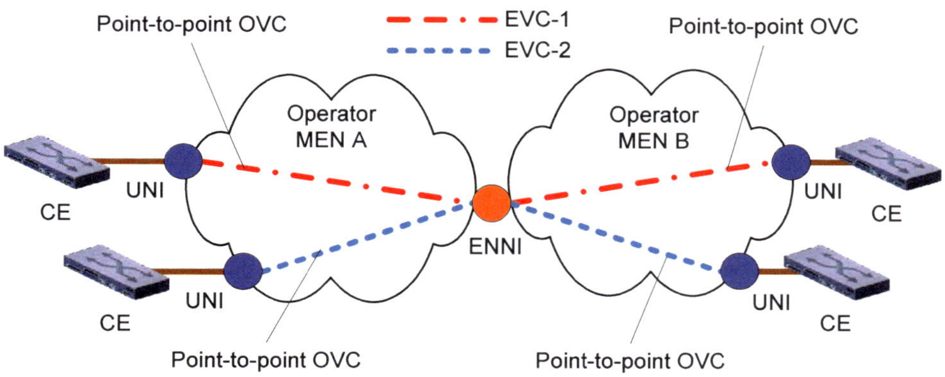

Figure 53: Two EPL Services

At the ENNI, principle concerns include:

- Distinguishing which ENNI frames belong to which OVC end point
- Propagating CoS (class of service) from one Operator MEN/CEN to the next
- Propagating Color (green/yellow drop-eligibility) from one Operator MEN/CEN to the next

Recall that service frames are assigned CoS and Color at UNIs per the SLA (service-level agreement) between the subscriber and the service provider:

- CoS is identified at UNIs in several ways: based on EVC, based on PCP bits in the CE-VLAN tag, based on DSCP in the IP header, or based on Layer 2 control protocol
- Color is assigned at UNIs based on Ingress bandwidth profile processing. If a frame is CIR/CBS-compliant, it is declared green and forwarded. Otherwise, if it is EIR/EBS-compliant, it is declared yellow and forwarded. Otherwise, it is dropped.

After CoS and Color are established at the ingress UNI, they typically remain fixed for the duration of the service frame's travel, UNI to UNI, through any number of Operator MEN/CENs.

MEF 26.1 requires ENNIs to support the IEEE 802.1ad (provider bridging) frame format for the purpose of propagating CoS, Color, and OVC end point mapping information between Operator MENs.

6.3.1 IEEE 802.1ad Ethernet Frames

IEEE 802.1ad-2005 extended the IEEE 802.1Q-2005 standard to support stacked VLAN tags as follows:

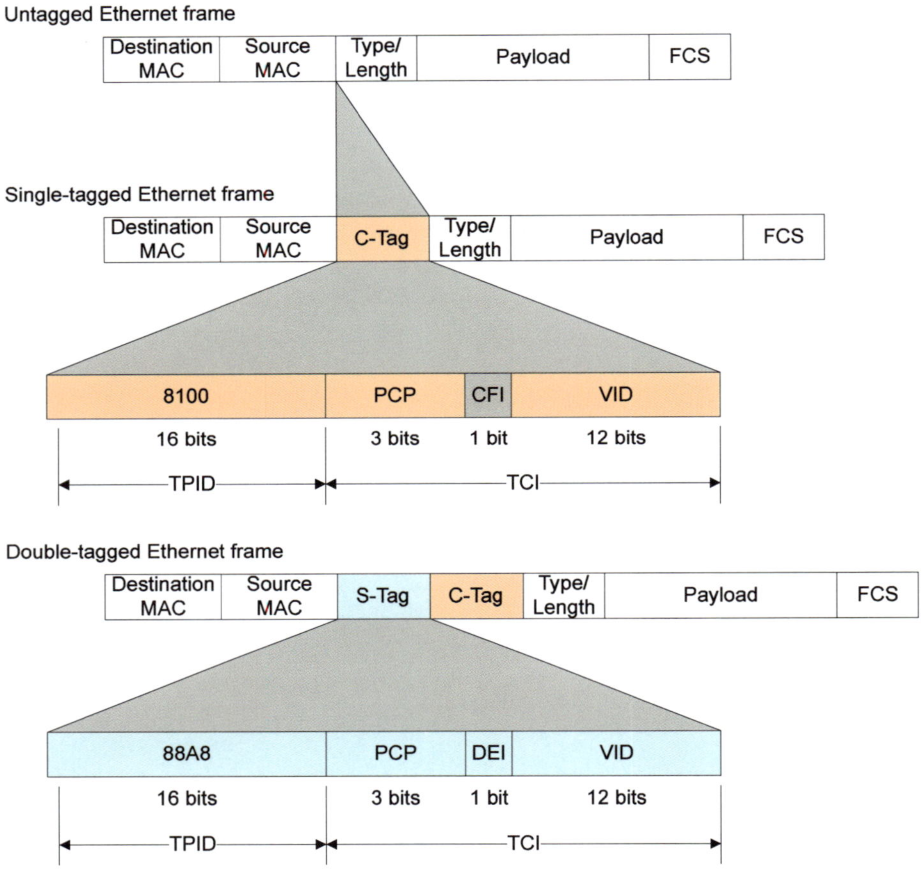

Figure 54: IEEE 802.1ad Ethernet Frame

Two types of VLAN tags are defined:

- **C-Tag**–Inner VLAN tag conforming to IEEE 802.1Q customer VLAN tag (the CE-VLAN tag)
- **S-Tag**–Outer VLAN tag that matches the IEEE 802.1Q customer VLAN tag but with the following differences:
    - TPID value is 88A8 (not 8100).
    - Drop-eligibility indicator (DEI) replaces CFI and is dedicated to indicating drop eligibility.

In MEF service applications, the C-Tag corresponds to the CE-VLAN tag, and the S-Tag is used to capture CoS, Color, and OVC end-point mapping information.

An S-Tag adds 4 bytes (32 bits) to an Ethernet frame.

Related Links
Size of Service Frame Compared to Payload
Layer 1 versus Layer 2 Ethernet Frames

6.3.2
Encoding Information in S-Tags

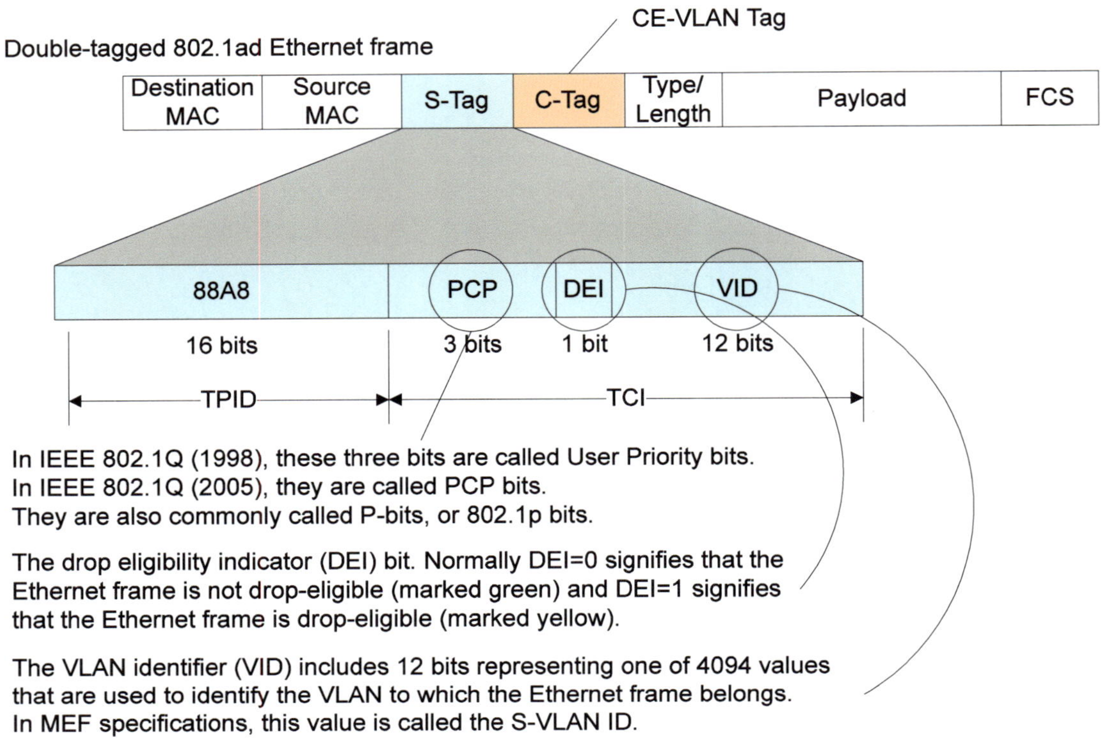

Figure 55: S-Tag Components

Note: *The 12-bit VID field allows 4096 values, but only 4094 values can be used for VLAN identification because VID values 0 and FFF are reserved for other uses.*

Per MEF 26.1, the following information is encoded in the S-Tag at ENNIs:

- **OVC End Point Identification**—Ingress ENNI frames are mapped to OVC end points based on S-VLAN ID value.
- **CoS identification**—Ingress ENNI frames are mapped to CoS based on S-Tag PCP value.

  Note: *Per MEF 26.1, for each OVC end point, each S-Tag PCP value (0–7) must be explicitly mapped to exactly one CoS. To explicitly discard frames, one CoS can be equated to discard.*

- **Color identification**—Color (green/yellow drop-eligibility) can be assigned to an ingress ENNI frame in either one of two ways:
  - Based on S-Tag DEI bit value
  - Based on S-Tag PCP value

The information contained in the ENNI frame S-Tag pertains to the Operator MEN/CEN receiving the frame, not the Operator MEN/CEN sending the frame. Thus, it is the responsibility of the transmitting Operator MEN/CEN to set S-Tag values appropriately for the receiving Operator MEN/CEN.

Related Links

CoS Identification at UNIs

6.3.3 ENNI Tagging Requirements

Per MEF 26.1, all ENNI frames must be in one of the following formats with C-Tags and S-Tags defined per IEEE 802.1ad (2005).

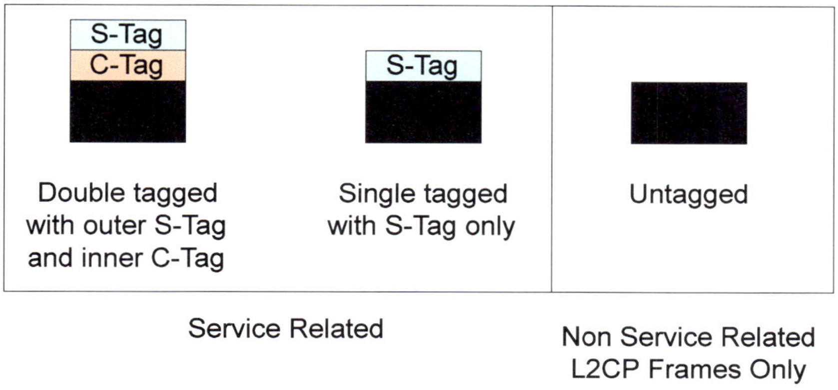

Figure 56: ENNI Frame Formats

Untagged frames are allowed at the ENNI only to support L2CPs, such as Link Aggregation Control Protocol (LACP), **which are not service specific**.

All frames associated with a service require an outer S-Tag. This includes service-associated L2CP frames, such as Spanning Tree Protocol (STP) bridge protocol data units (BPDUs).

The following figure shows tagging operations for two service frames as they progress through two Operator MENs. For clarity, only one direction of bidirectional traffic flow is shown.

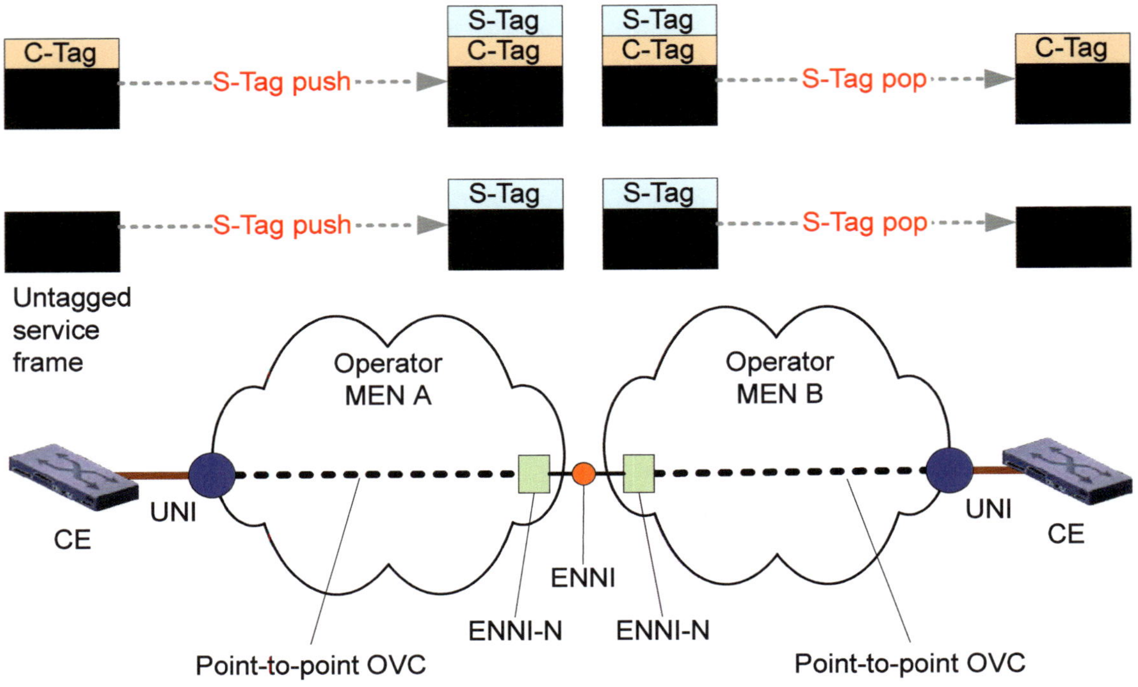

Figure 57: Service Frame Tagging at ENNI

Two operations are required:

- Operator MEN A pushes an S-Tag onto the service frame before transmitting it to Operator MEN B.
- Operator MEN B pops the S-Tag from the service frame before transmitting it to the CE.

The S-Tag is added (pushed) to capture information (CoS, Color, and S-VLAN ID) that is useful for service implementation, but the S-Tag must be removed (popped) before the service frame is transmitted to the subscriber.

Per MEF 26.1, S-Tags are only required to support service hand off at ENNIs. However, operators commonly push and pop S-Tags at UNIs, as shown in the following figure.

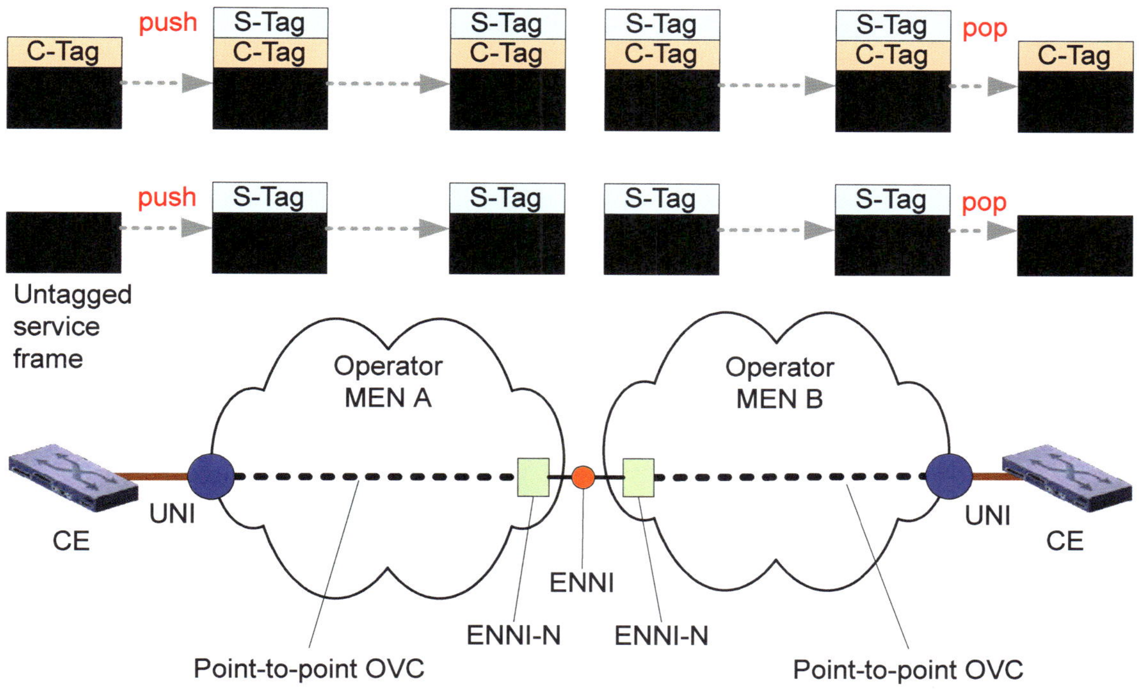

Figure 58: S-Tag Push/Pop at UNIs

Operators do this in order to use the S-Tag information (CoS, Color, and S-VLAN ID) for service implementation within the Operator MEN/CEN (not just at the ENNI).

Note: *In single-MEN/CEN service applications, service providers also commonly push and pop S-Tags at UNIs for the same reason: to mark each service frame with an S-VLAN ID, a CoS identifier, and a Color identifier.*

6.4
MEF Model for Service Across an Operator MEN/CEN

The MEF model for service across an Operator MEN/CEN is similar to the MEF model for the overall service, but it involves more service attributes and different stakeholders. The stakeholders are:

- **Service Provider**–The organization purchasing the service across the Operator MEN/CEN
- **Operator**–The organization providing the service across the Operator MEN/CEN

Just as MEF 6.1 uses service attributes to define an overall service, MEF 26.1 uses service attributes to define service across an Operator MEN/CEN. More service attributes are required, but the approach is basically the same: service across an Operator MEN/CEN is defined in terms of end point functionality (details of implementation are avoided).

In this document these services are called **operator services**, or **OVC-based services**, to distinguish them from UNI-to-UNI (EVC-based) Ethernet services.

Note: *This lesson describes how the MEF defines OVC-based services in general. The next lesson describes two MEF-standarized OVC-based services (the* ***Access EPL*** *and the* ***Access EVPL****), which the MEF has defined to support industry standardization of Ethernet access service.*

Related Links

MEF Model for Service Across an Operator MEN/CEN
E-Access Services

6.5
Operator Service Attributes

In this section:

Each operator service attribute is associated to one of five categories (rather than the three categories used for UNI-to-UNI service attributes):

ENNI service attributes

OVC service attributes

OVC End Point per ENNI service attributes

UNI service attributes

OVC per UNI service attributes

Attention: *Most of the operator service attributes are similar to those previously described (for the overall service) and will not be reviewed in detail within this lesson. The following tables list the attributes to give you a general idea of what is involved.* ***They are not meant to be studied in detail.***

Refer to MEF 26.1 for complete details on all operator service attributes.

6.5.1
ENNI Service Attributes

For each ENNI, there are two sets of ENNI Service Attributes, one for each Operator MEN (Operator CEN).

Table 9: ENNI Service Attributes (Reference: Table 2 in MEF 26.1)

Attribute Name	Summary Description	Possible Values
Operator ENNI Identifier	An identifier for the ENNI intended for management purposes	A string that is unique across the Operator MEN/CEN
Physical Layer	The physical layer of the links supporting the ENNI ***Note:*** *ENNIs are defined to be high bandwidth (1 Gb/s or 10 Gb/s).*	1000Base-SX, 1000Base-LX, 1000Base T, 10GBase-SR, 10GBase-LX4, 10GBase-LR, 10GBase-ER, 10GBase-SW, 10GBase-LW, 10GBase-EW

Table 9: ENNI Service Attributes (Reference: Table 2 in MEF 26.1) (Cont.)

Attribute Name	Summary Description	Possible Values
Frame Format	The format of the PDUs at the ENNI	Conforming to IEEE 802.1ad (2005)
Number of Links	The number of physical links in the ENNI	1 or 2
Protection Mechanism	The method for protection, if any, against a failure	Link Aggregation, None, or Other
ENNI Maximum Transmission Unit Size	The maximum length ENNI Frame in bytes are allowed at the ENNI	Integer ≥ 1526 bytes
End Point Map	The map that associates each S-Tagged ENNI Frame with an OVC End Point	A table with rows of the form <S-VLAN ID value, End Point Identifier, End Point Type>
Maximum Number of OVCs	The maximum number of OVCs that the Operator can support at the ENNI	Integer ≥ 1
Maximum Number of OVC End Points per OVC	The maximum number of OVC End Points that the Operator can support at the ENNI for an OVC.	Integer ≥ 1

Note: *The MEF provides no option to assign bandwidth profiles per ENNI.*

Related Links

Color-Aware Bandwidth Profiles at ENNI
Protection at ENNI
ENNI and OVC MTU Size
End Point Map

6.5.2 OVC Service Attributes

Table 10: OVC Service Attributes (Reference: Table 5 in MEF 26.1)

Attribute Name	Summary Description	Possible Values
OVC Identifier	Identifies the OVC for management purposes.	A string that is unique across the Operator MEN/CEN
OVC Type	Declares the OVC type.	Point-to-Point, Multipoint-to-Multipoint, or Rooted-Multipoint
OVC End Point List	Lists OVC End Points associated with the OVC.	A list of <OVC End Point Identifier, OVC End Point Role> pairs

Table 10: OVC Service Attributes (Reference: Table 5 in MEF 26.1) (Cont.)

Attribute Name	Summary Description	Possible Values
Maximum Number of UNI OVC End Points	Bounds the number UNIs on which OVC End Points terminate.	Integer ≥ 0
Maximum Number ENNI OVC End Points	Bounds the number ENNIs on which OVC End Points terminate.	Integer ≥ 1
OVC Maximum Transmission Unit Size	Declares the maximum size of frames supported over all OVC End Points.	Integer ≥ 1526 bytes
CE-VLAN ID Preservation	If Yes, the OVC preserves the value of the VLAN ID (VID) in the CE-VLAN tag.	Yes or No
CE-VLAN CoS Preservation	If Yes, the OVC preserves the value of the priority code point (PCP) bits in the CE-VLAN tag.	Yes or No
S-VLAN ID Preservation	If Yes, the OVC preserves the value of the VLAN ID (VID) in the S-Tag.	Yes or No
S-VLAN CoS Preservation	If Yes, the OVC preserves the value of the priority code point (PCP) bits in the S-Tag.	Yes or No
Color Forwarding	When Color Forwarding is Yes, the OVC cannot "promote" a frame from Yellow to Green.	Yes or No
Service Level Specification	Defines frame delivery performance definitions and objectives.	As described later in this lesson
Unicast Service Frame Delivery	Describes OVC delivery requirements for ingress frames with a unicast destination MAC address.	Deliver Unconditionally or Deliver Conditionally
Multicast Service Frame Delivery	Describes OVC delivery requirements for ingress frames with a multicast destination MAC address.	Deliver Unconditionally or Deliver Conditionally
Broadcast Service Frame Delivery	Describes OVC delivery requirements for ingress frames with a broadcast destination MAC address.	Deliver Unconditionally or Deliver Conditionally
Layer 2 Control Protocol Tunneling	The Layer 2 Control Protocols that are tunneled by the OVC	A list on Layer 2 Control Protocols

Note: *The initial MEF-CECP exam did not include OVC attributes* ***Service Level Specification*** *and* ***Layer 2 Control Protocol Tunneling****. They were introduced in MEF 26.1.*

Related Links

Service Level Specification (OVC Service Attribute)
Layer 2 Control Protocol Tunneling
OVC End Point Role
ENNI and OVC MTU Size
Color Forwarding

6.5.3 OVC End Point per ENNI Service Attributes

Table 11: OVC End Point per ENNI Service Attributes (Reference: Table 17 in MEF 26.1)

Attribute Name	Summary Description	Possible Values
OVC End Point Identifier	Identifies the OVC End Point for management purposes	A string that is unique across the Operator MEN/CEN
Trunk Identifiers	For a Trunk OVC End Point, specifies the S-VLAN ID values used on the ENNI to distinguish frames originating at a Root UNI and frames originating at a Leaf UNI	<Root S-VLAN ID value, Leaf S-VLAN ID value> for Trunk OVC End Points; Not Applicable to Root or Leaf OVC End Points
Class of Service Identifiers	Determines CoS for ingress ENNI frames mapped to the OVC End Point	Non-overlapping sets S-Tag PCP values; one for each CoS
Ingress Bandwidth Profile Per OVC End Point	Determines per-OVC-End-Point policing for ingress ENNI Frames mapped to the OVC End Point	None or one set of bandwidth profile values (CIR, CBS, EIR, EBS, CF, CM)
Ingress Bandwidth Profile Per ENNI Class of Service Identifier	Determines per-CoS policing for ingress ENNI Frames mapped to the OVC End Point	None or one set of bandwidth profile values (CIR, CBS, EIR, EBS, CF, CM) for each CoS
Egress Bandwidth Profile Per End Point	Determines per-OVC-End-Point policing for egress ENNI Frames mapped to the OVC End Point	None or one set of bandwidth profile values (CIR, CBS, EIR, EBS, CF, CM)
Egress Bandwidth Profile Per ENNI Class of Service Identifier	Determines per-CoS policing for egress ENNI Frames mapped to the OVC End Point	None or one set of bandwidth profile values (CIR, CBS, EIR, EBS, CF, CM) for each CoS

Note: *The initial MEF-CECP exam did not include the Trunk Identifiers attribute. It was introduced in MEF 26.1 to support rooted-multipoint OVCs.*

Note: *At ENNIs, bandwidth profiles (if used) must be color-aware.*

Related Links

Trunk Identifiers
Color-Aware Bandwidth Profiles at ENNI

6.5.4 UNI Service Attributes

The UNI service attributes for a service across an Operator MEN (Operator CEN) are identical to the UNI service attributes for the overall service.

Table 12: UNI Service Attributes (Reference: Table 3 in MEF 6.1)

Attribute Name	Requirement
UNI Identifier	Arbitrary text string to identify the UNI
Physical Medium	UNI Type 2 Physical Interface
Speed	10 Mb/s, 100 Mb/s, 10/100 Mb/s Auto-negotiation, 10/100/1000 Mb/s Auto-negotiation, 1 Gb/s, or 10 Gb/s
Mode	Full Duplex
MAC Layer	IEEE 802.3-2005
UNI MTU Size	≥ 1522 bytes
Service Multiplexing	Yes or No. MUST be No if *All to One Bundling* is *Yes*.
Bundling	Yes or No. MUST be No if *All to One Bundling* is *Yes*.
All to One Bundling	Yes or No. MUST be No if *Bundling* or *Service Multiplexing* is *Yes*.
CE-VLAN ID for untagged and priority tagged Service Frames	MUST specify *CE-VLAN ID for untagged and priority tagged Service Frames* in the range of 1-4094. This requirement does not apply for services with all to one bundling at the UNI.
Maximum number of EVCs	Integer ≥ 1
Ingress Bandwidth Profile Per UNI (optional)	If this attribute is supported, MUST specify CIR, CBS, EIR, EBS, CM, or CF. MUST NOT be combined with any other type of ingress bandwidth profile.
Egress Bandwidth Profile Per UNI (optional)	If this attribute is supported, MUST specify CIR, CBS, EIR, EBS, CM, or CF. MUST NOT be combined with any other type of egress bandwidth profile.
L2CP Processing	For each protocol, MUST specify one of the following: *Peer*, *Discard*, *Pass to EVC*, or *Peer and Pass to EVC*.

6.5.5 OVC per UNI Service Attributes

Note: *The term **OVC End Point per UNI** is equivalent to **OVC per UNI** . These terms are equivalent because an OVC can associated only one OVC End Point to a UNI.*

Table 13: OVC per UNI Service Attributes (Reference: Table 18 in MEF 26.1)

Attribute Name	Summary Description	Possible Values
UNI OVC Identifier	Identifies the OVC at UNI for management purposes	A string formed by the concatenation of the UNI Identifier and the OVC Identifier
OVC End Point Map	The CE-VLAN ID(s) that map to the OVC End Point at the UNI	A list of one or more CE-VLAN ID values

Table 13: OVC per UNI Service Attributes (Reference: Table 18 in MEF 26.1) (Cont.)

Attribute Name	Summary Description	Possible Values
Ingress Bandwidth Profile Per OVC End Point at a UNI	Determines policing for ingress frames mapped to the OVC End Point	None or one set of bandwidth profile values (CIR, CBS, EIR, EBS, CF, CM)
Ingress Bandwidth Profile Per Class of Service Identifier at a UNI	Determines per-CoS policing for ingress frames mapped to the OVC End Point	None or one set of bandwidth profile values (CIR, CBS, EIR, EBS, CF, CM) for each CoS
Egress Bandwidth Profile Per OVC End Point at a UNI	Determines policing for egress frames mapped to the OVC End Point	None or one set of bandwidth profile values (CIR, CBS, EIR, EBS, CF, CM)
Egress Bandwidth Profile Per Class of Service Identifier at a UNI	Determines per-CoS policing for egress frames mapped to the OVC End Point	None or one set of bandwidth profile values (CIR, CBS, EIR, EBS, CF, CM) for each CoS

6.6 Highlights of the MEF Model for Service Across an Operator MEN/CEN

In this section:

6.6.1 ENNI Frame Format

At an ENNI, all frames must be in one of the following formats, where C-Tags and S-Tags are as defined in IEEE 802.1ad (2005).

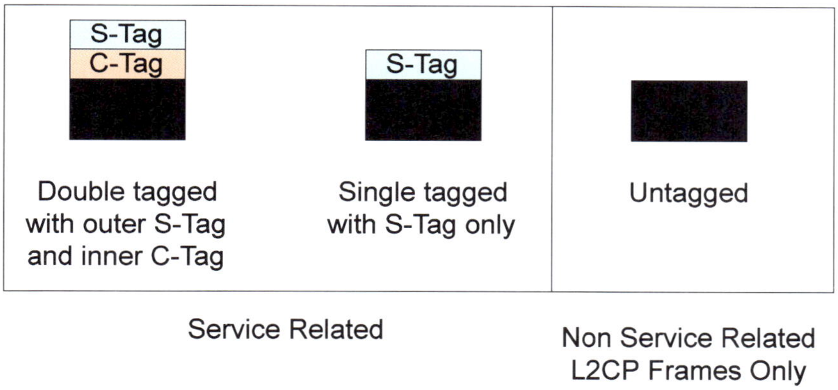

Untagged frames are allowed at the ENNI only to support L2CPs, such as link aggregation control protocol (LACP), which are not service specific.

All frames associated with a service require an outer S-Tag. This includes service-associated L2CP frames, such as spanning tree protocol (STP) bridge protocol data units (BPDUs).

6.6.2 ENNI and OVC MTU Size

Several operator service attributes specify MTU size:

Table 14: MTU Size Attributes

Attribute Category	Attribute Name	Requirement
ENNI Service Attribute	ENNI Maximum Transmission Unit Size	≥ 1526 bytes (required) ≥ 2000 bytes (recommended)
OVC Service Attribute	OVC Maximum Transmission Unit Size	≥ 1526 bytes (required) ≥ 2000 bytes (recommended)
UNI Service Attribute	UNI MTU Size	≥ 1522 bytes

ENNIs and OVCs are required to support frames at least as large as 1526 bytes, which is 4 bytes larger than the same requirement at UNIs (1522 bytes). The difference (4 bytes) accounts for the S-Tag required at ENNIs.

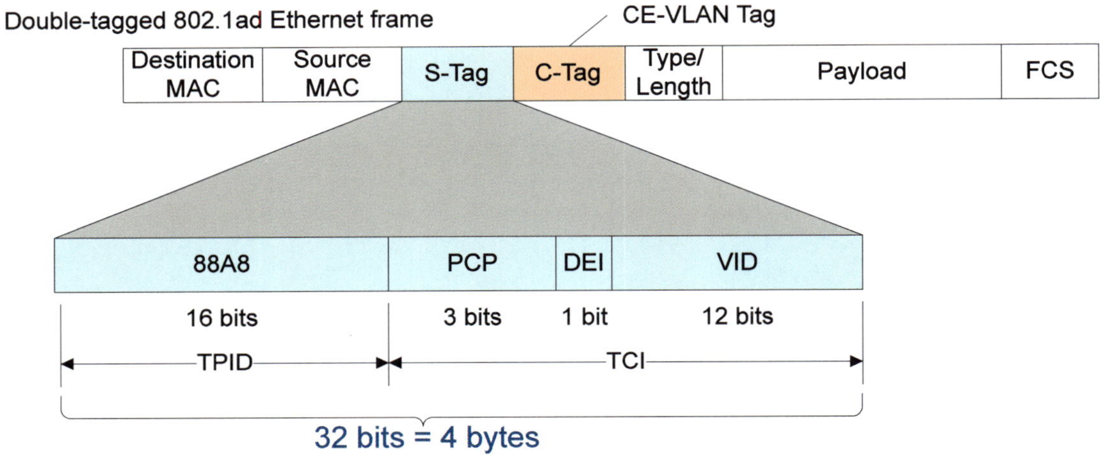

Related Links

MTU Size Service Attributes
ENNI Service Attributes
OVC Service Attributes

6.6.3 Protection at ENNI

If an ENNI includes two physical links, it must be capable of implementing link aggregation (per IEEE 802.1AX or as previously defined in IEEE 802.3ad or clause 43 of IEEE 802.3) with one port active and one standby, as follows:

Note: *MEF 26.1 cites clause 43 of IEEE 802.3 for link aggregation requirements.*

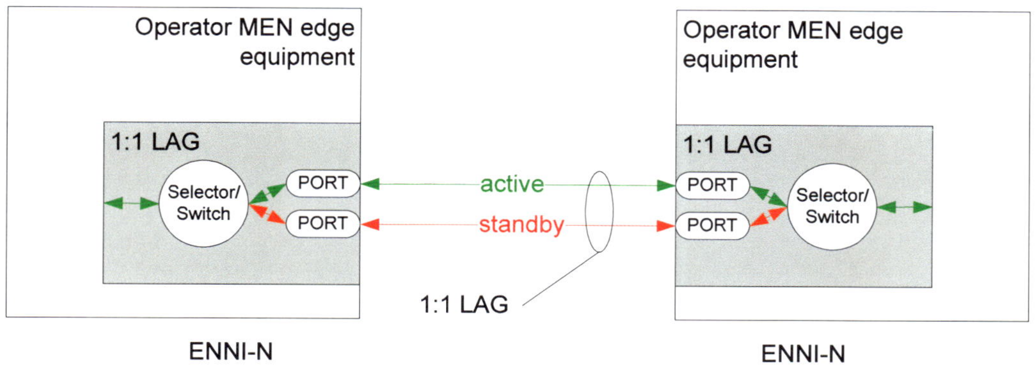

Figure 59: 1:1 LAG

ENNI protection is specified by the value of the **Protection Mechanism** ENNI service attribute:

- **None**—Applicable to ENNIs implemented with only one physical link
- **Link Aggregation**—Indicates that the ENNI is implemented using two physical links, one active and one standby, as shown in the preceding figure
- **Other**—Allows other unspecified protection mechanisms to be used by agreement

Per MEF 26.1, if Link Aggregation is used at the ENNI, it must be implemented using LACP (Link Aggregation Control Protocol) at each ENNI-N.

Related Links

ENNI Service Attributes

6.6.4 End Point Map

The ENNI service attribute **End Point Map** specifies how ingress ENNI frames are assigned to OVC end points based on S-VLAN ID.

The following figure shows three services (red, green, and blue) implemented over three Operator MENs. The green and blue services are both point-to-point. The red service is multipoint-to-multipoint. Three services are shown to help you visualize service multiplexing at the circled ENNI. At this ENNI, Ethernet frames from all three services arrive to Operator MEN B mixed together in the same traffic stream. The End Point Map specifies how incoming Ethernet frames are sorted and distributed to OVC end points, based on S-VLAN ID. The zoom-in detail shows only one direction of traffic flow. In the other flow direction, a different End Point Map would apply (and govern distribution of traffic incoming to Operator MEN A from Operator MEN B).

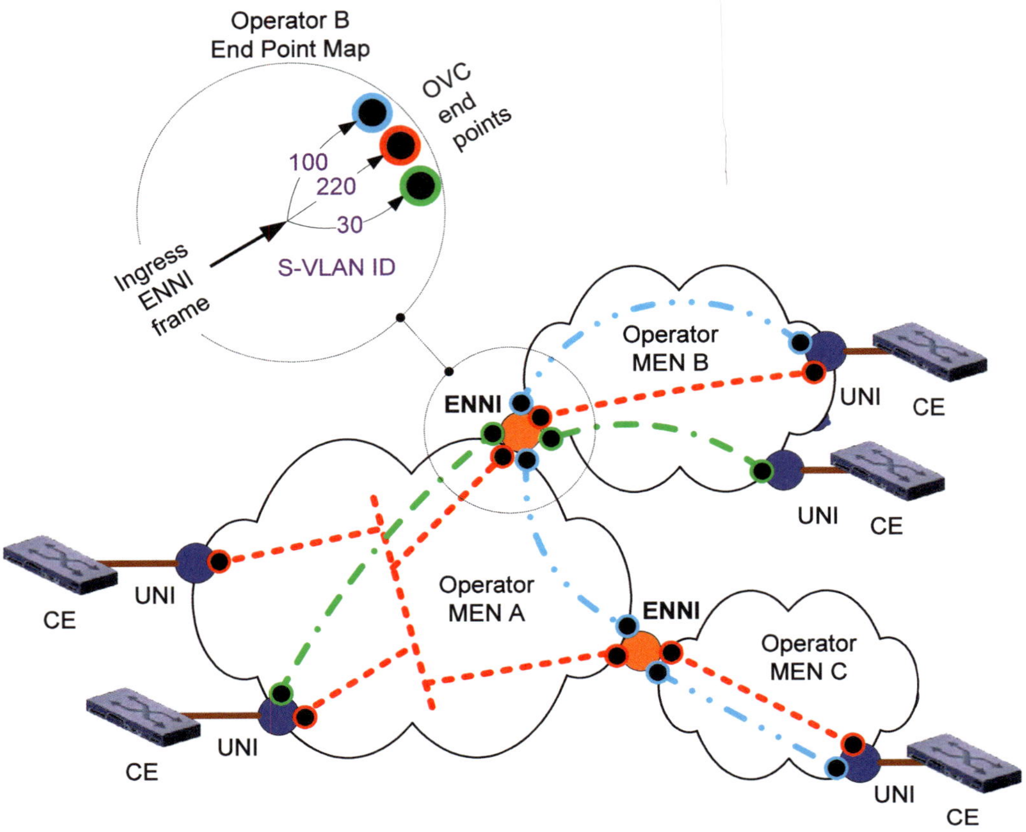

Figure 60: End Point Map

An S-VLAN ID can be mapped to no more than one OVC end point.

It is the responsibility of the transmitting Operator MEN/CEN to set S-VLAN ID values appropriately for the receiving Operator MEN/CEN.

Test your understanding: *Assign true or false to each of the following statements and explain why:* ***(a)*** *An End Point Map is not needed if the ENNI is port-based.* ***(b)*** *Two End Point Maps are required at every ENNI.* ***(c)*** *To implement services within an Operator MEN/CEN, the operator needs to know only one of the two End Point Maps at each ENNI.*

Answers: ***(a)*** *False. There is no such thing as a port-based ENNI.* ***(b)*** *True. One End Point Map is required for each Operator MEN/CEN.* ***(c)*** *False. At each ENNI, the operator uses both End Point Maps. The operator uses one End Point Map to assign ingress ENNI frames to OVC end points. The operator uses the other End Point Map to assign S-VLAN IDs to egress ENNI frames.*

Related Links

ENNI Service Attributes
End Point Map Bundling

6.6.5
Color-Aware Bandwidth Profiles at ENNI

At ENNIs, ingress and egress bandwidth profiles are assigned per OVC end point.

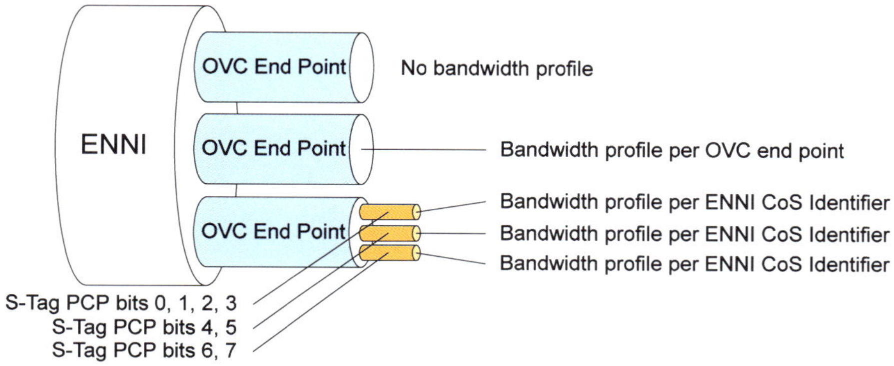

Figure 61: Bandwidth Profile Options at ENNI

For each OVC end point, there are three options:

1. Assign no bandwidth profile.
2. Assign one bandwidth profile applicable to all ENNI frames that map to the OVC end point.
3. Assign a set of bandwidth profiles (per CoS) to all ENNI frames that map to the OVC end point.

Note: *In the previous figure, S-Tag PCP bits are CoS identifiers.*

Note: *Bandwidth profiles cannot be assigned per ENNI.*

At ENNIs, bandwidth profiles are defined by six parameters, just as they are defined at UNIs. However, **at ENNIs, all bandwidth profiles must be color-aware**.

Table 15: Bandwidth Profile Parameters at ENNI

Parameter		Units of Measure or Values
CIR	Committed Information Rate	Bits per second
CBS	Committed Burst Size	Bytes
EIR	Excess Information Rate	Bits per second
EBS	Excess Burst Size	Bytes
CF	Coupling Flag	N or Y
CM	Color Mode	color-aware (color-blind not allowed at ENNI)

Note: *At ENNIs, color can be encoded in the DEI bit of the S-Tag or in the PCP bits of the S-Tag.*

The MEF does not require bandwidth profiles at ENNIs, but any bandwidth profiles used at an ENNI must be color-aware.

Color-awareness of ENNI bandwidth profiles helps minimize the risk of recoloring service frames and of dropping green service frames as follows:

Color-awareness ensures that yellow frames cannot be recolored to green, but more importantly, color-awareness ensures that yellow frames do not compete with green frames for CIR/CBS bandwidth. Color-awareness of an ingress bandwidth profile does not eliminate the possibility that a green frame will be recolored to yellow and/or dropped. It just reduces that possibility by ensuring that yellow frames do not compete with green frames for CIR/CBS bandwidth.

Note: *Recall that Color reflects the service frame's compliance to the SLA (service-level agreement) bandwidth profile (implemented by policing at the ingress UNI) and should not be changed during transport through the service, UNI-to-UNI.*

If bandwidth profiles are used at ENNIs, they should be carefully configured to ensure that green service frames are not dropped or recolored to yellow.

Related Links

ENNI Service Attributes
Color Mode
Color-Aware Bandwidth Profile Operation
OVC End Point per ENNI Service Attributes

6.6.6 Color Forwarding

The OVC service attribute *Color Forwarding* can be set to *Yes* or *No*. When Color Forwarding is Yes, the OVC cannot "promote" a frame from Yellow to Green.

Promoting a frame from Yellow to Green could have an undesired impact on EVC performance because newly promoted Green frames would compete for resources with frames that were marked Green at the ingress UNI.

If Color Forwarding is set to Yes, operators must ensure that:

- Color is encoded in the S-Tags of service frames prior to transmission across an ENNI
- Color is decoded from the S-Tags of service frames received from an ENNI

Related Links

OVC Service Attributes
Encoding Information in S-Tags
Color-Aware Bandwidth Profiles at ENNI

6.6.7
Hairpin Switching

Recall that hairpin switching refers to ENNI-N functionality that returns an ingress ENNI frame across the ENNI with changes that cause the frame to be associated to a different OVC upon return.

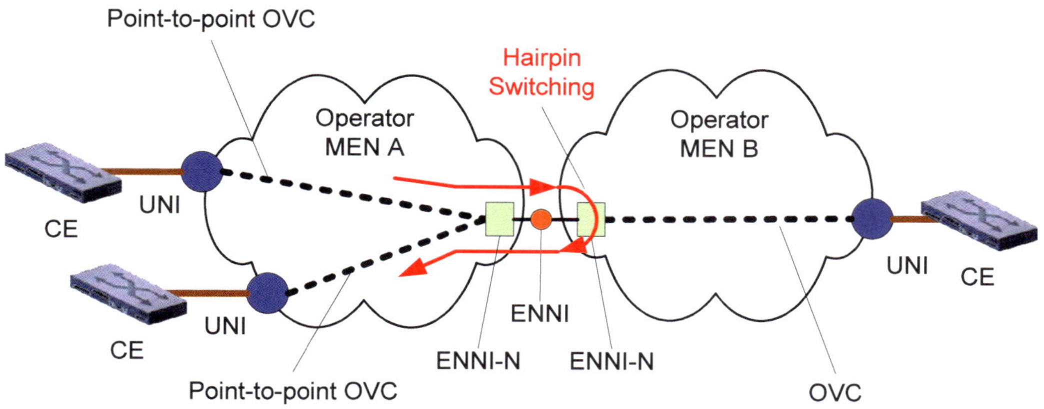

Figure 62: Example Hairpin Switch

In this example, hairpin switching is used to support the following EVP-LAN service.

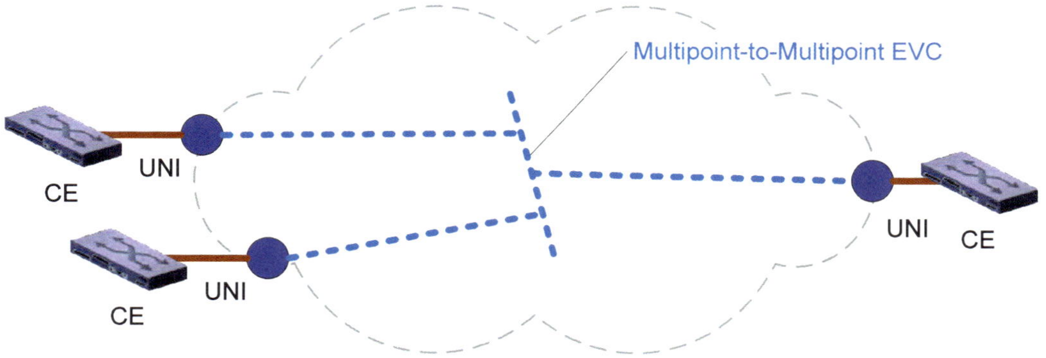

Figure 63: EVP-LAN Service for Example Hairpin Switch

This EVP-LAN service is implemented with three OVCs: two point-to-point OVCs in Operator MEN A and one multipoint-to-multipoint OVC in Operator MEN B.

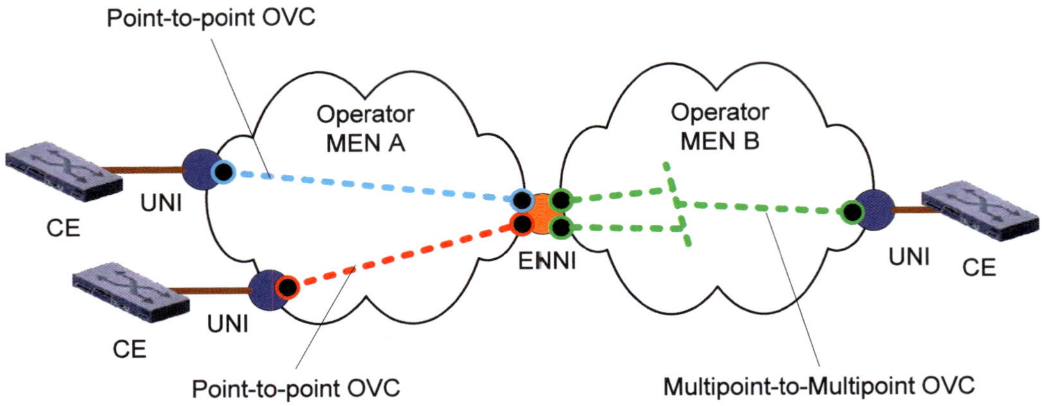

Figure 64: OVCs for Example Hairpin Switch

The multipoint-to-multipoint OVC in Operator MEN B includes two OVC end points at the ENNI.

Two End Point Maps (ENNI service attributes), one in each operator MEN, govern the flow of service frames between OVC end points, as follows.

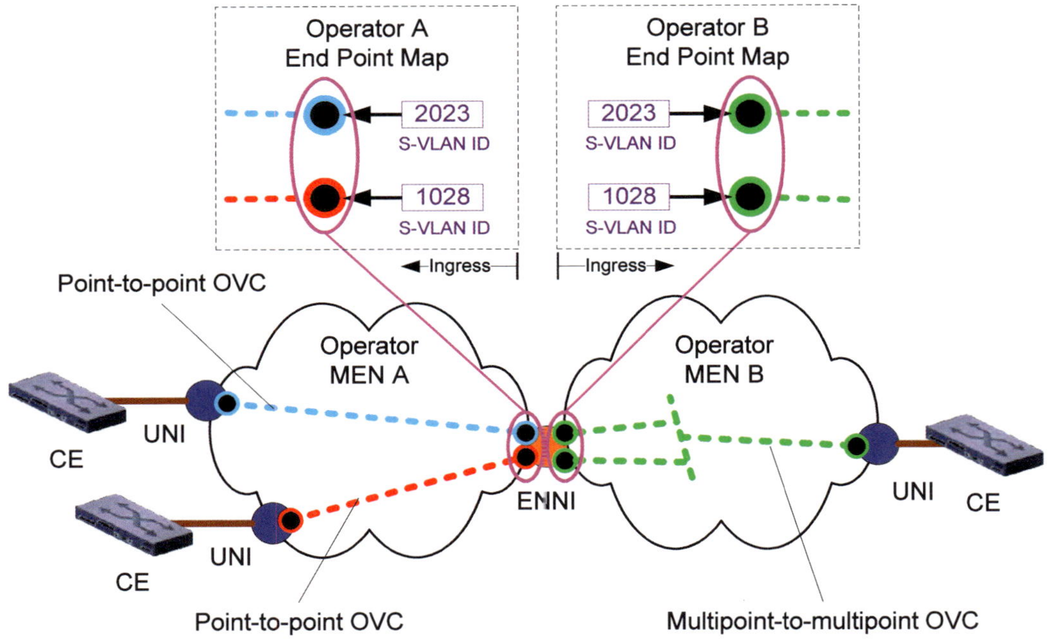

Figure 65: End Point Maps for Example Hairpin Switch

The End Point Maps, in effect, connect each point-to-point OVC to one endpoint of the multipoint-to-multipoint OVC, as follows:

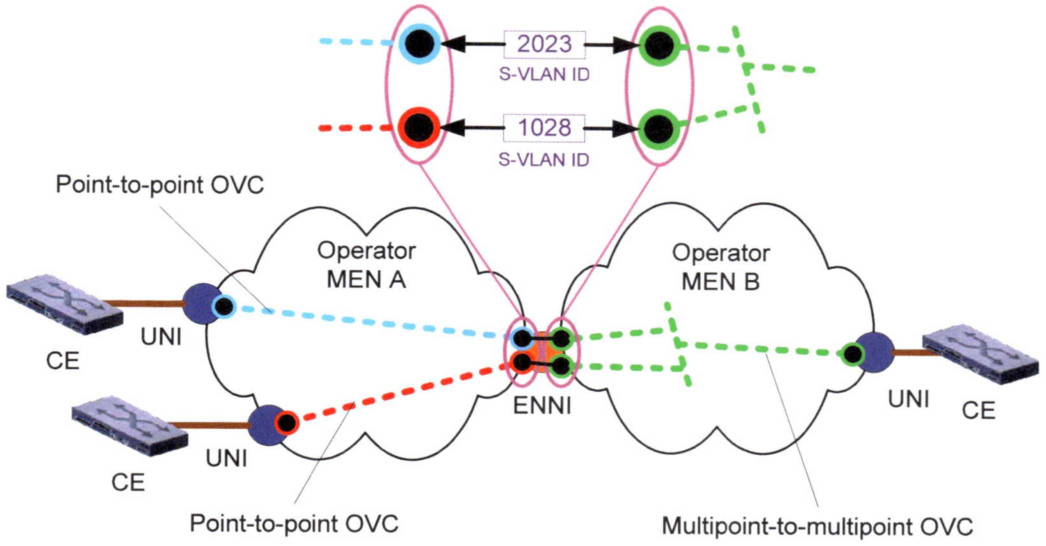

Figure 66: End Point Map Connections for Example Hairpin Switch

The multipoint-to-multipoint OVC now in effect now has two connections through the ENNI, one for each point-to-point OVC.

To implement a hairpin switch, the multipoint-to-multipoint OVC simply changes the S-VLAN ID on the ingress ENNI frame (from 2023 to 1028, in the example shown) and transmits it back through the ENNI.

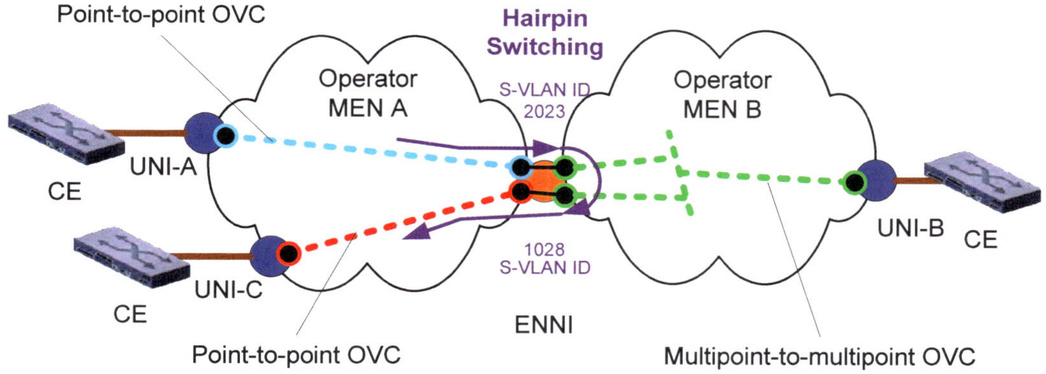

Figure 67: Example Hairpin Switch

Note: *The preceding example is based on Section 10.4, Example 3: Hairpin Switching in MEF 26.1.*

Improper use of hairpin switching can result in a data loop between two Operator MENs at a single ENNI, as shown in the following example.[13]

13 This example is based on *Section 10.5 Example 4: Data Loop at an ENNI with Hairpin Switching* in MEF 26.1.

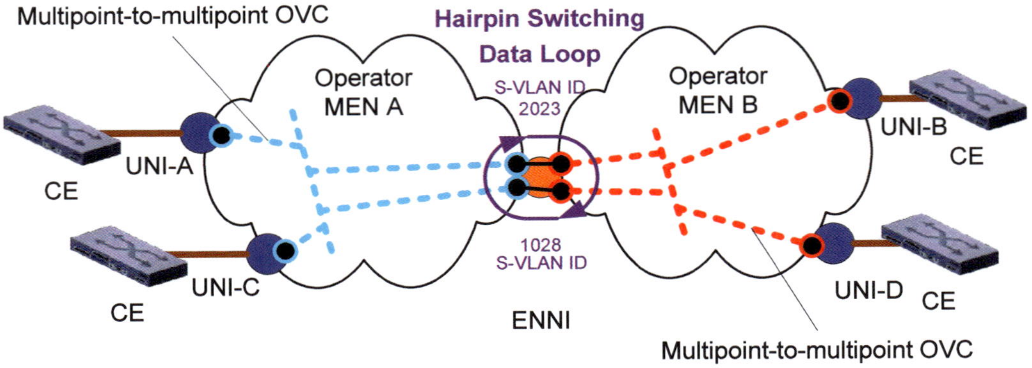

Figure 68: Example Data Loop at an ENNI with Hairpin Switching

The service provider and operators are responsible for ensuring that such loops do not occur.

In the previous example, hairpin switching is not needed and should not be used. The two multipoint-to-multipoint OVCs should interconnect at the ENNI through one pair of OVC end points as shown in the following figure.

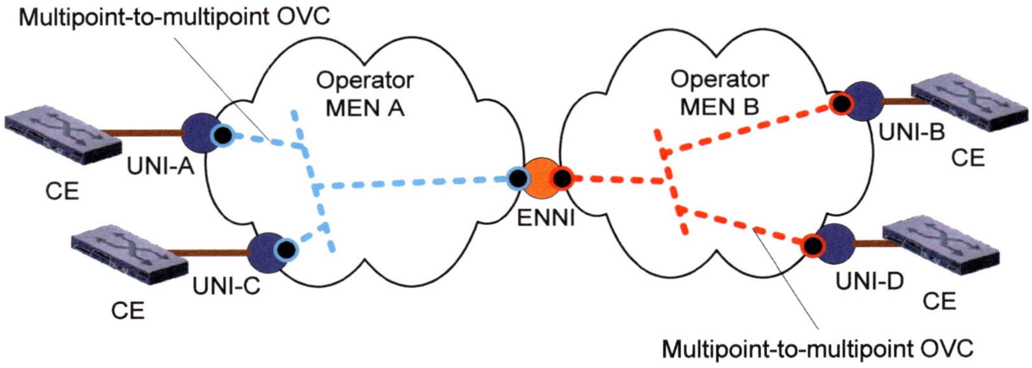

Related Links

Hairpin Switch

6.6.8
End Point Map Bundling

An End Point Map associates a set of S-VLAN ID values to a set of OVC end points. Typically, the number of S-VLAN ID values equals the number of OVC end points, and each OVC end point is assigned one S-VLAN ID value, as shown in the following example.

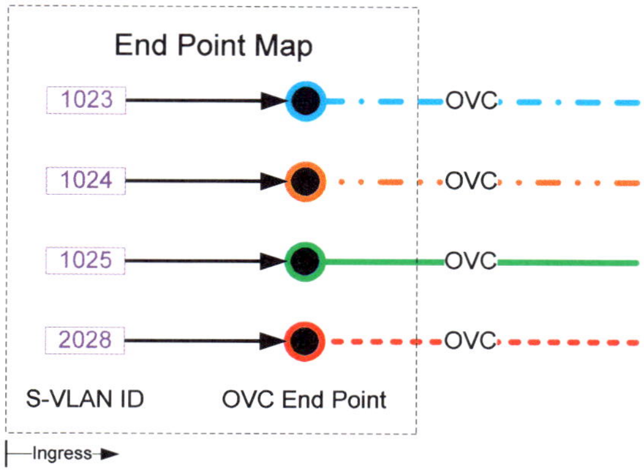

Figure 69: Basic End Point Map

However, MEF 26.1 also permits End Point Maps to assign multiple S-VLAN ID values to one or more OVC endpoints, as shown in the following example.

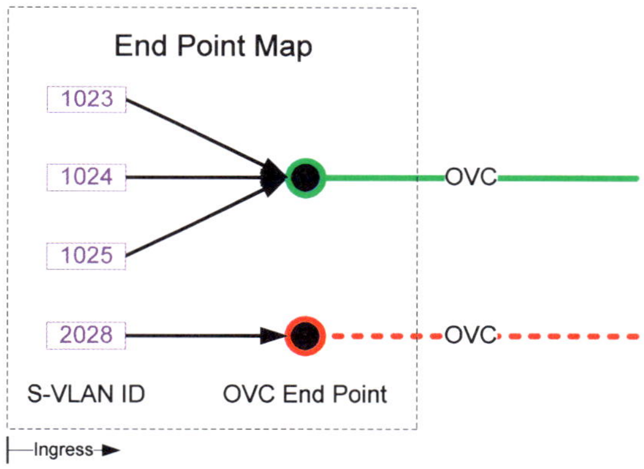

Figure 70: End Point Map with Bundling

When multiple S-VLAN ID values map to a single OVC End Point, the End Point Map is said to have Bundling.

End Point Map Bundling allows one OVC to support multiple services (S-VLAN ID values) as shown in the following example.

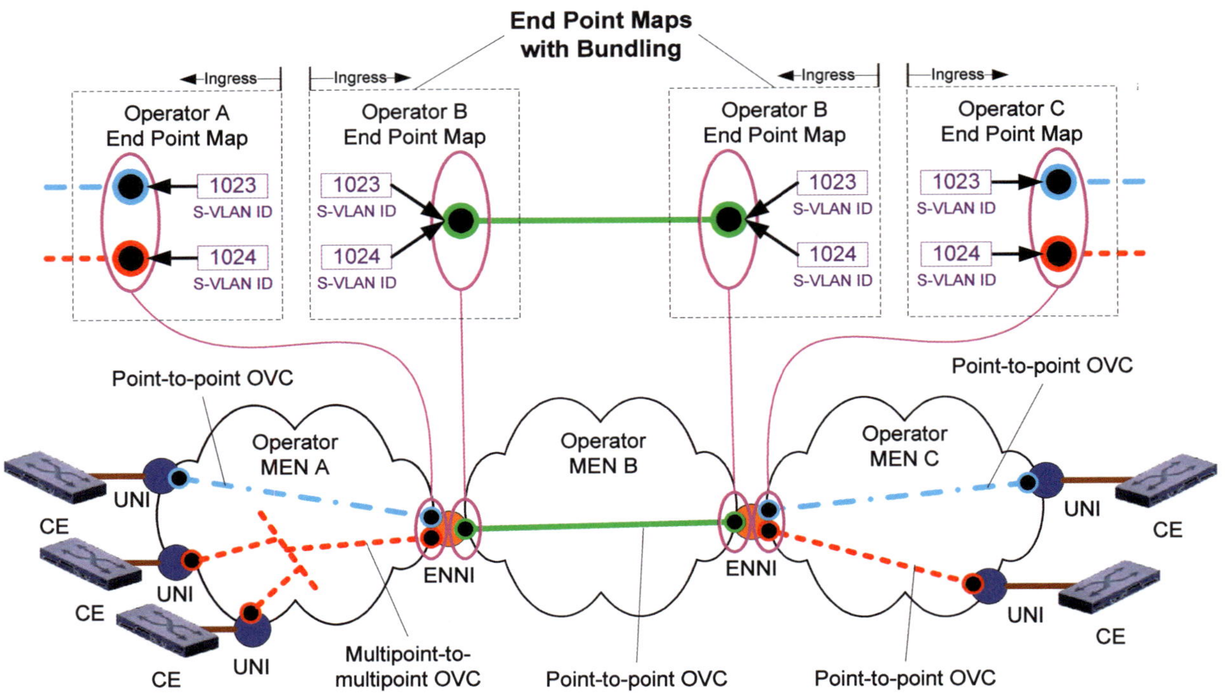

Figure 71: Example of End Point Map Bundling

In this example, the green OVC through Operator MEN B supports two services.

Per MEF 26.1, OVCs that support Bundling must meet the following requirements:

- The OVC must be point-to-point.
- The OVC must have its S-VLAN ID Preservation attribute set to Yes.
- The OVC must have its CE-VLAN ID Preservation attribute set to Yes.
- The OVC must have its CE-VLAN CoS Preservation attribute set to Yes.
- Both OVC end points must terminate at an ENNI.
- The OVC cannot support hairpin switching.

Note: *An OVC that supports Bundling can have its S-VLAN CoS Preservation attribute set to Yes or No.*

Related Links

Hairpin Switching
OVC Service Attributes
End Point Map

6.6.9
LOAM Requirement for ENNIs

ENNIs are required to support Link OAM (Operation, Administration and Management) as defined in MEF 26.1, based on clause 57 of IEEE 802.3. LOAM allows operator on both sides of the ENNI to monitor and diagnose link-level connectivity issues.

Link OAM functionality is covered later in this study guide, within a larger context that includes all types of links (not just ENNI links). For now, just note that LOAM support is an important requirement for ENNIs.

Related Links

Link OAM

6.6.10
Service Level Specification (OVC Service Attribute)

Note: *The initial MEF-CECP exam did not include OVC attribute* ***Service Level Specification****. It was introduced in MEF 26.1.*

OVC service attribute **Service Level Specification** specifies frame delivery performance for an OVC-based operator service using 8 performance objectives:

1. One-way Frame Delay (FD)
2. One-way Frame Delay Range (FDR)
3. One-way Mean Frame Delay (MFD)
4. Inter Frame Delay Variation (IFDV)
5. One-way Frame Loss Ratio (FLR)
6. One-way Availability (Availability)
7. One-way High Loss Intervals (HLI)
8. One-way Consecutive High Loss Intervals (CHLI)

Note: *Not Specified (N/S) is an acceptable value for any of these performance objectives.*

These performance objectives are similar to those defined for EVC Performance Attributes, except that frame delivery performance is defined EI-to-EI, rather than UNI-to-UNI.

Note: *An EI (External Interface) is either a UNI or an ENNI.*

Note: *The MEF addresses the problem of measuring OVC performance using SOAM Performance Management (as described in a later lesson). The focus here is on defining OVC performance.*

Related Links

EVC Performance Attributes: Metrics, Objectives, and Parameters
SOAM Performance Management

OVC versus EVC Performance Objectives

When an EVC is composed of OVCs, the service provider is faced with the problem of defining performance objective values for each OVC as well as for the EVC.

Conceptually, EVC performance results from the concatenation of the OVC performances. However, precise relationships between OVC and EVC performance can be hard to pin down.

The problem can be viewed from two perspectives:

- **Concatenation**– Determine EVC performance objectives, given OVC performance objectives
- **Sectionalization**–Determine OVC performance objectives, given EVC performance objectives

In either case, the objective is to come up with a complete set of performance objective values (for each OVC and for the EVC) that is *consistent* in the sense that the OVC performance objectives will ensure that EVC performance objectives are met.

As a simple example, consider an EPL service supporting one class of service over two operator MENs. Mean frame delay (MFD) might be specified as follows:

- MFD = 80 ms (for the EVC)
- MFD = 10 ms (for OVC 1)
- MFD = 50 ms (for OVC 2)

The methods, techniques, and negotiations needed to arrive at acceptable performance objectives are beyond the scope of MEF specifications and this study guide.

6.6.11 Layer 2 Control Protocol Tunneling

Note: *The initial MEF-CECP exam did not include OVC attribute* ***Layer 2 Control Protocol Tunneling****. It was introduced in MEF 26.1.*

OVC service attribute **Layer 2 Control Protocol Tunneling** specifies a list of Layer 2 Control Protocols that are tunneled by the OVC.

MEF 26.1 defines two types of L2CP frames:

- **L2CP Service Frame**–A service frame[14] with a L2CP MAC address (one of the MAC addresses shown in the following table)
- **L2CP ENNI Frame**–An ENNI frame[15] with a L2CP MAC address (one of the MAC addresses shown in the following table)

14 A service frame is an Ethernet frame transmitted across the UNI.
15 An ENNI frame is an Ethernet frame transmitted across the ENNI.

Table 16: List of Standardized Layer 2 Control Protocols (Table 30 in MEF 6.1)

Destination MAC Addresses	Layer 2 Control Protocol
01-80-C2-00-00-00 through 01-80-C2-00-00-0F	Bridge Block of protocols
01-80-C2-00-00-20 through 01-80-C2-00-00-2F	GARP Block of protocols

Note: *Table 1 in MEF 10.2 provides a similar list of standard L2CP MAC addresses, but that table includes the "All Bridges Protocol" MAC address 01-80-C2-00-00-10, which is not included here because, as explained in MEF 6.1, " it has been depreciated in 802.1Q-2005" and "In the unlikely event that a customer may use this MAC DA, MEF services are expected to treat them as normal service frames."*

Per MEF 26.1, when a *L2CP Service Frame* or *L2CP ENNI Frame* is tunneled:

1. The frame must be delivered to all OVC End Points, other than the ingress OVC End Point.
2. The frame format at egress must conform to requirements listed in the following table.

Table 17: Format Relationships for Tunneled L2CP Service and ENNI Frames (Table 16 in MEF 26.1)

Ingress Interface	Egress Interface	Egress Frame Format
UNI	UNI	Identical to the ingress frame
UNI	ENNI	Same as ingress frame, but with S-Tag added
ENNI	UNI	Same as ingress frame, but with S-Tag removed
ENNI	ENNI	Same as ingress frame, but with S-Tag possibly changed

Note: *If the egress frame differs from the ingress frame, the Frame Check Sequence in the egress frame is recalculated.*

6.7 Rooted-Multipoint OVCs

In this section:

Note: *The initial MEF-CECP exam did not include rooted-multipoint OVCs. They were introduced in MEF 26.1.*

Rooted-multipoint OVCs are used to support rooted-multipoint EVCs for E-Tree services.

Each OVC end point in a rooted-multipoint OVC is assigned an OVC end point role (Root, Leaf, or Trunk) that is related to UNI designations (Root and Leaf) in the associated rooted-multipoint EVC:

- OVC end point role **Root** – Indicates that all frames ingressing to the OVC end point originated at a Root UNI and can be forwarded to any OVC end point (Root, Leaf, or Trunk)
- OVC end point role **Leaf**–Indicates that all frames ingressing to the OVC end point originated at a Leaf UNI and must not be forwarded to a Leaf OVC end point
- OVC end point role **Trunk**–Indicates that frames ingressing to the OVC end point have mixed origins (some originate at a Root UNI and others originate at a Leaf UNI)

In MEF 26.1, *OVC end point role* is described as "a property of an OVC end point that determines the forwarding behavior between it and other OVC end points that are associated with the OVC end point by an OVC."

OVC end point roles appear in OVC service attribute *OVC End Point List*, which is a list of <OVC End Point Identifier, OVC End Point Role> pairs.

Related Links

OVC Service Attributes
OVC End Point Role

6.7.1 Applications with Root and Leaf OVC End Points Only

Example Application A

A rooted-multipoint EVC (for an E-Tree service) is implemented with 3 OVCs: a rooted-multipoint OVC in Operator MEN A, a multipoint-to-multipoint OVC in Operator MEN B, and a point-to-point OVC in Operator MEN C.

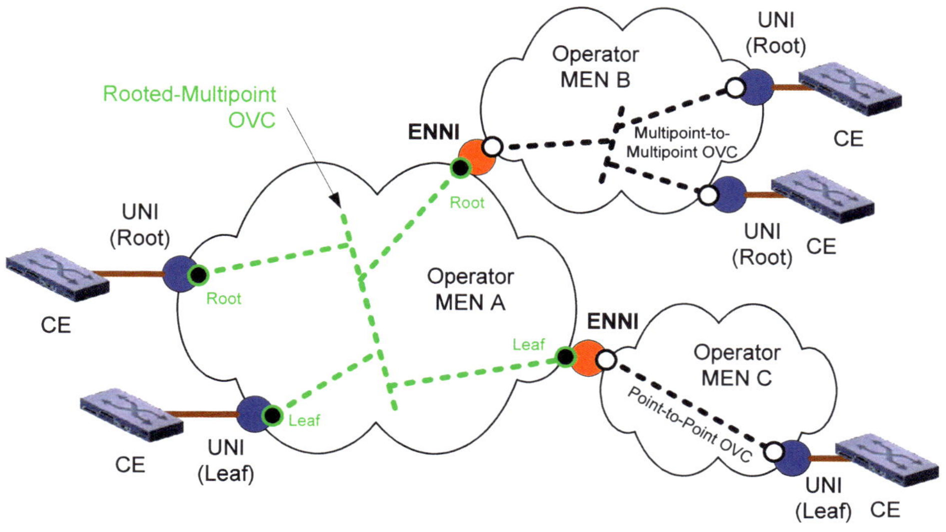

Observations:

1. The concatenation of OVCs results in a rooted-multipoint EVC that connects 5 UNIs (3 Root UNIs and 2 Leaf UNIs).
2. In Operator MEN B and Operator MEN C, OVC end point roles are not shown because the OVCs are not type rooted-multipoint.[16]
3. In Operator MEN A, OVC end point roles are assigned as follows:
   - At the UNIs, OVC end point roles simply match the UNI designation (Leaf or Root).
   - At the ENNIs, OVC end point roles match the designation of all UNIs that connect to it through the ENNI.

16 For point-to-point and multipoint-to-multipoint OVCs, each OVC end point is assigned role *Root*, but that assignment is trivial. It is done for consistency only.

Test your understanding: *Suppose the UNI in Operator MEN C was a Root UNI (instead of a Leaf UNI). Which OVC end point role would you assign to the associated OVC end point (in Operator MEN A): Root, Leaf, or Trunk?*

Answer: *Root (to match the designation of all UNIs that connect to it through the ENNI)*

Test your understanding: *Suppose that the two UNIs in Operator MEN B were both Leaf UNIs (instead of Root UNIs). Which OVC end point role would you assign to the associated OVC end point (in Operator MEN A): Root, Leaf, or Trunk?*

Answer: *None of the suggested answers is correct because this scenario is invalid. The OVC in MEN B cannot support any Leaf UNIs because it is a multipoint-to-multipoint OVC (not a rooted-multipoint OVC). However, if the OVC in MEN B were a rooted-multipoint OVC (with Leaf end point roles at the two UNIs and Root end point role at the ENNI), then the associated end point (in Operator MEN A) would be Leaf.*

Test your understanding: *Under what conditions can a rooted-multipoint OVC connect to a multipoint-to-multipoint OVC through an ENNI?*

Answer: *All UNIs that connect to the rooted-multipoint OVC through the multipoint-to-multipoint OVC must be Root UNIs.*

Test your understanding: *A point-to-point OVC supports a rooted-multipoint EVC and connects to a UNI. What is the UNI designation?*

Answer: *The UNI can be either a Root UNI or a Leaf UNI. The example includes such a point-to-point OVC in Operator MEN C.*

Example Application B

A rooted-multipoint EVC (for an E-Tree service) is implemented with 4 OVCs: a rooted-multipoint OVC in Operator MEN A, 2 point-to-point OVCs in Operator MEN B, and 1 point-to-point OVC in Operator MEN C.

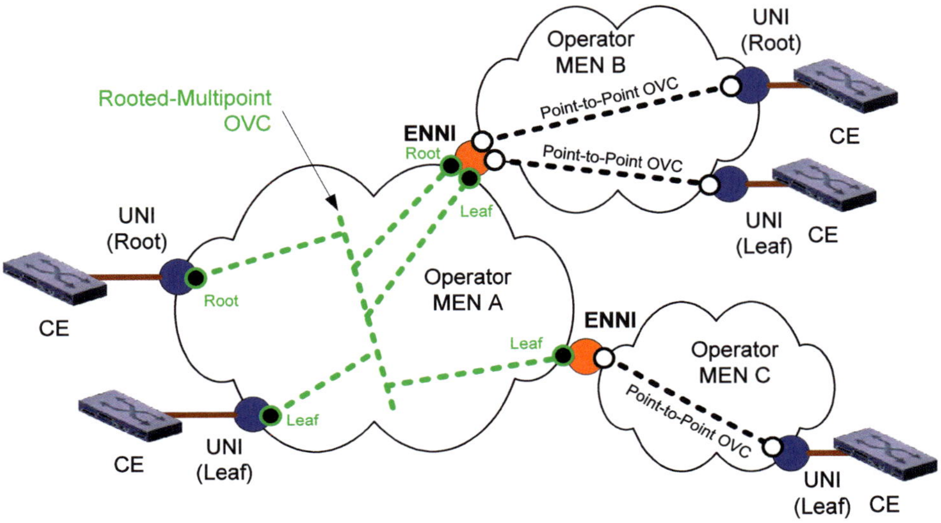

Observations:

1. The concatenation of OVCs results in a rooted-multipoint EVC that connects 5 UNIs (2 Root UNIs and 3 Leaf UNIs).
2. In Operator MEN B and Operator MEN C, OVC end point roles are not shown because the OVCs are not type rooted-multipoint.[17]
3. In Operator MEN A, OVC end point roles are assigned as follows:
   - At the UNIs, OVC end point roles simply match the UNI designation (Leaf or Root).
   - At the ENNIs, OVC end point roles match the designation of all UNIs that connect to it through the ENNI.
4. Hairpin switching in Operator MEN A is required to support traffic between the two UNIs in Operator MEN B.

17 For point-to-point and multipoint-to-multipoint OVCs, each OVC end point is assigned role *Root*, but that assignment is trivial. It is done for consistency only.

Related Links

Hairpin Switching

6.7.2 Applications with Trunk OVC End Points

OVC end point role Trunk is used when the source of ingress frames is ambiguous: when a frame ingressing to an OVC end point might have originated at a Root UNI or at a Leaf UNI. This situation only occurs in applications that involve more than one rooted-multipoint OVC.

The following example includes two rooted-multipoint OVCs

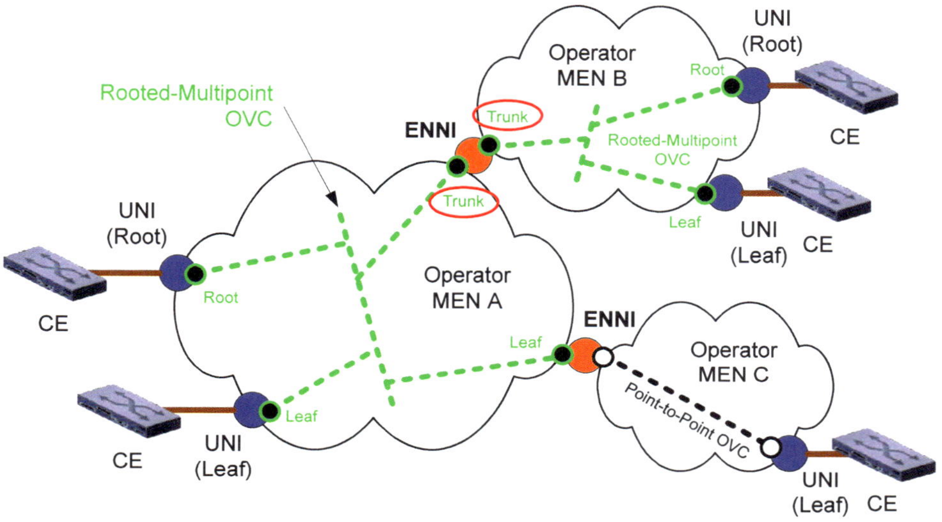

In this example, the OVC end points at the top ENNI cannot be assigned role Leaf or Root because an ingress frame can originate from either a Root or Leaf UNI. Such OVC end points are assigned the OVC end point role Trunk.

Assignment of role Trunk to such OVC end points does not fully resolve frame forwarding issues. It is only part of the solution.

To understand the problem better, consider the rooted-multipoint EVC formed by concatenation of the OVCs.

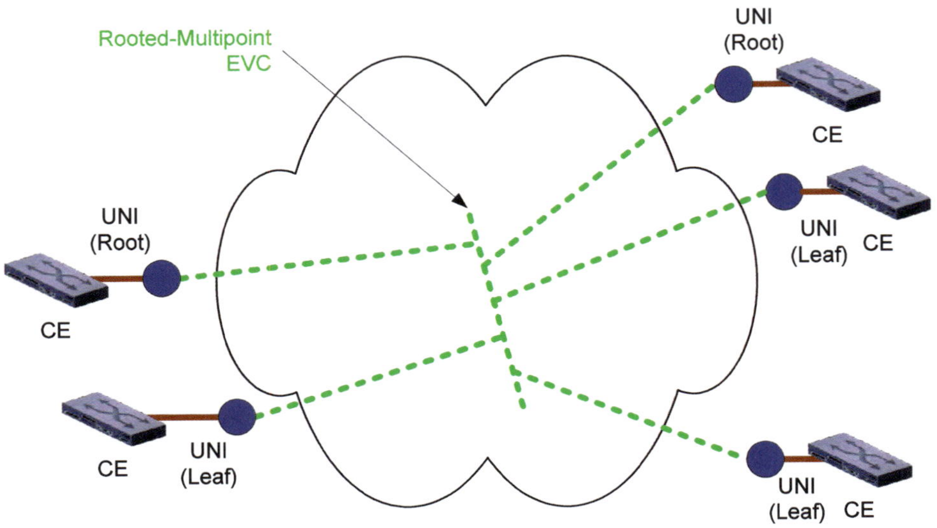

Each frame within a rooted-multipoint EVC is governed by a forwarding requirement that is determined by its UNI of origin:

Frame Source	Forwarding Requirement
Leaf UNI	The frame must not be forwarded to another leaf UNI. It may be forwarded to any root UNI.
Root UNI	The frame may be forwarded to any leaf UNI or root UNI.

If an E-Tree service is implemented over multiple operator MENs, a mechanism is needed to propagate forwarding requirements through ENNIs.

If only one rooted-multipoint OVC is involved, this issue is fully resolved by assigning each OVC end point the role of Leaf or Root (as previously described):

- OVC end point role **Root** – Indicates that all frames ingressing to the OVC end point originated at a Root UNI and can be forwarded to any OVC end point (Root, Leaf, or Trunk)

- OVC end point role **Leaf**–Indicates that all frames ingressing to the OVC end point originated at a Leaf UNI and must not be forwarded to another Leaf OVC end point

However, this system for propagating forwarding requirements across an ENNI does not work for Trunk OVC end points, which have to support frames originating from both types of UNIs: Leaf UNIs and Root UNIs.

- OVC end point role **Trunk**–Indicates that frames ingressing to the OVC end point have mixed origin (some originating at a Root UNI and others originating at a Leaf UNI)

Trunk Identifiers

When the OVC end point role is Trunk, *Trunk Identifiers* are assigned to resolve the issue of propagating forwarding requirements across the ENNI.

The following two values constitute the *OVC End Point per ENNI* service attribute named **Truck Identifiers**:

- **Root S-VLAN ID value**–This value identifies ingress frames that originated at a **Root UNI**.
- **Leaf S-VLAN ID value**–This value identifies ingress frames that originated at a **Leaf UNI**.

Note: *Trunk identifiers are only applicable to OVC end points with role Trunk.*

Because Trunk Identifiers define S-VLAN ID values for Ingress frames, it is the responsibility of the transmitting Operator MEN/CEN to set them appropriately.

Note: *The method by which frame forwarding requirements are preserved as frames are forwarded through an Operator MEN/CEN depends upon the technology used within the Operator MEN/CEN and is beyond the scope of MEF specifications.*

When the OVC end point role is Trunk, the ENNI service attribute **End Point Map** associates two S-VLAN ID values to the OVC end point. When the OVC end point role is Leaf or Root, the ENNI End Point Map associates just one S-VLAN ID value to the OVC end point.

Related Links

End Point Map
OVC End Point per ENNI Service Attributes

Example Application C

A rooted-multipoint EVC (for an E-Tree service) is implemented with 3 OVCs: a rooted-multipoint OVC in Operator MEN A, a rooted-multipoint OVC in Operator MEN B, and a point-to-point OVC in Operator MEN C.

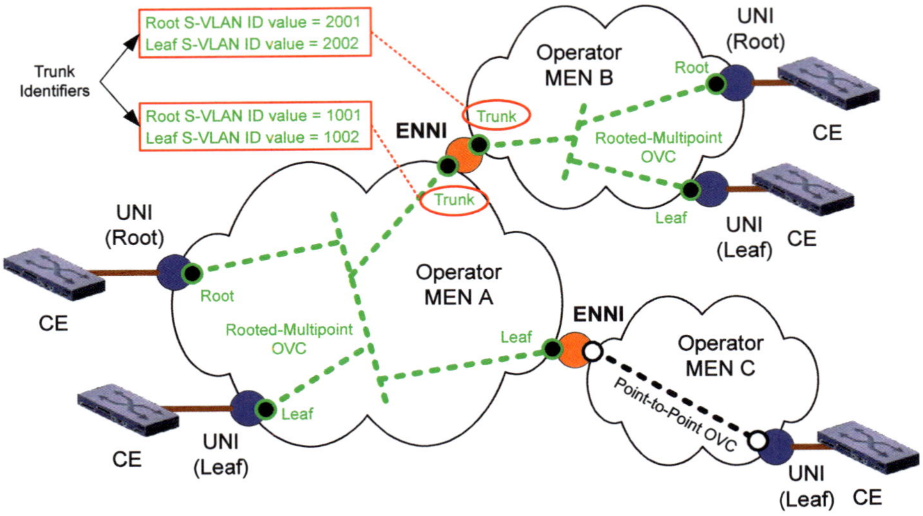

Observations:

1. The concatenation of OVCs results in a rooted-multipoint EVC that connects 5 UNIs (2 Root UNIs and 3 Leaf UNIs).
2. In Operator MEN C, OVC end point roles are not shown because the OVC is not type rooted-multipoint.[18]
3. In Operator MEN A, OVC end point roles are assigned as follows:
   - At the UNIs, OVC end point roles simply match the UNI designation (Leaf or Root).
   - At the ENNI connecting to Operator MEN C, the OVC end point role is set to Leaf, matching the UNI that connects to it through the ENNI.
   - At the ENNI connecting to Operator MEN B, the OVC end point role is set to Trunk because frames (ingressing from Operator MEN B) can originate from either a Root UNI or Leaf UNI.
4. In Operator MEN B, OVC end point roles are assigned as follows:
   - At the UNIs, OVC end point roles simply match the UNI designation (Leaf or Root).
   - At the ENNI, the OVC end point role is set to Trunk because frames (ingressing from Operator MEN A) can originate from either a Root UNI or Leaf UNI.
5. In Operator MEN A, the following Trunk Identifiers are assigned to the Trunk OVC end point:
   - Root S-VLAN ID value = 1001
   - Leaf S-VLAN ID value = 1002
6. In Operator MEN B, the following Trunk Identifiers are assigned to the Trunk OVC end point:
   - Root S-VLAN ID value = 2001
   - Leaf S-VLAN ID value = 2002

18 For point-to-point and multipoint-to-multipoint OVCs, each OVC end point is assigned role *Root*, but that assignment is trivial. It is done for consistency only.

Test your understanding: *A frame with broadcast MAC address (FF-FF-FF-FF-FF-FF) is transmitted from the CE into Operator MEN C. When it gets to Operator MEN B, the frame includes an S-Tag. What is the value of S-VLAN ID?*

Answer: *2002 (The frame originated at a Leaf UNI, so it should enter Operator MEN B with the Leaf S-VLAN ID value that is associated with the Trunk OVC Endpoint in Operator MEN B.)*

Test your understanding: *A frame with broadcast MAC address (FF-FF-FF-FF-FF-FF) and S-VLAN ID = 1001 enters Operator MEN A at the Trunk OVC end point. To which UNI(s) is it forwarded?*

Answer: *The frame is forwarded to both UNIs in Operator MEN A and to the UNI in Operator MEN C. S-VLAN ID = 1001 indicates that the frame originated at a Root UNI. It can therefore be forwarded to any Leaf UNI or Root UNI.*

Test your understanding: *Suppose that the Root UNI in Operator MEN B is changed into a Leaf UNI (so that both UNIs in Operator MEN B are Leaf UNIs) and that the role of the associated OVC end point is also changed from Root to Leaf (because all ingress frames now originate at a Leaf UNI).*

1. *Are Trunk OVC end points still required at the ENNI?*
2. *If not, what OVC end point roles can be used in place of Trunk?*

Answer:

1. *No. Trunk OVC end points can still be used, but they are no longer required because the OVC end points no longer have to support two forwarding states.*
2. *As shown in the following figure:*
    - *The Trunk OVC end point in Operator MEN B can be changed to a Root OVC end point (because it now only has to support ingress frames that originate at a Root UNI).*
    - *The Trunk OVC end point in Operator MEN A can be changed into a Leaf OVC end point (because it now only has to support ingress frames that originate at Leaf UNIs).*

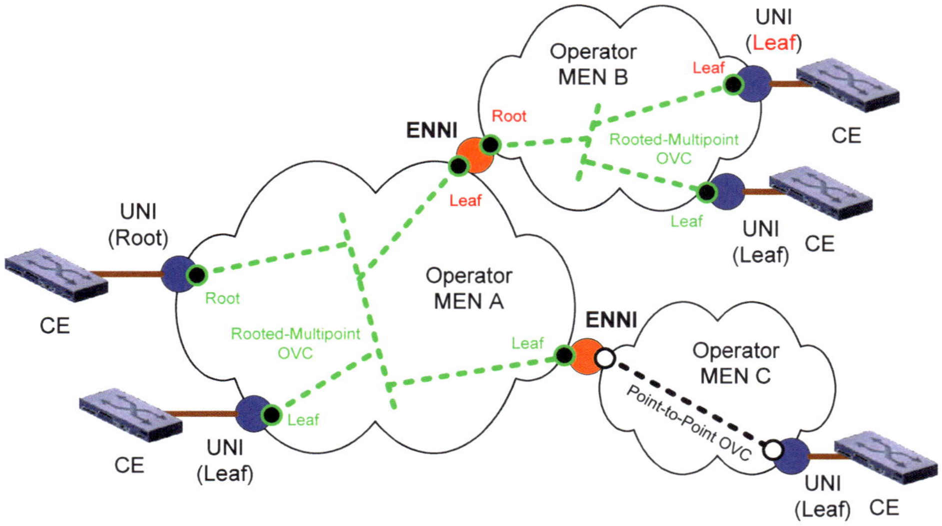

Note: *This solution is a more elegant (because it eliminates Trunk Identifiers), but it only works if all UNIs in Operator MEN B are Leaf UNIs.*

Example Application D

A rooted-multipoint EVC (for an E-Tree service) is implemented with 3 OVCs: a rooted-multipoint OVC in Operator MEN A, a bundled OVC in Operator MEN B, and a rooted-multipoint OVC in Operator MEN C.

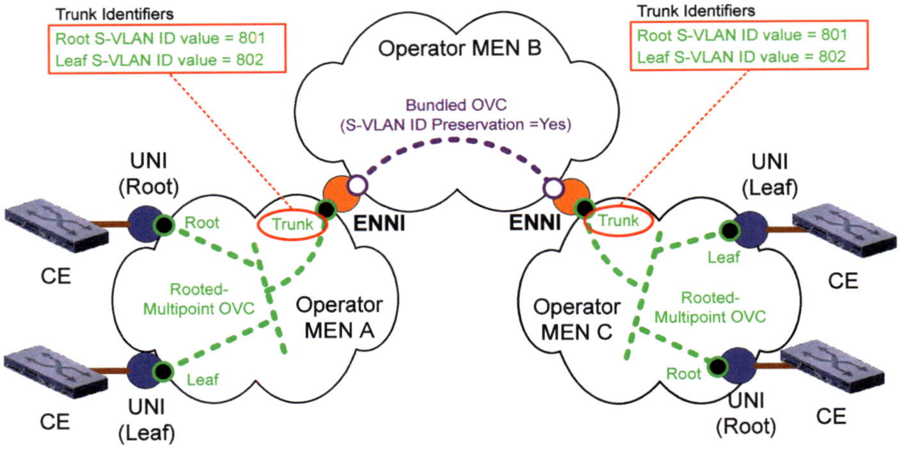

Observations:

1. The concatenation of OVCs results in a rooted-multipoint EVC that connects 4 UNIs (2 Root UNIs and 2 Leaf UNIs).
2. In Operator MEN B, OVC end point roles are not shown because the OVC is not type rooted-multipoint.[19]
3. In Operator MEN A and Operator MEN C, OVC end point roles are assigned as follows:
   - At the UNIs, OVC end point roles simply match the UNI designation (Leaf or Root).
   - At the ENNI, the OVC end point role is set to Trunk because frames (ingressing from Operator MEN B) can originate from either a Root UNI or Leaf UNI.
4. In Operator MEN A and Operator MEN C, the following Trunk Identifiers are assigned to the Trunk OVC end points:
   - Root S-VLAN ID value = 801
   - Leaf S-VLAN ID value = 802

19 For point-to-point and multipoint-to-multipoint OVCs, each OVC end point is assigned role *Root*, but that assignment is trivial. It is done for consistency only.

Test your understanding: *In this example, Trunk Identifier values (Root S-VLAN ID, Leaf S-VLAN) in Operator MEN A match those in Operator MEN C. Do they have to match? If so, why?*

Answer: *Yes, they have to match because the OVC in Operator MEN B is a bundled OVC (supporting multiple S-VLAN ID values), and bundled OVCs are required to have S-VLAN ID Preservation = Yes.*

Related Links

End Point Map Bundling

6.8
Review

1. Which entity delivers Ethernet frames between ENNIs and UNIs/ENNIs?
   - a. Operator Virtual Connection
   - b. Ethernet Virtual Circuit
   - c. Intermediate Virtual Connection
   - d. Operator Virtual Circuit
   - e. Ethernet Virtual Connection
2. Per MEF 26.1, which three values are recognized for the *Protection Mechanism* ENNI service attribute? (Choose three.)
   - a. LACP
   - b. None
   - c. Link Aggregation
   - d. E-LMI
   - e. Other
3. The *OVC End Points* of an operator virtual connection (OVC) must associate to:
   - a. Exactly two UNIs and any number of ENNIs
   - b. Exactly one ENNI and any number of UNIs
   - c. UNIs only
   - d. At least one ENNI
   - e. Only ENNIs
4. Two or more operators can:
   - a. connect through an ENNI
   - b. be involved in an OVC
   - c. be responsible for an EVC

   d. be involved in an EVC
   e. be responsible for an OVC

5. Which three frame formats are permitted at an ENNI? (Choose three.)
   a. Untagged
   b. C-Tag
   c. S-Tag and no C-Tag
   d. S-Tag and C-Tag
   e. Stacked S-Tags

6. How many EPL services can an ENNI support?
   a. One
   b. Two
   c. Up to 4094
   d. About 16 million
   e. Unlimited

7. Which statement about the External Network-to-Network Interface (ENNI) is true?
   a. The subscriber and service provider agree on its exact physical location so that responsibilities are clearly defined.
   b. It can only support one EPL service.
   c. Service provider is responsible for the service from the UNI to the ENNI, and then the organization in charge of the Operator MEN/CEN is responsible.
   d. It is a physical demarcation point between two Operator MEN/CENs.

8. An OVC requires which one of the following?
   a. A UNI
   b. Two or more UNIs
   c. Two or more EVCs
   d. An ENNI
   e. Two or more ENNIs

9. An *OVC End Point Map* has *bundling*. Which statement is true about the associated OVC?

   a. *S-VLAN ID Preservation* can be *Yes* or *No*.

   b. *CE-VLAN ID Preservation* can be *Yes* or *No*.

   c. *CE-VLAN CoS Preservation* can be *Yes* or *No*.

   d. *S-VLAN ID CoS Preservation* can be *Yes* or *No*.

10. In which category of operator service attributes is *Broadcast Service Frame Delivery*?

    a. OVC Service Attributes

    b. PVC End Point per ENNI Service Attributes

    c. ENNI Service Attributes

    d. UNI Service Attributes

    e. OVC per UNI Service Attributes

11. An *OVC End Point Map* has *bundling*. Which statement is <u>not</u> true about the associated OVC?

    a. The OVC cannot support hairpin switching.

    b. All end points of the OVC have bundling.

    c. *S-VLAN ID Preservation* is set to *Yes*.

    d. At least one end point of the OVC terminates at a UNI.

    e. The OVC is point-to-point.

12. Two Operator CENs connect through an ENNI. What is the advantage of using an ENNI with Link OAM?

    a. It provides link protection using two physical links, one active and one standby

    b. It allows the ENNI to support hairpin switching.

    c. It allows both operators to monitor and diagnose ENNI connectivity issues

    d. It allows two different types of physical layers to be connected at the ENNI

    e. It allows class of service (CoS) propagation through IEEE 802.1ad Ethernet Frames

13. Which OVC service attribute is <u>incompatible</u> with hairpin switching?

    a. CE-VLAN CoS Preservation = Yes

    b. S-VLAN ID Preservation = Yes

c. S-VLAN CoS Preservation = Yes

d. CE-VLAN ID Preservation = Yes

14. Which statement is not true about the service provider in the MEF 26.1 service model?

a. The service provider is responsible for the service to the subscriber.

b. The service provider is responsible for coordinating operators.

c. The service provider may or may not be an operator.

d. Two or more operators can collaboratively function as the service provider.

15. The smallest maximum transmission unit (MTU) size allowed for an operator virtual connection (OVC) is:

a. 1522 bytes

b. 1526 bytes

c. 4094 bytes

d. 4096 bytes

e. None of the above

16. How many EVCs can be supported by one OVC?

a. Exactly one

b. Up to 4094 if the OVC is point-to-point between two ENNIs

c. Up to 4094 if UNI is configured with *Service Multiplexing = Yes*

d. Exactly one if the UNI is configured with *All to One Bundling = Yes*

17. Which three of the following statements are true? (Choose three.)

a. At ENNIs, drop-eligibility can be encoded in the S-Tag DEI bit or in the S-Tag PCP value.

b. At ENNIs, all frames require C-Tags.

c. At an ENNI, both Operators are responsible for setting S-Tag values appropriately before transmitting frames across the ENNI.

d. Ingress ENNI frames are mapped to OVC end points based on S-VLAN ID value.

e. At ENNIs, CoS may be encoded in the C-Tag or in the S-Tag.

18. Which of the following is <u>not</u> an ENNI requirement?

- **a.** ENNIs are required to support E-LMI.
- **b.** At ENNIs, all bandwidth profiles must be color-aware.
- **c.** ENNIs must support MTU sizes at least as large as 1526 bytes.
- **d.** If an ENNI includes two physical links, it must be capable of implementing link aggregation.
- **e.** ENNIs are required to support Link OAM.

19. Which of the following is an ENNI service attribute?

- **a.** Bundling
- **b.** End Point Map
- **c.** Service Multiplexing
- **d.** All-to-one bundling

20. The *End Point Map* attribute:

- **a.** is inactive if the ENNI is provisioned with all-to-one bundling
- **b.** specifies how ingress ENNI frames are assigned to OVC end points based on S-VLAN ID
- **c.** lists the CE-VLAN IDs that map to the EVC
- **d.** is an *OVC End Point per ENNI* service attribute

21. A rooted-multipoint OVC has 5 OVC end points: 2 are designated Root, 2 are designated Leaf, and 1 is designated Trunk. A broadcast frame ingresses at a Leaf OVC end point. How many frames does the OVC deliver if OVC attribute *Broadcast Service Frame Delivery* is configured to *Deliver Unconditionally*?

- **a.** 1
- **b.** 2
- **c.** 3
- **d.** 4
- **e.** 5

22. Why is the minimum MTU size requirement at ENNIs (1526 bytes) four bytes larger than the minimum MTU size requirement at UNIs (1522 bytes)?

    a. Because ENNIs do not support all-to-one bundling.

    b. To allow hairpin switching.

    c. To account for the S-Tag.

    d. Because color-aware bandwidth profiles are required at ENNIs.

23. An EVPL service is implemented across three Operator CENs. How many UNIs, ENNIs, EVCs, and OVCs are involved?

    a. 2 UNIs, 2 ENNIs, 3 EVCs, and 1 OVC

    b. 2 UNIs, 3 ENNIs, 1 EVC, and 3 OVCs

    c. 2 UNIs, 2 ENNIs, 1 EVC, and 3 OVCs

    d. 2 UNIs, 3 ENNIs, 1 EVC, and 2 OVCs

24. An EPL service spans three Operator MENs and each OVC is configured with *S-VLAN ID Preservation = Yes*. Which of the following statements is true?

    a. The S-Tag VLAN identifier will match at both UNIs.

    b. The S-Tag VLAN identifier is preserved within each OVC, but can be translated at the ENNIs.

    c. The S-Tag VLAN identifier will match at both ENNIs.

    d. The S-Tag VLAN identifier will match at both UNIs and both ENNIs.

25. An EP-LAN service spans two Operator CENs that connect through one ENNI. At all UNIs, ingress bandwidth profiles are configured with CIR=0, CBS=0, EIR=20 Mb/s, and EBS= 2,000 bytes. How should the service be configured to ensure that yellow frames are not "promoted" to green?

    a. At all OVC end points, configure ingress and egress bandwidth profiles with *Color Mode = Color Aware*.

    b. At those OVC end points that connect to the ENNI (not those connecting to UNIs), configure ingress and egress bandwidth profiles with *Color Mode = Color Aware*.

    c. For both OVCs, configure *Color Forwarding = Yes*.

    d. At those OVC end points that connect to UNIs (not those connecting to the ENNI), configure ingress bandwidth profiles with *Color Mode = Color Aware*. Also, do not configure any bandwidth profiles at the OVC end points that connect to the ENNI.

26. Which statement about CoS identification at ENNIs is <u>not</u> true?

- **a.** CoS is determined by S-Tag PCP bit values.
- **b.** At a given OVC end point, an S-Tag PCP bit value can map to more than one CoS.
- **c.** For each OVC end point, a CoS must be associated with each S-Tag PCP value.
- **d.** At a given OVC end point, two different S-Tag PCP bit values can map to the same CoS.
- **e.** The S-Tag supports up to eight classes of service.

27. A service is implemented over 3 Operator MEN/CENs. Which three statements are correct? (Choose three.)

- **a.** All three operators are responsible to the subscriber for the service.
- **b.** One of the operators can be the service provider.
- **c.** The subscriber has a business relationship with each operator.
- **d.** Each Operator MEN/CEN includes at least one ENNI-N.
- **e.** At least two of the operators have no business relationship with the subscriber.

28. An ENNI is to support many EVCs, including some EVCs configured with *EVC MTU Size* as large as 1800 bytes. What is the smallest value of the ENNI MTU Size that will support these EVCs?

- **a.** 1526 bytes
- **b.** 1800 bytes
- **c.** 1804 bytes
- **d.** 2000 bytes

29. A point-to-point OVC-based service can be configured with OVC attribute *S-VLAN ID Preservation* set to *Yes* or *No*. Which two statements are correct? (Choose two.)

- **a.** If *S-VLAN ID Preservation* = *Yes*, the OVC-based service involves exactly 1 S-VLAN ID value.
- **b.** If *S-VLAN ID Preservation* = *Yes*, the OVC-based service can involve up to 4094 S-VLAN ID values.
- **c.** If *S-VLAN ID Preservation* = *No*, the OVC-based service involves exactly 2 S-VLAN ID values.
- **d.** If *S-VLAN ID Preservation* = *No*, the OVC-based service involves 1 or 2 S-VLAN ID values.

Related Links

Answers: Extending MEF Services over Multiple Operator MEN/CENs

7

MEF-Standardized Classes of Service

In this chapter:

Note: *The initial MEF-CECP exam did not include this material. It was introduced in MEF 23.1.*

EVC performance specification (or OVC service level specification), as previously described, requires the service provider and the subscriber (or the service provider and the operator) to reach a bilateral agreement with respect to many CoS details, including the following (for each CoS involved in the service):

CoS performance objectives–values of performance metrics[20] that represent agreed performance commitments
CoS performance parameters–values, such as time intervals, required to derive and specify CoS performance objectives
CoS ID–Class of Service Identifier. The mechanism and values used to identify CoS at UNIs (and ENNIs).
Color ID–Color Identifier. The mechanism and values used to identify frame color at UNIs (and ENNIs).

20 Performance metrics include Availability, High Loss Intervals (HLI), Consecutive High Loss Intervals (CHLI), Frame Loss Ration (FLR), Frame Delay (FD), Mean Frame Delay (MFD), Frame Delay Range (FDR), and Inter-Frame Delay Variation (IFDV).

So far, we have learned how these four items are used to define classes of service (what is allowed, how details are specified). We now look at some specific classes of service.

This lesson describes twelve classes of service defined in MEF 23.1 that conform to previously described requirements, but are much more prescribed. The purpose of these MEF-standardized classes of service is to establish a common baseline to help subscribers, service providers, and operators easily agree on CoS-related service attributes. MEF-standardized classes of service do not constrain CoS options offered by a service provider or operator. They are meant to help establish a common reference for communicating CoS performance requirements and capabilities.

Development of an industry-wide menu of standardized classes of service makes good business sense from all perspectives (subscriber, service provider, or operator) because it simplifies the buying and selling of Ethernet services. The twelve classes of service described in this lesson represent a step in that direction.

The MEF-standardized classes of service described in this lesson can be applied in any MEF service application. They appear explicitly in mobile backhaul CoS requirements and in CE 2.0 certification.

7.1
Definitions and Requirements

In this section:

7.1.1
CoS Labels and Performance Tiers

Two values are used to identify each of the MEF-standardized classes of service:

- **CoS Label** – H, M, or L
- **Performance Tier** – PT1, PT2, PT3, or PT4

Each combination of two values (for example, M and PT3) identifies one of 12 classes of service defined in MEF 23.1.

Informally, the CoS Labels (L, M, H) indicate relative performance level (Low, Medium, High). Performance Tiers (PT1, PT2, PT3, PT4) indicate "field of use" or "area of applicability," which approximately aligns to network size, transmission distance, and/or propagation delay as shown in the following table.

Performance Tier	Field of Use (Area of Applicability)		
	Network Size	Transmission Distance	Propagation Delay
PT1	Metro	< 250 km	< 2 ms
PT2	Regional	< 1200 km	< 8 ms
PT3	Continental	< 7000 km	< 44 ms
PT4	Global	< 27500 km	< 172 ms

For each (CoS Label, Performance Tier) pair, MEF 23.1 specifies class of service details. As the naming convention suggests, these details were derived to represent realistic performance levels (Low, Medium, High) in different networking contexts (Metro, Regional, Continental, Global). However, these associations are informal; they place no limitations on how these MEF-standardized classes of service are used in practice.

MEF 23.1 also includes the following table, which maps applications to values of CoS Label and Performance Tier. Again, this table is only meant to provide helpful information; it does not limit how MEF-standardized classes of service are actually used in practice.

CoS Label	H				M				L			
Performance Tier	**1**	**2**	**3**	**4**	**1**	**2**	**3**	**4**	**1**	**2**	**3**	**4**
VoIP	X	X	X	X								
VoIP & videoconf signaling					X	X	X	X				
Videoconf data					X	X	X	X				
IPTV data					X	X	X					
IPTV control					X	X	X					
Streaming media									X	X	X	X
Interactive gaming	X	X			X	X						
SANs synch replication					X							
SANs asynch replication					X							
Network attached storage									X	X	X	X
Text & graphics terminals									X	X	X	X
T.38 fax over IP					X	X	X	X				
Database hot standby					X							
Database WAN replication					X							
Database client/server									X	X	X	X
Financial/Trading	X											
CCTV					X	X	X	X				
Telepresence	X	X	X									
Circuit Emulation	X											
Mobile BH H	X											
Mobile BH M					X							
Mobile BH L									X			

Figure 72: Application Mapping for Derivation of CoS Performance Objectives (Table 36 in MEF 23.1)

7.1.2 Ingress Bandwidth Profile Constraints (per CoS Label)

Per MEF 23.1, CoS Labels impose the following constraints on ingress bandwidth profiles.

CoS Label	Ingress Bandwidth Profile Constraints	
H	CIR > 0	EIR ≥ 0 (typically EIR=0)
M	CIR > 0	EIR ≥ 0
L	CIR ≥ 0	EIR ≥ 0

Key Observation: *CoS Labels M and H require CIR > 0.*

Note: *Although CoS Label H allows EIR > 0, EIR is typically set to zero for CoS Label H. As stated in MEF 26.1, EIR>0 is allowed for CoS Label H to support "some situations for certain applications such as Mobile Backhaul."*

Related Links

Bandwidth Profiles

7.1.3
CoS Label and Color Identification Mechanisms

Mechanisms for CoS Label Identification

Per MEF 23.1, CoS Label value (L, M, H) can be assigned to ingress frames using any of the usual mechanisms:

CoS Label Identification Mechanisms	
At UNIs	**At ENNIs**
• Based on EVC (or OVC end point) • Based on the value of PCP bits in C-Tag • Based on the value of the differentiated services code point (DSCP) in the IP header • Based on Layer 2 Control Protocol	• Based on OVC end point • Based on the value of PCP bits in S-Tag

Mechanisms for Color Identification

Color value (Green, Yellow), if needed[21], can be assigned to ingress frames based on DEI bit value (in S-Tag), PCP bit value (in C-Tag or S-Tag), or DSCP value (in IP header).

Related Links

CoS Identification at UNIs
Encoding Information in S-Tags
Color-Aware Bandwidth Profiles at ENNI
Color Forwarding

7.1.4
CoS Label and Color Identification Using PCP Bits, DEI Bit, and/or EVC/OVC End Point

Note: *Values shown in the following tables are not example values. They are required per MEF 23.1.*

If CoS Labels are identified using PCP bits, PCP bits map to CoS Label and Color values as follows:

21 Color identification is used for color forwarding and color-aware ingress bandwidth profiles.

CoS Label	MEF-Specified PCP Bit Value			
	In C-Tag (or in S-Tag without DEI supported)		In S-Tag with DEI Supported	
	Color Green	Color Yellow	Color Green	Color Yellow
H	5	Not Supported	5 (with DEI=0)	Not Supported
M	3	2	3 (with DEI=0)	3 (with DEI=1)
L	1	0	1 (with DEI=0)	1 (with DEI=1)

If the S-Tag DEI bit is used to encode color, then PCP bits encode only CoS Label value.

If CoS Labels are assigned per EVC or per OVC end point, PCP bits can encode Color as follows:

MEF-Specified PCP Bit Value in C-Tag	
Color Green	Color Yellow (CoS Label ≠H)
5, 3, or 1	2 or 0 **Note:** *Color Yellow is not supported for CoS Label H.*

In this case, PCP bits do not determine CoS Label value (L, M, H). They only determine Color.

7.1.5 CoS Label and Color Identification Using DSCP Values and/or EVC/OVC End Point

Note: *Values shown in the following tables are not example values. They are required, per MEF 23.1.*

Note: *For reference, the following tables include IP per-hop behavior (PHB) for each differentiated services code point (DSCP) value. In the differentiated services (DiffServ) computer networking architecture, DSCP values determine PHB.*

If CoS Labels are identified using DSCP values, DSCP values map to CoS Label and Color values as follows:

CoS Label	MEF-Specified PHB (DSCP Value)	
	Color Green	Color Yellow
H	EF (46)	Not Supported
M	AF31 (26)	AF32 (28) or AF33 (30)
L	AF11 (10)	AF12 (12), AF13 (14), or DF(0)

If CoS Labels are assigned per EVC or per OVC end point, DSCP values can encode Color as follows:

MEF-Specified PHB (DSCP Value)	
Color Green	**Color Yellow (CoS Label ≠H)**
EF (46), AF31 (26), or AF11 (10)	AF32 (28), AF33 (30), AF12 (12), AF13 (14), or DF(0) ***Note:*** *Color Yellow is not supported for CoS Label H*

In this case, DSCP values do not determine CoS Label value (L, M, H). They only determine Color.

7.1.6 CoS Label Identification for L2CP Frames

Per MEF 23.1, subscriber L2CP frames:

- Can be assigned to any CoS Label value that is supported by the EVC/OVC
- Should be assigned CoS Label M if M is supported. Otherwise, CoS Label H if H is supported. Otherwise, CoS Label L.

Note: *As explained in MEF 23.1, "the M CoS Label is chosen for L2CP whenever available, based on its superior loss performance, and a desire to keep it separate from real-time applications."*

7.1.7 Key Observations About CoS Label and Color Identification

- Ingress frames are mapped to CoS Label values (L, M, H) using the usual mechanisms (based on EVC, OVC end point, PCP bits, DSCP value, or L2CP).
- If PCP bits encode CoS Label only, values 5, 3, and 1 correspond to H, M, and L, respectively.
- If PCP bits encode CoS Label and Color, values 5, 3, 2, 1, and 0 correspond to H (Green), M (Green), M (Yellow), L (Green), and L (Yellow), respectively.
- Color Yellow is not supported for CoS Label H.
- L2CP frames can be assigned to any CoS Label value that is supported by the EVC/OVC. For this purpose, M is preferred over H/L, and H is preferred over L.

7.1.8 Performance Metrics for CoS Labels

EVC/OVC performance for a CoS Label is defined using the usual set of 8 performance metrics:

1. One-way Frame Delay (FD)
2. One-way Frame Delay Range (FDR)
3. One-way Mean Frame Delay (MFD)
4. Inter Frame Delay Variation (IFDV)
5. One-way Frame Loss Ratio (FLR)
6. One-way Availability (Availability)
7. One-way High Loss Intervals (HLI)
8. One-way Consecutive High Loss Intervals (CHLI)

Per MEF 23.1 an SLS that is based on a CoS Label must include at least one of either FD or MFD and at least one of either FDR or IFDV.

Related Links

EVC Performance Service Attributes
CPOs (per CoS Label and Performance Tier)

7.1.9
CoS Parameter Values (per CoS Label)

The following table from MEF 23.1 defines CoS Parameter values for CoS Labels H, M, and L.

Performance Metric	Parameter Name	Parameter Values for CoS Label H	Parameter Values for CoS Label M	Parameter Values for CoS Label L
FD	Percentile (P_d)	≥ 99.9th	≥ 99th	≥ 95th
	Time Interval (T)	≤ Month	≤ Month	≤ Month
MFD	Time Interval (T)	≤ Month	≤ Month	≤ Month
IFDV	Percentile (P_v)	≥ 99.9th	≥ 99th or N/S[1]	N/S
	Time Interval (T)	≤ Month	≤ Month or N/S[1]	N/S
	Pair Interval (Δt)	≥ 1sec	≥ 1sec or N/S[1]	N/S
FDR	Percentile (P_r)	≥ 99.9th	≥ 99th or N/S[1]	N/S
	Time Interval (T)	≤ Month	≤ Month or N/S[1]	N/S
FLR	Time Interval (T)	≤ Month	≤ Month	≤ Month
Availability	TBD	TBD	TBD	TBD
High Loss Interval	TBD	TBD	TBD	TBD
Consecutive High Loss Interval	TBD	TBD	TBD	TBD

[1] Parameters are N/S only when CPO is N/S
Note: each parameter value > 0

The inequalities in this table indicate that service providers and operators may provide more stringent values than the maximum and minimum values shown.

7.1.10 CPOs (per CoS Label and Performance Tier)

MEF 23.1 defines CoS Performance Objective (CPO) values for point-to-point EVCs and OVCs, as follows:

Table 18: CPOs for Point-to-Point EVCs and OVCs (From Tables 6-9 in MEF 23.1)

CoS Label	H				M				L			
Performance Tier	PT1	PT2	PT3	PT4	PT1	PT2	PT3	PT4	PT1	PT2	PT3	PT4
FD (ms)	≤10	≤25	≤77	≤230	≤20	≤75	≤115	≤250	≤37	≤125	≤230	≤390
MFD (ms)	≤7	≤18	≤70	≤200	≤13	≤30	≤80	≤220	≤28	≤50	≤125	≤240
IFDV (ms)	≤3	≤8	≤10	≤32	≤8*	≤40*			N/S	N/S	N/S	N/S
FDR (ms)	≤5	≤10	≤12	≤40	≤10*	≤50*			N/S	N/S	N/S	N/S
FLR (%)	≤0.01		≤0.025	≤0.05	≤0.01		≤0.025	≤0.05	≤0.1			
Availability	TBD	TBD	TBD	TBD	TBD	TBD	TBD	TBD	TBD	TBD	TBD	TBD
HLI	TBD	TBD	TBD	TBD	TBD	TBD	TBD	TBD	TBD	TBD	TBD	TBD
CHLI	TBD	TBD	TBD	TBD	TBD	TBD	TBD	TBD	TBD	TBD	TBD	TBD

* Compliant services can leave this value unspecified.
N/S indicates Not Specified.

Additionally:

- The inequalities in this table indicate that service providers and operators may provide more stringent values than the maximum values shown.
- CPO values are not required for both FD and MFD (at least one must be specified).
- CPO values are not required for both IFDV and FDR (at least one must be specified).
- All CPO values apply to Green frames only.
- MEF 23.1 does not define CPO values for multipoint-to-multipoint (or rooted-multipoint) EVCs and OVCs.

Related Links

EVC Performance Service Attributes
Service Level Specification (OVC Service Attribute)
Performance Metrics for CoS Labels

7.2 Review

1. How many classes of service does MEF 23.1 define for Performance Tier 2?

    a. 2 classes of service

    b. 3 classes of service

    c. 4 classes of service

    d. 6 classes of service

    e. 12 classes of service

2. An *Ingress Bandwidth Profile Per ENNI Class of Service Identifier* is required at an OVC end point to police ingress ENNI frames marked with *CoS Label M*. Which one of the following ingress bandwidth profiles satisfies all MEF requirements?

    a. CIR=0, CBS=0 bytes, EIR=200 Mb/s, EBS=2000 bytes, CF=Y, CM=Y

    b. CIR=200 Mb/s, CBS=2000 bytes, EIR=200 Mb/s, EBS=2000 bytes, CF=N, CM=N

    c. CIR=200 Mb/s, CBS=2000 bytes, EIR=200 Mb/s, EBS=2000 bytes, CF=N, CM=Y

    d. CIR=0, CBS=0 bytes, EIR=200 Mb/s, EBS=2000 bytes, CF=Y, CM=N

3. A point-to-point service uses *CoS Label H*. Which CPO must be specified in the SLS?

    a. Both Frame Delay (FD) and Frame Delay Range (FDR)

    b. Both Mean Frame Delay (MFD) and Inter Frame Delay Variation (IFDV)

    c. Frame Loss Ratio (FLR)

    d. High Loss Intervals (HLI)

    e. Consecutive High Loss Intervals (CHLI)

4. According to MEF 23.1, if CoS Label and Color are encoded in S-Tag PCP bits, PCP bit value 3 indicates...

    a. M, Yellow

    b. M, Green

    c. H, Green

    d. Nothing

5. Which two statements about Class of Service Performance Objectives (CPOs) are correct? (Choose two.)
   - **a.** CPOs for *CoS Label L* are more demanding than CPOs for *CoS Label H*.
   - **b.** CPOs for *Performance Tier 4* are more demanding than CPOs for *Performance Tier 3*.
   - **c.** CPOs for *CoS Label M* are more demanding than CPOs for *CoS Label L*.
   - **d.** CPOs for *Performance Tier 1* are more demanding than CPOs for *Performance Tier 3*.
6. How many performance tiers does MEF 23.1 define for CoS Label H?
   - **a.** 3 performance tiers
   - **b.** 4 performance tiers
   - **c.** 6 performance tiers
   - **d.** 12 performance tiers
7. For each of the following services, all service frames map to *CoS Label L*. Which service is subject to CPO values defined in MEF 23.1?
   - **a.** *Performance Tier 1* EP-LAN service with ingress bandwidth profile CIR > 0.
   - **b.** *Performance Tier 4* EVPL service with ingress bandwidth profile CIR > 0.
   - **c.** *Performance Tier 1* EPL service with ingress bandwidth profile CIR = 0.
   - **d.** *Performance Tier 4* EVP-LAN service with ingress bandwidth profile CIR = 0.
8. According to MEF 23.1, if CoS Label and Color is encoded in S-Tag PCP bits, PCP bit value 4 indicates...
   - **a.** M, Yellow
   - **b.** M, Green
   - **c.** H, Green
   - **d.** Nothing

Related Links

Answers: MEF-Standardized Classes of Service

8

E-Access Services

In this chapter:

Note: *The initial MEF-CECP exam did not include this material. It was introduced in MEF 33.*

8.1
Overview

In this section:

The previously described general framework for implementing Carrier Ethernet services across multiple Operator MENs supports the development of OVC-based services (services deployed by an operator and sold to the service provider). But this framework is very open, leaving it up to the service providers and operators to work out many service details.

In defining E-Access services, the MEF goes one step further: to define industry-standard OVC-based Ethernet services for Ethernet access applications.

Ethernet access is a critical topic in Carrier Ethernet because carrier networking has evolved into a tiered industry with separate businesses and networks operating at local, metro, regional, continental, and global levels. Large carriers that are well positioned to offer wholesale Ethernet services generally have to work with many access providers (providers that own and operate the first/last mile network connections) to acquire Ethernet access to subscriber sites. Access providers that are well positioned to offer Ethernet access services commonly work with more than one large carrier.

Today, global Ethernet service providers may have hundreds of interconnect agreements with many different access providers. Agreements that are unique and time consuming to procure will delay service deployments. E-Access services are designed to address this issue, making the Ethernet access business manageable for Ethernet service providers and Ethernet access providers alike. E-Access services allow access providers to mass market Ethernet access services similarly to how they currently sell standardized T1 or E1 access circuits. They also greatly simplify the purchasing process for large Ethernet service providers, allowing them to purchase standardized Ethernet access services similarly to how they have purchased standardized T1 or E1 access circuits for decades from any access provider they partner with.

An E-Access service can be used in two ways:

1. **In combination with other OVC-based service**—to support EVC-based service, providing UNI-to-UNI connectivity between customer sites

2. **Alone**—to connect a customer site to an IP network

8.1.1
EVC-Based Service Application

In this application a service provider uses E-Access services to reach out-of-franchise subscriber locations.

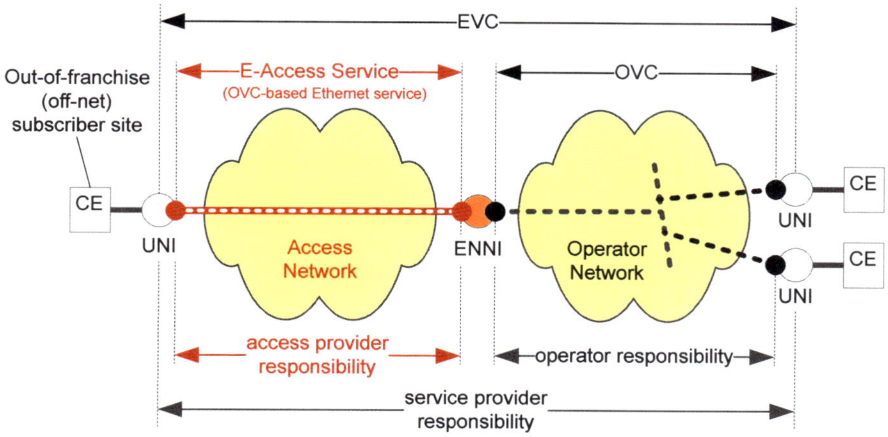

The business model has two principle stakeholders:

- **Ethernet Access Provider**–The organization that operates the access network and offers the E-Access service to the service provider
- **Service Provider**–The organization that purchases the E-Access service to reach an out-of-franchise (off-net) subscriber site in order to deliver the end-to-end service to the subscriber

Note: *The subscriber has no business relationship with the Ethernet access provider and may be unaware of the E-Access service.*

8.1.2
IP Network Access Application

In this application, an access provider uses E-Access services to connect customer sites to an IP network, for example, the network of an Internet service provider.

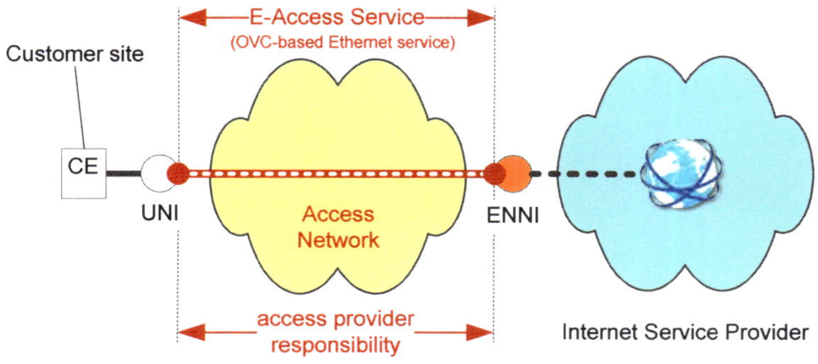

MEF 33 defines two E-Access services: the **Access EPL** (Access Ethernet Private Line) and the **Access EVPL** (Access Ethernet Virtual Private Line). Both provide a point-to-point OVC between a UNI and an ENNI.

8.1.3
Access EPL

The Access EPL (Access Ethernet Private Line) is a port-based point-to-point E-Access service.

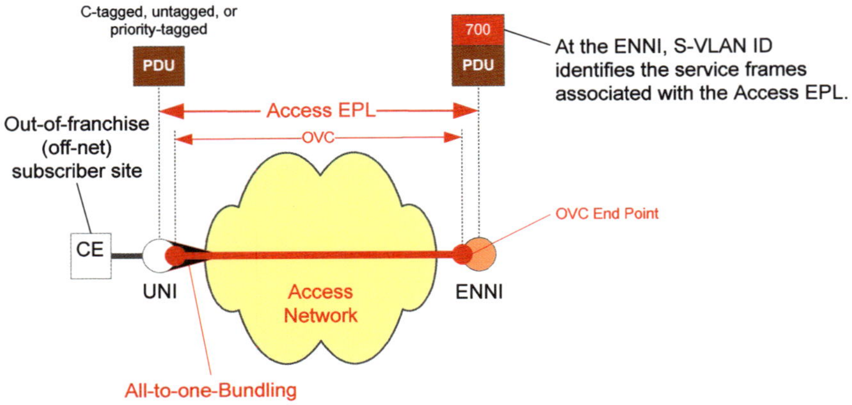

All frames at the UNI map to the service. At the ENNI, service frames are identified by S-VLAN ID.

Note: *The UNI of an Access EPL implicitly supports all-to-one bundling.*

8.1.4
Access EVPL

The Access EVPL (Access Ethernet Virtual Private Line) is a VLAN-based point-to-point E-Access service.

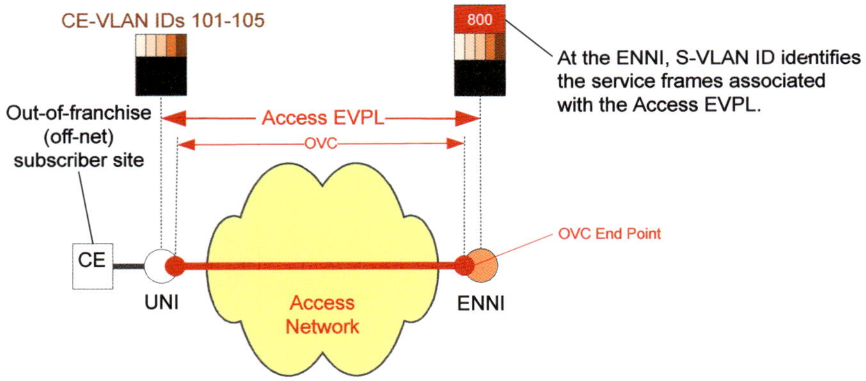

At the UNI, frames are mapped to the service based on CE-VLAN ID. At the ENNI, service frames are identified by S-VLAN ID.

Note: *The Access EVPL UNI implicitly supports bundling and service multiplexing. The OVC does not support CE-VLAN ID translation.*

8.2
E-Access Service Attributes

In this section:

MEF 33 defines E-Access services using a subset of MEF 26.1 service attributes (attributes used in MEF 26.1 to define general OVC-based operator services).

The following tables list MEF 26.1 service attributes and indicate how they are used and modified in MEF 33 to define E-Access services. These tables are meant for overview purposes only. Scanning them, you should notice that many attributes are unchanged from the general case and changes are often just the straight-forward consequence of specialization to point-to-point, port-based, or VLAN-based service.

Attributes shown in **bold** are explained later in this lesson.

Related Links

Operator Service Attributes

8.2.1
ENNI Service Attributes for E-Access

General OVC-Based Operator Service Attribute (MEF 26.1)		E-Access Service Attribute (MEF 33)	
Name	**Possible Values**	**Access EPL**	**Access EVPL**
Operator ENNI Identifier	A string that is unique across the Operator MEN/CEN	No change	No change
Physical Layer	1000Base-SX, 1000Base-LX, 1000Base T, 10GBase-SR, 10GBase-LX4, 10GBase-LR, 10GBase-ER, 10GBase-SW, 10GBase-LW, 10GBase-EW	No change	No change
Frame Format	Conforming to IEEE 802.1ad (2005)	No change	No change
Number of Links	1 or 2	No change	No change
Protection Mechanism	Link Aggregation, None, or Other	No change	No change
ENNI Maximum Transmission Unit Size	Integer ≥ 1526 bytes	No change	No change

General OVC-Based Operator Service Attribute (MEF 26.1)		E-Access Service Attribute (MEF 33)	
Name	Possible Values	Access EPL	Access EVPL
End Point Map	A table with rows of the form <S-VLAN ID value, End Point Identifier, End Point Type>	No change	No change
Maximum Number of OVCs	Integer ≥ 1	No change	No change
Maximum Number of OVC End Points per OVC	Integer ≥ 1	No change	No change

Related Links

ENNI Service Attributes

8.2.2 OVC Service Attributes for E-Access

General OVC-Based Operator Service Attribute (MEF 26.1)		E-Access Service Attribute (MEF 33)	
Name	Possible Values	Access EPL	Access EVPL
OVC Identifier	A string that is unique across the Operator MEN/CEN	No change	No change
OVC Type	Point-to-Point, Multipoint-to-Multipoint, Rooted-multipoint	Must be Point-to-Point	Must be Point-to-Point
OVC End Point List	A list of OVC End Point Identifiers	Exactly 2: one at UNI and one at ENNI	Exactly 2: one at UNI and one at ENNI
Maximum Number of UNI OVC End Points	Integer ≥ 0	Must be 1	Must be 1
Maximum Number ENNI OVC End Points	Integer ≥ 1	Must be 1	Must be 1
OVC Maximum Transmission Unit Size	Integer ≥ 1526 bytes	No change	No change
CE-VLAN ID Preservation	**Yes or No**	**Must be Yes**	**Must be Yes**
CE-VLAN CoS ID Preservation	**Yes or No**	**Must be Yes**	**Must be Yes**
S-VLAN ID Preservation	Yes or No	Not applicable (only one ENNI in the service)	Not applicable (only one ENNI in the service)
S-VLAN CoS Preservation	Yes or No	Not applicable (only one ENNI in the service)	Not applicable (only one ENNI in the service)
Color Forwarding	Yes or No	Should be Yes	Should be Yes

General OVC-Based Operator Service Attribute (MEF 26.1)		E-Access Service Attribute (MEF 33)	
Name	**Possible Values**	**Access EPL**	**Access EVPL**
Service Level Specification	Same as for general OVC-based service. Eight performance metrics are specified, where Not Specified (N/S) is an acceptable value: 1. One-way Frame Delay (FD) 2. One-way Frame Delay Range (FDR) 3. One-way Mean Frame Delay (MFD) 4. Inter Frame Delay Variation (IFDV) 5. One-way Frame Loss Ratio (FLR) 6. One-way Availability (Availability) 7. One-way High Loss Intervals (HLI) 8. One-way Consecutive High Loss Intervals (CHLI)		
Unicast Service Frame Delivery	Deliver conditionally or unconditionally	Must deliver unconditionally	Deliver conditionally or unconditionally
Multicast Service Frame Delivery	Deliver conditionally or unconditionally	Must deliver unconditionally	Deliver conditionally or unconditionally
Broadcast Service Frame Delivery	Deliver conditionally or unconditionally	Must deliver unconditionally	Deliver conditionally or unconditionally
Layer 2 Control Protocol Tunneling	A list on Layer 2 Control Protocols that are tunnelled.	Processing of L2CP frames is for further study. Until defined in a future revision of MEF 33, processing of L2CP frames is agreed to by the two parties involved in the Access Service.	

Related Links

OVC Service Attributes
Service Level Specification (OVC Service Attribute)

8.2.3
OVC End Point per ENNI Service Attributes for E-Access

General OVC-Based Operator Service Attribute (MEF 26.1)		E-Access Service Attribute (MEF 33)	
Name	**Possible Values**	**Access EPL**	**Access EVPL**
OVC End Point Identifier	A string that is unique across the Operator MEN/CEN	No change	No change
Class of Service Identifiers	**Non-overlapping sets of S-Tag PCP values; one for each CoS**	**All frames map to one CoS (regardless of S-Tag PCP value)**	**All frames map to one CoS (regardless of S-Tag PCP value)**

General OVC-Based Operator Service Attribute (MEF 26.1)		E-Access Service Attribute (MEF 33)	
Name	**Possible Values**	**Access EPL**	**Access EVPL**
Ingress Bandwidth Profile Per OVC End Point	**None or one set of bandwidth profile values (CIR, CBS, EIR, EBS, CF, CM)**	**Supported per special requirements, as explained later in this lesson**	**Supported per special requirements, as explained later in this lesson**
Ingress Bandwidth Profile Per ENNI Class of Service Identifier	**None or one set of bandwidth profile values (CIR, CBS, EIR, EBS, CF, CM) for each CoS**	**Not supported**	**Not supported**
Egress Bandwidth Profile Per End Point	**None or one set of bandwidth profile values (CIR, CBS, EIR, EBS, CF, CM)**	**Not supported**	**Not supported**
Egress Bandwidth Profile Per ENNI Class of Service Identifier	**None or one set of bandwidth profile values (CIR, CBS, EIR, EBS, CF, CM) for each CoS**	**Not supported**	**Not supported**

Related Links

OVC End Point per ENNI Service Attributes
Bandwidth Profiles for E-Access Service

8.2.4
UNI Service Attributes for E-Access

General OVC-Based Operator Service Attribute (MEF 26.1)		E-Access Service Attribute (MEF 33)	
Name	**Possible Values**	**Access EPL**	**Access EVPL**
UNI Identifier	Arbitrary text string to identify the UNI	No change	No change
Physical Medium	UNI Type 2 Physical Interface	No change	No change
Speed	10 Mb/s, 100 Mb/s, 10/100 Mb/s Auto-negotiation, 10/100/1000 Mb/s Auto-negotiation, 1 Gb/s, or 10 Gb/s	No change	No change
Mode	Full Duplex	No change	No change
MAC Layer	IEEE 802.3-2005	No change	No change
UNI MTU Size	≥ 1522 bytes	No change	No change
Service Multiplexing	**Yes or No. MUST be No if *All to One Bundling* is *Yes*.**	**Not applicable**	**Not applicable**

General OVC-Based Operator Service Attribute (MEF 26.1)		E-Access Service Attribute (MEF 33)	
Name	**Possible Values**	**Access EPL**	**Access EVPL**
Bundling	**Yes or No. MUST be No if *All to One Bundling* is *Yes*.**	**Not applicable**	**Not applicable**
All to One Bundling	**Yes or No. MUST be No if *Bundling* or *Service Multiplexing* is *Yes*.**	**Not applicable**	**Not applicable**
CE-VLAN ID for untagged and priority tagged Service Frames	MUST specify *CE-VLAN ID for untagged and priority tagged Service Frames* in the range of 1-4094. This requirement does not apply for services with all to one bundling at the UNI.	Not applicable	No change
Maximum number of EVCs	Integer ≥ 1	Must be 1	Integer ≥ 1
Ingress Bandwidth Profile Per UNI (optional)	**If this attribute is supported, MUST specify CIR, CBS, EIR, EBS, CM, or CF. MUST NOT be combined with any other type of ingress bandwidth profile.**	**Must not specify**	**Must not specify**
Egress Bandwidth Profile Per UNI (optional)	**If this attribute is supported, MUST specify CIR, CBS, EIR, EBS, CM, or CF. MUST NOT be combined with any other type of egress bandwidth profile.**	**Must not specify**	**Must not specify**
L2CP Processing	For each protocol, MUST specify one of the following: *Peer*, *Discard*, *Pass to EVC*, or *Peer and Pass to EVC*.	Processing of L2CP frames is for further study. Until defined in a future revision of MEF 33, processing of L2CP frames is agreed to by the two parties involved in the Access Service	

Related Links

UNI Service Attributes

8.2.5 OVC per UNI Service Attributes for E-Access

General OVC-Based Operator Service Attribute (MEF 26.1)		E-Access Service Attribute (MEF 33)	
Name	**Possible Values**	**Access EPL**	**Access EVPL**
UNI OVC Identifier	A string formed by the concatenation of the UNI Identifier and the OVC Identifier	No change	No change
OVC End Point Map	A list of one or more CE-VLAN ID values	All service frames map to a single OVC.	A list of one or more CE-VLAN ID values

General OVC-Based Operator Service Attribute (MEF 26.1)		E-Access Service Attribute (MEF 33)	
Name	**Possible Values**	**Access EPL**	**Access EVPL**
Ingress Bandwidth Profile Per OVC End Point at a UNI	None or one set of bandwidth profile values (CIR, CBS, EIR, EBS, CF, CM)	Supported per special requirements, as explained later in this lesson	Supported per special requirements, as explained later in this lesson
Ingress Bandwidth Profile Per Class of Service Identifier at a UNI	None or one set of bandwidth profile values (CIR, CBS, EIR, EBS, CF, CM) for each CoS	Not supported	Not supported
Egress Bandwidth Profile Per OVC End Point at a UNI	None or one set of bandwidth profile values (CIR, CBS, EIR, EBS, CF, CM)	Not supported	Not supported
Egress Bandwidth Profile Per Class of Service Identifier at a UNI	None or one set of bandwidth profile values (CIR, CBS, EIR, EBS, CF, CM) for each CoS	Not supported	Not supported

Related Links

OVC per UNI Service Attributes
Bandwidth Profiles for E-Access Service

8.3 Highlights of E-Access Service Definitions

In this section:

8.3.1 Bundling and Multiplexing for E-Access Services

E-Access services (unlike general OVC-based services) do not include UNI service attributes **Bundling**, **Service Multiplexing**, or **All to One Bundling**.

MEF 33 explains that these attributes are omitted because they are not relevant to the agreement between the Access Provider and the Service Provider. However, their values (were they to be included) would be as shown in the following table.

UNI Service Attribute	Access EPL	Access EVPL
Service Multiplexing	No	Yes
Bundling	No	Yes/No
All to One Bundling	Yes	No

Observe that the values shown for the Access EPL are consistent with the service being port-based at the UNI.

The values shown for the Access EVPL are consistent with all MEF 33 requirements. In particular:

- The service being VLAN-based at the UNI
- The service supporting the association of multiple CE-VLAN IDs to the OVC at the UNI[22]
- The service requiring OVC service attribute *CE-VLAN ID Preservation=Yes*

22 To support more than one CE-VLAN ID, UNI attribute Bundling must be Yes.

Related Links

Bundling versus Service Multiplexing
Service Multiplexing, Bundling, and VLAN Tag Preservation Service Attributes

8.3.2
CE-VLAN ID Preservation for E-Access Services

MEF 33 requires OVC service attribute **CE-VLAN ID Preservation** = **Yes** for both kinds of E-Access service. For the Access EPL, this requirement is expected because the Access EPL is a port-based service. For the Access EVPL, this requirement represents a constraint compared to other VLAN-based services. Other VLAN-based services (EVPL, EVP-LAN, and EVP-Tree services) support CE-VLAN ID translation if UNI attribute *Bundling=No*. The Access EVPL, however, does not support CE-VLAN ID translation.

Note: *Elimination of CE-VLAN ID translation support simplifies E-Access services, making them easier to manage.*

Test your understanding: *Can an Access EVPL support an EVC-based service that includes CE-VLAN ID translation? Explain why or why not.*

Answer: *Yes. If the Access EVPL is deployed in combination with a general OVC-based operator service that includes CE-VLAN ID translation, the resulting EVC-based service would include CE-VLAN ID translation.*

8.3.3
CoS Identification for E-Access Services

MEF 33 simplifies CoS identification for E-Access services compared to general OVC-based services, as summarized in the following table.

Location	CoS Identification Mechanisms	
	General OVC-Based Operator Service	Access EPL/EVPL
UNI	CoS identification based on EVC	All frames map to a single CoS
	CoS identification based on value of PCP bits in IEEE 802.1Q customer VLAN tag (CE-VLAN tag)	
	CoS identification based on the value of the differentiated services code point (DSCP) in the IP header	
	CoS identification based on Layer 2 Control Protocol	
ENNI	CoS identification based on S-Tag PCP bits	All frames map to a single CoS

All frames associated with an E-Access service instance map to the same CoS and get the same QoS treatment in the access network.

Note: *Allowing only one CoS simplifies E-Access services, making them easier to manage.*

Related Links

CoS Identification at UNIs
OVC End Point per ENNI Service Attributes

8.3.4 CE-VLAN CoS ID Preservation for E-Access Services

MEF 33 requires OVC service attribute **CE-VLAN CoS ID Preservation** = **Yes** for both kinds of E-Access service.

For the Access EPL, this requirement is expected because the Access EPL is a port-based service.

For the Access EVPL, this requirement represents a constraint compared to other VLAN-based services. Other VLAN-based services (EVPL, EVP-LAN, and EVP-Tree services) support CE-VLAN CoS ID translation. The Access EVPL, however, does not support CE-VLAN CoS ID translation.

Note: *Elimination of CE-VLAN CoS ID translation support simplifies E-Access services, making them easier to manage.*

Test your understanding: *Can an Access EVPL support an EVC-based service that includes CE-VLAN CoS ID translation? Explain why or why not.*

Answer: *Yes. If the Access EVPL is deployed in combination with a general OVC-based operator service that includes CE-VLAN CoS ID translation, the resulting EVC-based service would include CE-VLAN CoS ID translation.*

8.3.5 Performance Objectives for E-Access Services

Up to 8 performance objectives (FLR, Availability, HLI, CHLI, FD, FDR, MFD, and IFDV) are specified for E-Access services using the OVC service attribute **Service Level Specification**, just as they are for general OVC-based services.

When an E-Access service is used to support an EVC-based service, performance objectives for the E-Access service are chosen to align with EVC performance objectives (as explained previously for general OVC-based services).

If an access provider offers a CE 2.0 certified E-Access service, performance objectives for that service will meet or exceed MEF-standardized CoS performance objectives (CPO's) for a particular CoS Label and Performance Tier.

Related Links

OVC Service Attributes for E-Access
Service Level Specification (OVC Service Attribute)
OVC versus EVC Performance Objectives
CPOs (per CoS Label and Performance Tier)

8.3.6 Bandwidth Profiles for E-Access Service

MEF 33 simplifies bandwidth profile support for E-Access services compared to general OVC-based services, as summarized in the following table.

Bandwidth Profile Attribute		Support	
Classification	Name	General OVC-Based Operator Service	Access EPL/ EVPL
UNI	Ingress Bandwidth Profile Per UNI (optional)	If this attribute is supported, MUST specify CIR, CBS, EIR, EBS, CM, or CF. MUST NOT be combined with any other type of ingress bandwidth profile.	Must not specify
	Egress Bandwidth Profile Per UNI (optional)	If this attribute is supported, MUST specify CIR, CBS, EIR, EBS, CM, or CF. MUST NOT be combined with any other type of egress bandwidth profile.	Must not specify
OVC per UNI	Ingress Bandwidth Profile Per OVC End Point	None or one set of bandwidth profile values (CIR, CBS, EIR, EBS, CF, CM)	Supported as specified in MEF 33
	Ingress Bandwidth Profile Per Class of Service Identifier at a UNI	None or one set of bandwidth profile values (CIR, CBS, EIR, EBS, CF, CM) for each CoS	Not supported
	Egress Bandwidth Profile Per OVC End Point at a UNI	None or one set of bandwidth profile values (CIR, CBS, EIR, EBS, CF, CM)	Not supported
	Egress Bandwidth Profile Per Class of Service Identifier at a UNI	None or one set of bandwidth profile values (CIR, CBS, EIR, EBS, CF, CM) for each CoS	Not supported
OVC End Point per ENNI	Ingress Bandwidth Profile Per OVC End Point	None or one set of bandwidth profile values (CIR, CBS, EIR, EBS, CF, CM)	Supported as specified in MEF33
	Ingress Bandwidth Profile Per ENNI Class of Service Identifier	None or one set of bandwidth profile values (CIR, CBS, EIR, EBS, CF, CM) for each CoS	Not supported
	Egress Bandwidth Profile Per End Point	None or one set of bandwidth profile values (CIR, CBS, EIR, EBS, CF, CM)	Not supported
	Egress Bandwidth Profile Per ENNI Class of Service Identifier	None or one set of bandwidth profile values (CIR, CBS, EIR, EBS, CF, CM) for each CoS	Not supported

Key observations:

1. Access EPL/EVPL services only support **ingress** bandwidth profiles (no egress bandwidth profiles).
2. Access EPL/EVPL services only support **per-OVC-end-point** ingress bandwidth profiles (no per-UNI or per-CoS-identifier bandwidth profiles).
3. Access EPL/EVPL services support two ingress bandwidth profiles (one at the UNI and one at the ENNI).
4. Access EPL/EVPL services are required to support ingress bandwidth profiles according to special rules, specified in MEF 33, which are described next.

Related Links

Bandwidth Profiles
Bandwidth Profile Service Attributes
UNI Service Attributes
OVC per UNI Service Attributes
OVC End Point per ENNI Service Attributes
OVC End Point per ENNI Service Attributes for E-Access
OVC per UNI Service Attributes for E-Access

Special Ingress Bandwidth Profile Requirements

Ingress bandwidth profiles for Access EPL and Access EVPL services are defined in the usual way, by specifying values for the six parameters listed in following table.

Table 19: Bandwidth Profile Parameters

Parameter		Units of Measure or Values	Requirements
CIR	Committed Information Rate	Bits per second	CIR ≥ 0
CBS	Committed Burst Size	Bytes	If CIR > 0, then CBS ≥ maximum MTU over all EVCs to which the bandwidth profile applies
EIR	Excess Information Rate	Bits per second	EIR ≥ 0
EBS	Excess Burst Size	Bytes	If EIR > 0, then EBS ≥ maximum MTU over all EVCs to which the bandwidth profile applies
CF	Coupling Flag	N or Y	n/a
CM	Color Mode	color-blind or color-aware	n/a

However, MEF 33 adds the following requirements, which ensure that services offered by all Ethernet access providers meet a common baseline standard for capability.

1. CIR must *be configurable* to each of the following 37 values:
   1, 2, 3, 4, 5, 6, 7, 8, 9, 10, 20, 30, 40, 50, 60, 70, 80, 90, 100, 200, 300, 400, 500, 600, 700, 800, 900 Mb/s,
   plus 1, 2, 3, 4, 5, 6, 7, 8, 9, 10 Gb/s
   Subject to the following limits:
   Up to 70% of **UNI** speed (for **Ingress Bandwidth Profile Per OVC End Point at UNI**)
   Up to 70% of **ENNI** speed (for **Ingress Bandwidth Profile Per OVC End Point at ENNI**)

Note: *The service may support other values of CIR, but these values must be supported.*

Note: *This requirement concerns capability, not the values of CIR used for a particular service instance.*

Note: *As explained in Appendix A of MEF 33, CIR values are limited to 70% of UNI/ENNI speed because bandwidth profile algorithms do not count interframe overhead (interframe GAP + preamble bits), which consume UNI/ENNI bandwidth. Setting CIR too close to the bit rate of the physical layer of the UNI/ENNI has consequences, which may appear only under certain traffic conditions.*

Example 1: *If UNI speed is 100 Mb/s, CIR in **Ingress Bandwidth Profile Per OVC End Point at UNI** must be configurable to 1, 2, 3, 4, 5, 6, 7, 8, 9, 10, 20, 30, 40, 50, 60, and 70 Mb/s.*

Example 2: *If ENNI speed is 10 Gb/s, **Ingress Bandwidth Profile Per OVC End Point at ENNI** must be configurable to 1, 2, 3, 4, 5, 6, 7, 8, 9, 10, 20, 30, 40, 50, 60, 70, 80, 90, 100, 200, 300, 400, 500, 600, 700, 800, and 900 Mb/s, plus 1, 2, 3, 4, 5, 6, and 7 Gb/s.*

2. **Ingress Bandwidth Profile Per OVC End Point at UNI** must allow configuration of
   EIR=0, EBS=0, CF=N, CM="**color-blind**".

   Note: *The service may support other values of EIR, EBS, CF, and CM, but these values must be supported.*

   Note: *This requirement concerns capability, not the values of EIR, EBS, CF, and CM used for a particular service instance.*

3. **Ingress Bandwidth Profile Per OVC End Point at ENNI** must allow configuration of
   EIR=0, EBS=0, CF=N, CM="**color-aware**".

   Note: *The service may support other values of EIR, EBS, CF, and CM, but these values must be supported.*

   Note: *This requirement concerns capability, not the values of EIR, EBS, CF, and CM used for a particular service instance.*

4. CBS must be ≥ 12176 bytes for both bandwidth profiles:
   Ingress Bandwidth Profile Per OVC End Point at UNI
   Ingress Bandwidth Profile Per OVC End Point at ENNI

   Note: *This requirement concerns the actual value of CBS used in any service instance.*

5. If **Ingress Bandwidth Profile Per OVC End Point at UNI** has CIR>0 and EIR=0, egress frames at ENNI must be marked green (via S-Tag).

Related Links

About Line Rate and CIR

8.4 E-Access Service Applications

In this section:

8.4.1 E-Access Services Applied in EVC-Based Services

E-Access services are designed to support any EVC-based service (EPL, EVPL, EP-LAN, EVP-LAN, EP-Tree, or EVP-Tree service), regardless of how the EVC-based service is defined.

Access EPLs and Access EVPLs, on their own, have limited capability compared to EVC-based services:

- No support for multipoint-to-multipoint functionality
- No support for rooted-multipoint functionality
- No support for multiple CoS
- No support for CE-VLAN ID translation
- No support for CE-VLAN CoS ID translation

However, these limitations do not restrict their ability to support EVC-based services that include these capabilities. Conceptually, Access EPL and Access EVPLs are role players, designed to provide point-to-point connectivity to out-of-franchise (off-net) customer sites. To support general EVC-based services, Access EPL/EVPLs are combined with general OVC-based operator services, which have general capability.

Example 1

A subscriber requires an EP-LAN service supporting 3 CoS levels (Low, Medium, and High) between five sites, three of which are off-net. To support this service, the service provider:

1. Purchases three Access EPLs, each supporting all traffic with 1 CoS (High), from two different Ethernet access providers.

2. Deploys a multipoint-to-multipoint OVC-based operator service within their own network supporting three CoS (Low, Medium, and High).

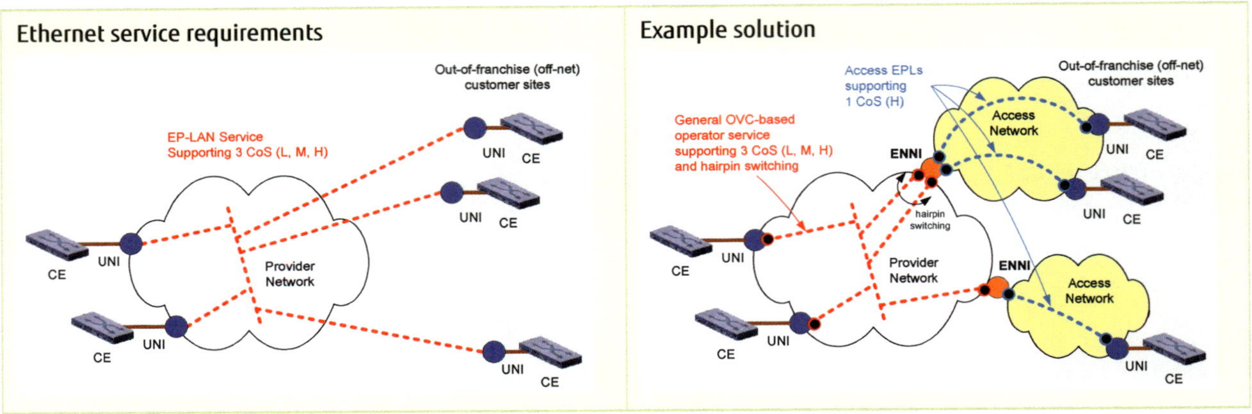

The Access EPLs are purchased with CoS level High in order to support all three CoS levels (Low, Medium, and High) that are supported by the EP-LAN service, in particular the most demanding CoS level (High).

Note: *In this example, the general OVC-based operator service requires a multipoint-to-multipoint OVC with multi-CoS support and hairpin switching support.*

Example 2

A subscriber requires two services involving four sites, two of which are off-net: an EVP-Tree service with CE-VLAN ID translation and an EVPL service supporting CE-VLAN IDs 100–120.

To support these services, the service provider:

1. Purchases four Access EVPLs.
2. Deploys a point-to-point OVC-based operator service within their own network to support the EVPL service.
3. Deploys a rooted-multipoint OVC-based operator service within their own network to support the EVP-Tree service.

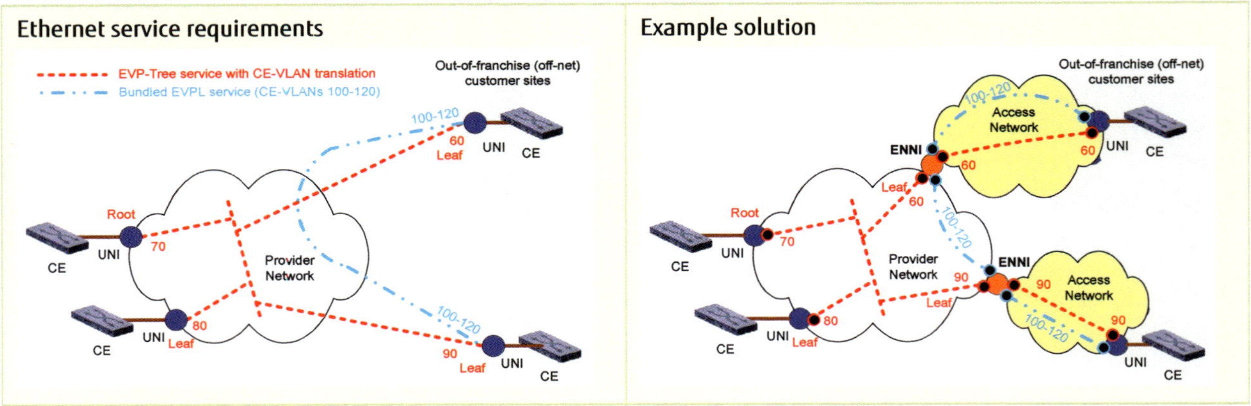

Note: *In this example, the rooted-multipoint OVC provides all CE-VLAN ID translation support.*

8.4.2
E-Access Services Applied for IP Network Services

E-Access Support for Internet Access

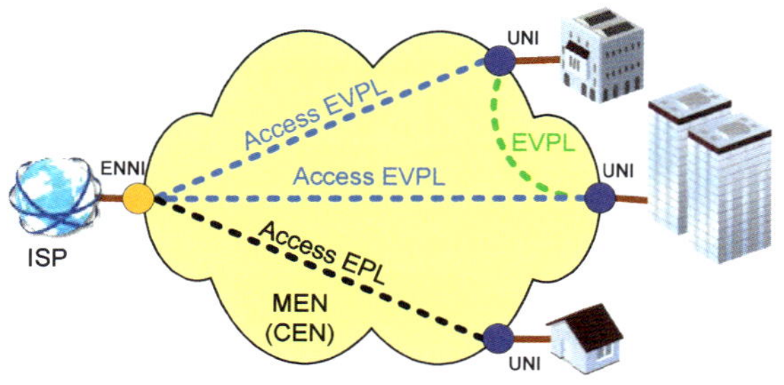

Access-EPL and Access-EVPL services support this application using one ISP interface (ENNI) to support multiple independent IP service connections for different subscribers. At the ENNI, each Access EPL/EVPL is identified by S-VLAN ID.

Note: *A single ENNI can support up to 4094 Access EPL and/or Access EVPL service connections.*

Note: *Each Access EVPL can support any set of CE-VLAN ID values at the ENNI. CE-VLAN ID values need not be distinct, as would be required for EVPLs multiplexing to an ISP UNI.*

8.5
E-Access Compared to E-Line

Key Similarities:

- Access EPL, Access EVPL, EPL, and EVPL services are all point-to-point.
- Access EPL and EPL are both port based at UNI(s).
- Access EVPL and EVPL are both VLAN-based at UNI(s).

Key Differences:

Access EPL / EVPL	EPL / EVPL
OVC-based	EVC-based
UNI-to-ENNI	UNI-to-UNI
ENNI supports multiplexing of Access EPLs and/or Access EVPLs	UNI configured with Service Multiplexing = Yes supports multiple EVPLs UNI configured with Service Multiplexing = No supports one EPL
No support for CE-VLAN ID translation	EVPL supports CE-VLAN ID translation if UNIs are configured with Bundling = No
Supports one CoS only	Supports multiple CoS

8.6
Review

1. Both the Access EVPL and the Access EPL... (Choose two.)

    a. Allow CE-VLAN ID translation

    b. Associate each service instance to one S-VLAN ID value at the ENNI

    c. Are point-to-point OVCs

    d. Connect one ENNI and one UNI

2. Two access providers establish an ENNI connection between their networks. What types of EVC-based services can they offer subscribers using only Access EPL and Access EVPL services? Assume that one of the access providers takes the role of service provider.

    a. E-Line only

    b. E-Line and E-LAN only

    c. E-Line and E-Tree only

    d. E-Line, E-LAN, and E-Tree

3. An E-Access service must support *CM=color-blind* for which bandwidth profile?

    a. Ingress bandwidth profile per OVC end point at UNI

    b. Egress bandwidth profile per OVC end point at UNI

    c. Ingress bandwidth profile per OVC end point at ENNI

    d. Egress bandwidth profile per OVC end point at ENNI

4. Access EVPL differs from EVPL in that...

    a. Access EVPL supports multiple CoS, while EVPL supports only one CoS.

    b. Access EVPL is VLAN-based, while EVPL is port-based.

    c. Access EVPL is OVC-based, while EVPL is EVC-based.

    d. Access EVPL supports CE-VLAN ID Preservation, while EVPL does not.

5. An Ethernet access provider offers Access EPL service with a 1 Gb/s UNI. The value of CIR in service attribute ***Ingress Bandwidth Profile Per OVC End Point*** at the UNI must be configurable to...

    a. 500 Mb/s and 1 Gb/s

    b. 400 Mb/s and 800 Mb/s

    c. 20 Mb/s and 300 Mb/s

    d. 250 Mb/s and 500 Mb/s

6. Which service supports CE-VLAN ID translation?

    a. Access EVPL

    b. Access EPL

    c. EVPL

    d. EPL

7. E-Access services and E-Line services both include at least one...

    a. UNI

    b. ENNI

    c. OVC

    d. EVC

8. A service provider has been using four Access EPLs to support an EP-LAN for a subscriber. The subscriber now wants to convert the EP-LAN service to an EP-Tree service. How are the Access EPLs impacted by this change?

    a. They can no longer be used.

    b. The must be converted to Access EVPLs.

    c. They must now filter traffic.

    d. They are not impacted.

9. Both Access EVPL and Access EPL...

    a. Support hairpin switching

    b. Can connect a customer site to an Internet service provider

c. Support multiple CoS

d. Are EVC-based services

10. If an E-Access service is applied to reach an off-net customer site in an EVC-based service application, the EVC-based service...

   a. Can support only one class of service

   b. Cannot support CE-VLAN ID translation

   c. Must be point-to-point

   d. Must involve at least one ENNI

11. As a service provider you have agreed to provide an EPL service supporting IP packets up to 2000 bytes in length. However, one customer site is off-net, so you need to purchase an Access EPL service to reach that customer site. The access provider only offers Access EPLs with ingress bandwidth profiles at the ENNI and only supports the following values of CBS. Which value do you choose if you want to minimize CBS?

   a. CBS = 2000 bytes

   b. CBS = 2010 bytes

   c. CBS = 2020 bytes

   d. CBS = 2030 bytes

   e. CBS = 2040 bytes

12. Which of the following statements is false?

   a. A UNI can support up to 4094 Access EVPLs.

   b. A UNI can support up to 4094 Access EPLs.

   c. An ENNI can support up to 4094 Access EPLs.

   d. An ENNI can support up to 4094 Access EVPLs.

13. Which two services are used to reach off-net customer sites? (Choose two.)

   a. Access EVPL

   b. EVPL

   c. Access EPL

   d. EPL

14. Access EPL is similar to EPL in that... (Choose two.)

- **a.** Both support multiple classes of service.
- **b.** Both support untagged frames and C-Tagged frames at the UNI.
- **c.** Both support CE-VLAN ID translation.
- **d.** Both support CE-VLAN CoS ID preservation.
- **e.** Both are OVC-based services.

15. An E-Access service must support *CM=color-aware* for which bandwidth profile?

- **a.** Ingress bandwidth profile per OVC end point at UNI
- **b.** Egress bandwidth profile per OVC end point at UNI
- **c.** Ingress bandwidth profile per OVC end point at ENNI
- **d.** Egress bandwidth profile per OVC end point at ENNI

Related Links

Answers: E-Access Services

9

Ethernet OAM

In this chapter:

Operations, administration, and maintenance (OAM) is a vital component of the Service Management attribute of Carrier Ethernet.

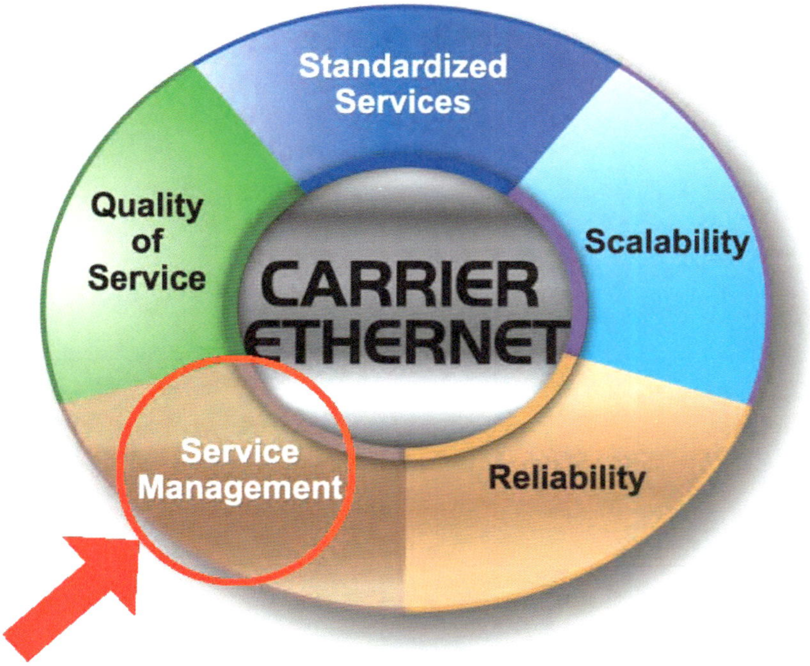

MEF specifications are designed to provide Carrier-Class OAM at all levels of service application: locally, regionally, and globally.

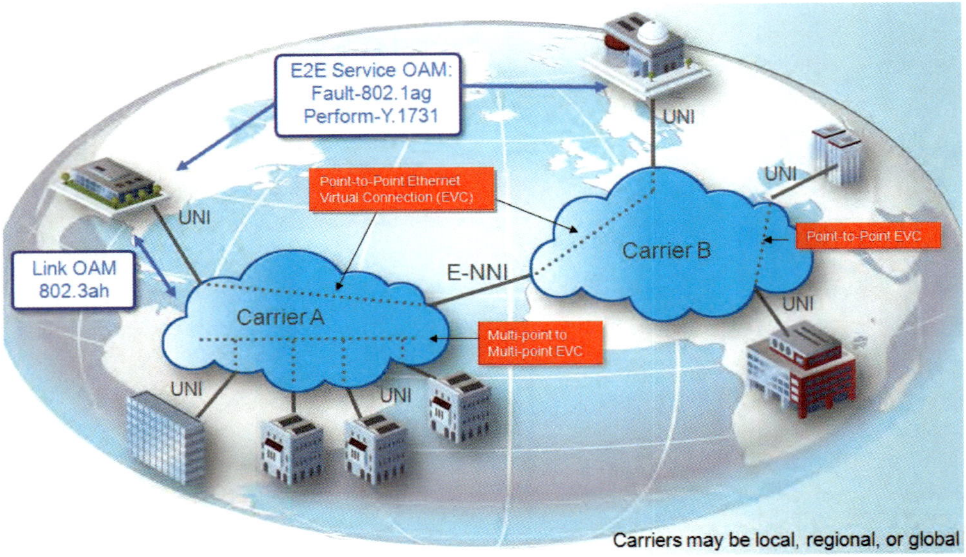

Reproduced with permission of the Metro Ethernet Forum

OAM can be broken down into two categories:

- **Link OAM (LOAM)**–Applies to the physical link interconnecting two devices (conforms to IEEE 802.3ah)
- **Service OAM (SOAM)**–Applies to the end-to-end Carrier Ethernet service (conforms to MEF 17, ITU-T Y.1731, and IEEE 802.1Q[23])

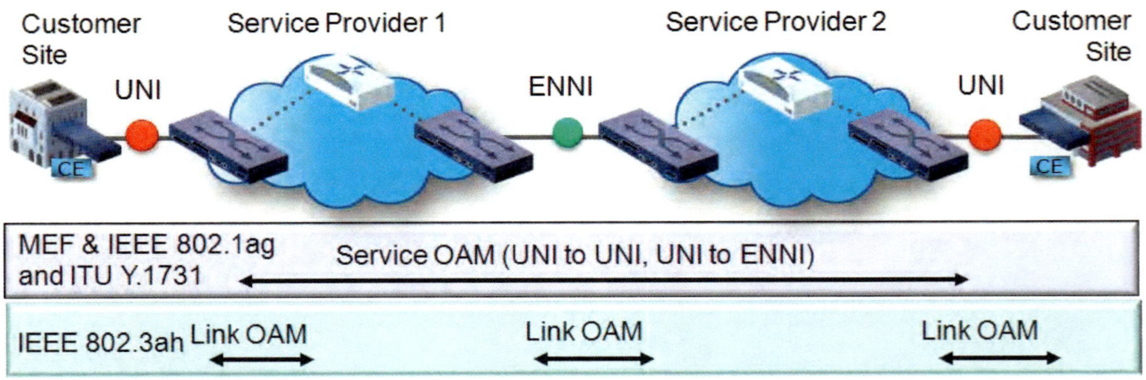

Reproduced with permission of the Metro Ethernet Forum

Figure 73: LOAM and SOAM

23 IEEE 802.1ag-2007 (the original source of IEEE SOAM standards, referenced in older MEF specifications) has been incorporated into and superseded by IEEE 802.1Q-2011.

Although LOAM can be applied to any physical link, LOAM support is especially important at UNIs and ENNIs, where link issues typically occur. LOAM functions only at the physical link level and cannot be extended to the service level.

SOAM provides per-service OAM functionality, end-to-end, across the network and distributes that functionality between administrative organizations (subscribers, service providers, and operators).

LOAM and SOAM are independent (have no effect on each other). However, at the link level, LOAM and SOAM can have some overlapping functionality.

LOAM and SOAM are both important for a comprehensive fault management solution.

9.1
Link OAM

MEF requirements for LOAM are based on clause 57 (OAM) of the IEEE 802.3ah standard.

LOAM is used to monitor and troubleshoot individual Ethernet links.

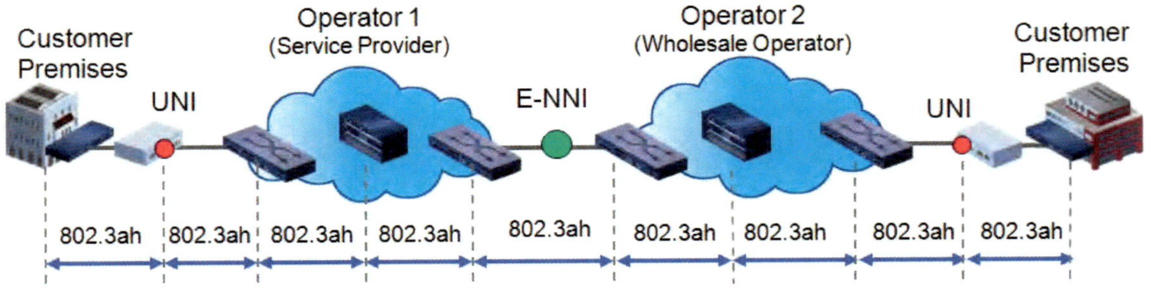

Reproduced with permission of the Metro Ethernet Forum

Although LOAM can be applied to any Ethernet link, it is most relevant to links between administrative organizations (at UNIs and ENNIs). LOAM allows operators to troubleshoot link issues with access from only one side of the link.

Important: *LOAM support is required for Type 2.2 UNIs (per MEF 20) and at ENNIs (per MEF 26.1).*

LOAM functionality includes:

Link Discovery–After activation, LOAM goes into discovery mode to identify the LOAM capability of its peer on the other side of the link.

Link Monitoring–After LOAM is operating, each side periodically sends a LOAM frame to its peer and expects to periodically receive a LOAM frame from its peer. LOAM frames are normally sent at 1-second intervals (per the IEEE 802.3ah standard). If one side stops receiving LOAM frames from its peer, it raises an alarm condition (locally) and sends a remote defect indication (RDI) to its peer (encoded within outgoing LOAM frames).

Remote Failure Indication–RDI from a peer indicates that the peer is not receiving frames sent.

Loopback–LOAM loopback triggers creation of a port-based loopback in the peer across the link. The loopback redirects all Ethernet traffic (except LOAM control packets) back to the local NE.

Caution: *LOAM loopback is service disrupting. For this reason (per MEF 26.1), if LOAM is enabled at an ENNI-N, the LOAM loopback capability should be disabled.*

Related Links

LOAM Requirement for Type 2.2 UNIs
LOAM Requirement for ENNIs

9.2
Service OAM Overview

In this section:

Service OAM (SOAM) requirements are based on IEEE® 802.1Q[24], ITU-T Y.1731, and ITU-T G.8021 standards.

MEF 17 and MEF 30.1 define the SOAM framework and requirements for Carrier Ethernet.

MEF 17 and MEF 35 specify the SOAM requirements for **Performance Management (PM)**, including the measurement of frame loss and frame delay for the calculation of performance metrics.

MEF 30.1 specifies SOAM requirements for **Connectivity Fault Management (CFM)**, including continuity check message (CCM), remote defect indication (RDI), alarm indication signal (ETH-AIS), lock signal (ETH-LCK), test signal (ETH-Test), loopback, and link trace.

SOAM functionality is based on the exchange of Ethernet frames between SOAM components provisioned within network elements (NEs) across the network.

Related Links

SOAM Requirement for Type 2 UNIs

24 IEEE 802.1ag-2007 (the original source of IEEE SOAM standards, referenced in older MEF specifications) has been incorporated into and superseded by IEEE 802.1Q-2011.

9.2.1
The MEF Service Life Cycle

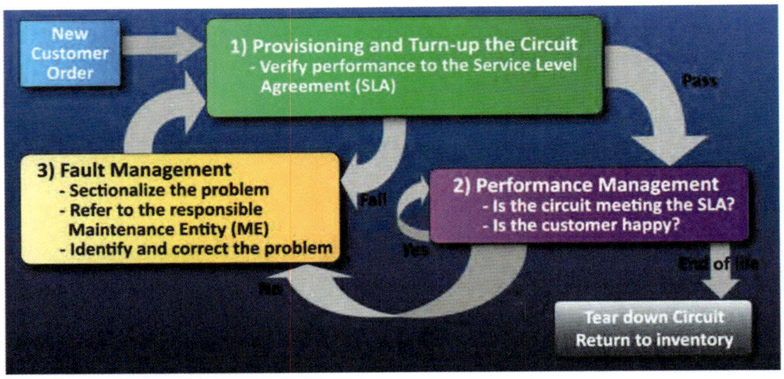

Figure 74: The MEF Service Life Cycle

The MEF identifies Performance Management (PM) and Fault Management (FM) as critical components of a Carrier Ethernet service life cycle. Both are based on SOAM.

9.2.2
SOAM Domains

An SOAM domain is an Ethernet network or subnetwork that is of interest to, or the responsibility of, an administrative entity, such as the subscriber, the service provider, or an operator.

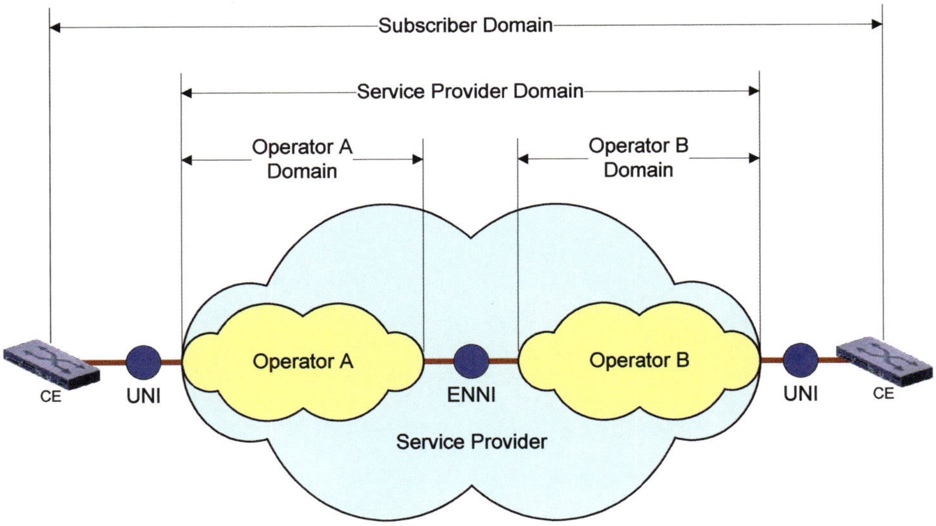

Figure 75: SOAM Domains

SOAM domains are defined, as required, subject to the following constraints:

- SOAM domains form a hierarchy with the subscriber domain at the highest level.
- Lower-level domains nest within higher-level domains.
- SOAM domains cannot overlap.

If service fails, the hierarchy of SOAM domains provides a framework for locating the failure and identifying the administrative organization that is responsible for fixing the fault.

SOAM provides each organization with the information it needs to perform its function without sharing privileged internal information with other organizations.

9.2.3 SOAM Frames

The term **SOAM frame** refers to any Ethernet frame injected into the network to support a SOAM function. SOAM frames are defined, as required, per SOAM function and have function-specific names, as shown in the following table.

SOAM Area	SOAM Function	SOAM Frame Name
Connectivity Fault Management	Connectivity Monitoring ***Note:*** *RDI (Remote Defect Indication) is encoded in CCM frames.*	CCM (Continuity Check Message)
		AIS (Alarm Indication Signal)
		LCK (Locked Signal)
		Test (Test Signal)
	Loopback	LBM (Loopback Message)
		LBR (Loopback Reply)
	Link Trace	LTM (Link Trace Message)
		LTR (Link Trace Reply)
Performance Management	Single-Ended Delay Measurement	DMM (Delay Measurement Message)
		DMR (Delay Measurement Reply)
	Single-Ended Synthetic Frame Loss Measurement	SLM (Synthetic Loss Message)
		SLR (Synthetic Loss Reply)
	Dual-Ended Delay Measurement	1DM (One-Way Delay Measurement)
	Single-Ended Service Frame Loss Measurement	LMM (Loss Measurement Message)
		LMR (Loss Measurement Reply)

Note: *The terms **OAM frame**, **SOAM PDU**, and **SOAM frame** are used interchangeably in MEF 17, MEF 30.1, and MEF 35. The term **SOAM frame** is used in this study guide.*

9.2.4 SOAM Components

The following table provides a summary of SOAM component definitions as they appear in MEF 30.1. These definitions follow ITU-T Y.1731 terminology. The last column indicates equivalent terminology from IEEE 802.1Q[25].

Attention: *If you are not already familiar with SOAM components, the definitions in the following table may be hard to follow. More friendly descriptions follow.*

Table 20: SOAM Component Definitions (per MEF 30.1 and ITU-T Y.1731)

Component		Definition	IEEE 802.1Q Term
MEP	MEG End Point	An actively managed SOAM entity associated with a specific service instance that can generate and receive SOAM frames and track any responses. It is an end point of a single MEG and is an endpoint of a separate ME for each of the other MEPs in the same MEG.	Equivalent to a Maintenance association End Point (MEP) in IEEE 802.1Q
MIP	MEG Intermediate Point	A SOAM entity, associated with a single maintenance domain, that can generate SOAM frames, but only in response to received SOAM frames.	Equivalent to a Maintenance domain Intermediate Point (MIP) in IEEE 802.1Q
ME	Maintenance Entity	A point-to-point relationship between two MEPs within a single MEG.	Equivalent to a Maintenance Entity (ME) in IEEE 802.1Q.
MEG	Maintenance Entity Group	A set of MEs that exists in the same administrative boundary with the same MEG Level and MEG ID.	Equivalent to a Maintenance Association (MA) in IEEE 802.1Q

MEPs and MIPs

SOAM functionality is based on the exchange of SOAM frames between MEPs and MIPs provisioned within NEs across the network.

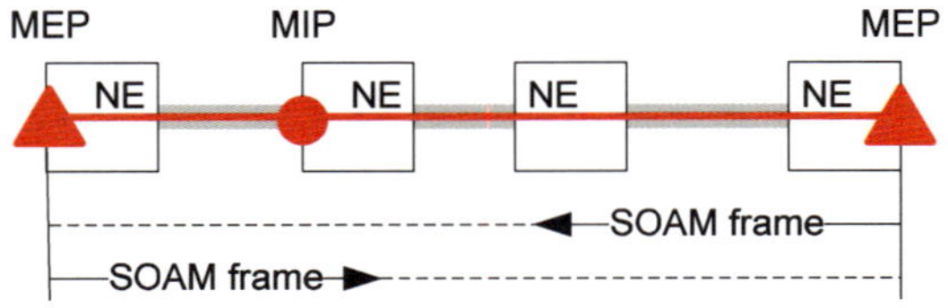

Figure 76: MEPs Exchanging SOAM Frames

25 IEEE 802.1ag-2007 (the original source of IEEE SOAM standards, referenced in older MEF specifications) has been incorporated into and superseded by IEEE 802.1Q-2011.

A **MEP (MEG End Point)** is a provisioned component that can initiate and terminate SOAM frames/processes and can also react to SOAM frames.

A **MIP (MEG Intermediate Point)** is a provisioned component that can only react to SOAM frames (for example, to support loopback or link trace). A MIP cannot initiate an SOAM process, but it can generate an SOAM frame in response to a received SOAM frame.

MEPs are always provisioned in pairs. One pair, plus any number of optional MIPs, is dedicated to a segment of the path of Carrier Ethernet service flow. The MEPs define endpoints for the path segment. MIPs, if used, are positioned between the MEPs.

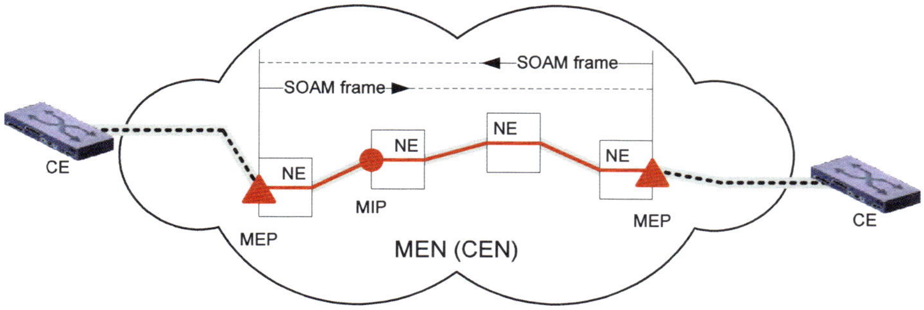

Figure 77: MEPs and MIPs on a Segment of the Path of Carrier Ethernet Service Flow

SOAM frames route through the network like service frames, using the same Ethernet-layer infrastructure and routing mechanisms. However, unlike service frames, SOAM frames are subject to processing at MEPs and MIPs.

Maintenance Entity (ME)

The ME is a higher-level SOAM component, associated to a segment of the path of Carrier Ethernet service flow.

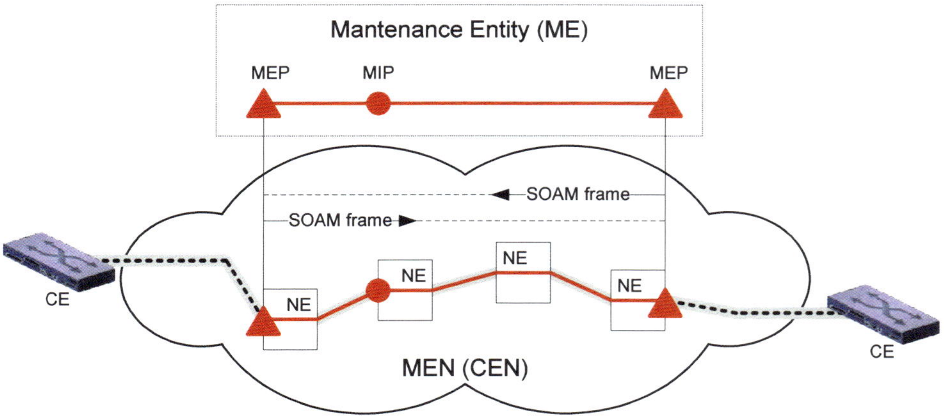

Figure 78: Maintenance Entity (ME)

Every ME includes exactly two MEPs, defining endpoints for the ME, plus any number of (optional) MIPs.

Maintenance Entity Group (MEG)

The MEG is the highest-level SOAM component. It associates a set of MEs to a particular service and SOAM domain. The following figure illustrates a MEG supporting the service provider domain for a multipoint-to-multipoint EVC with three UNIs. This MEG includes three MEs (one for each combination of UNIs).

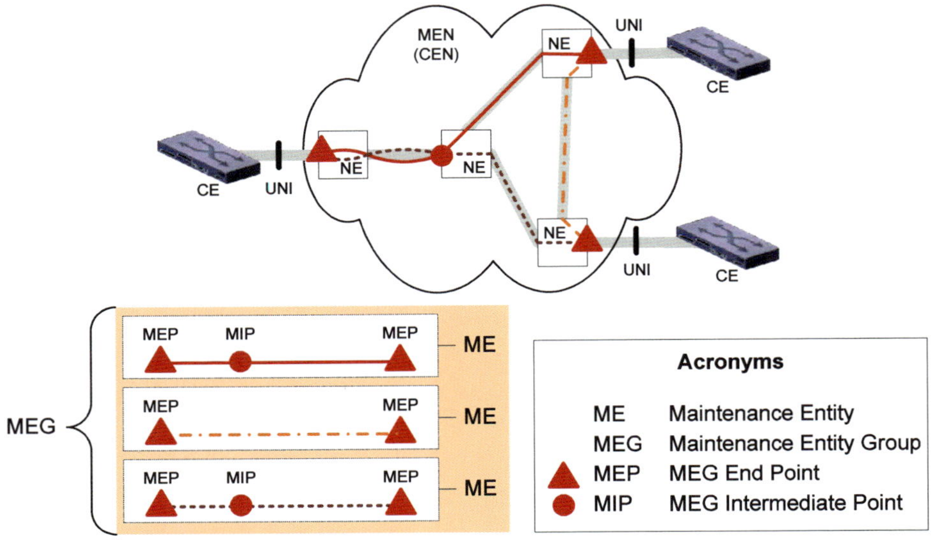

Figure 79: Example of a MEG Associating Three MEs

The MEG associates the entire group of components (MEs, MEPs, and MIPs) to a particular service and SOAM domain. A Carrier Ethernet service can have many MEGs.

A MEG for a point-to-point EVC contains a single ME, as shown in the following figure.

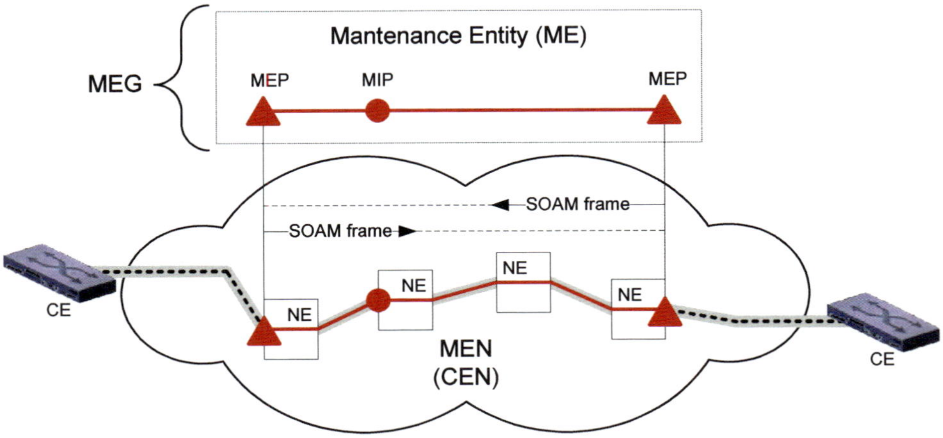

Figure 80: Example of MEG for a Point-to-Point EVC

The MEG for a multipoint-to-multipoint EVC with n UNIs contains n*(n-1)/2 MEs.

Test your understanding: ***(a)*** *Which SOAM components can initiate an SOAM process?* ***(b)*** *How many MIPs can an ME contain?* ***(c)*** *How many MEPs can an ME contain?* ***(d)*** *Can a point-to-point service have more than one ME?*

Answers: ***(a)*** *Only the MEP.* ***(b)*** *Any number.* ***(c)*** *2.* ***(d)*** *Yes. Any service, including any point-to-point service, can contain many MEGs. Each MEG in a point-to-point service has only one ME, but there can be many MEGs, so there can be many MEs.*

Up MEPs and Down MEPs

Every MEP is located at a specific network element (NE) interface and is provisioned to be either an Up MEP or a Down MEP.

If the MEP's peer lies in the direction outward from the interface (away from the NE), it is provisioned as a **Down MEP**.

If the MEP's peer lies in the opposite direction (inward toward the NE), it is provisioned as an **Up MEP**.

The two MEPs comprising an ME can be any combination of Up and Down MEPs: two Up MEPs, two Down MEPs, or one of each type, as shown in the following figure.

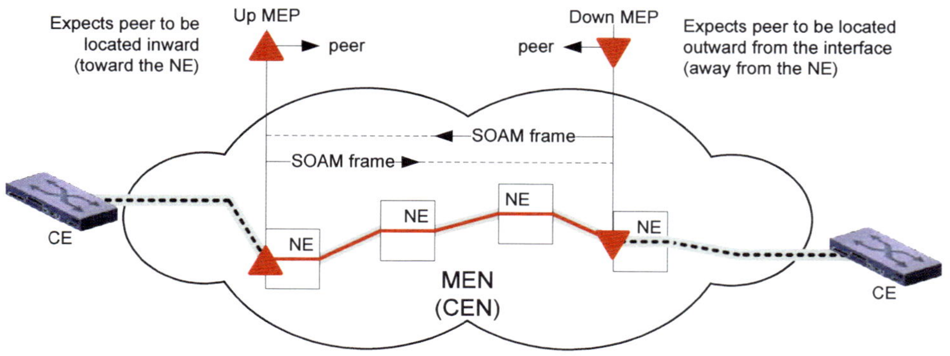

Figure 81: ME with One Up MEP and One Down MEP

9.2.5 SOAM MEG Levels

Per ITU-T Y.1731, SOAM supports eight MEG levels (0–7), which are used to differentiate SOAM traffic per administrative organization (subscriber, service provider, and operator).

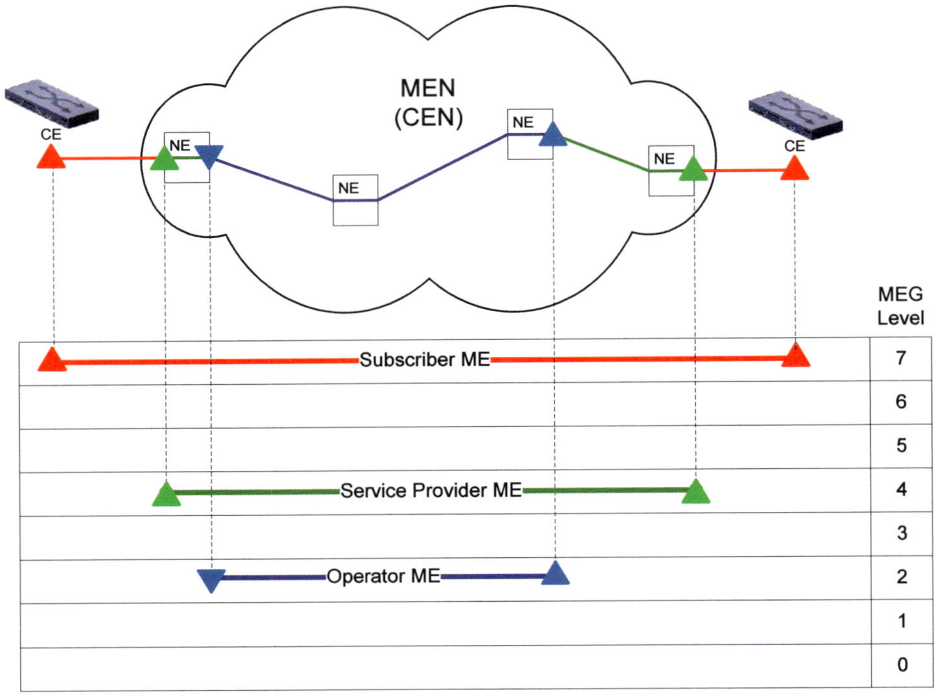

Figure 82: SOAM MEG Levels

Lower-level MEGs are nested within higher-level MEGs.

Important: *With proper provisioning, each SOAM frame is confined to the MEG in which it originates. Lower-level MEGs allow SOAM frames for higher-level MEGs to pass through transparently (they do not react to, or change, higher-level frames; they simply forward them).*

Suggested MEs and Default MEG Levels

MEG levels are allocated between administrative organizations (subscriber, service provider, and operators) based on mutual agreement.

MEF 17 suggests the following assignments as a baseline[26]:

- Subscriber uses MEG Level 7, 6, or 5.
- Service Provider uses MEG Level 3 or 4.
- Operators use MEG Levels 2, 1, or 0.

MEF 30.1 further refines suggested ME allocation as illustrated in the following figure.

26 This baseline allocation of MEG levels follows default MEG level assignments specified in ITU-T Y.1731.

Note: *This is only an example. Not all MEs are required for every application.*

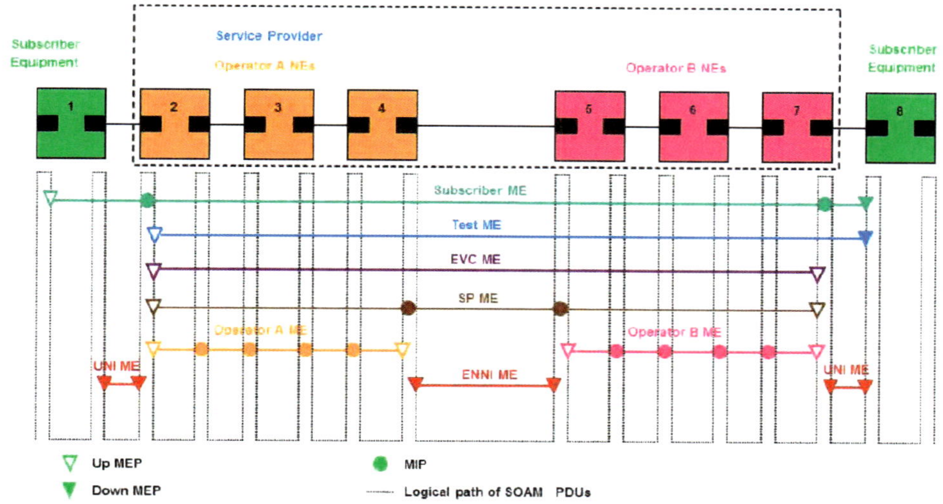

Figure 83: Example SOAM MEs (Figure 1 in MEF 30.1)

Per MEF 30.1, default MEG levels must conform to the following table.

Table 21: Default MEG Levels (from Table 3 in MEF 30.1)

MEG	Suggested Usage	Default MEG Level
Subscriber MEG	Subscriber monitoring of a Carrier Ethernet service	6
Test MEG	Service Provider isolation of subscriber-reported problems	5
EVC MEG	Service Provider monitoring of EVC	4
Service Provider MEG	Service Provider Monitoring of Service Provider network	3
Operator MEG	Network Operator monitoring OVC	2
UNI MEG	Service Provider monitoring of a UNI	1
ENNI MEG	Network Operators' monitoring of an ENNI	1

The default allocation of the MEG Level space can be changed by mutual agreement of the parties involved (subscribers, service providers, and operators).

Test your understanding: *For a given Carrier Ethernet service, can a service provider and an operator use the same MEG level?*

Answer: *In general no. But as shown in the preceding table, UNI MEGs (used by the service provider) and ENNI MEGs (used by operators) can be at the same MEG level.*

9.3 SOAM Connectivity Fault Management

In this section:

SOAM connectivity fault management (CFM) is implemented per IEEE 802.1Q[27], ITU-T Y.1731 and ITU-T G.8021.

9.3.1 Continuity Check Message (CCM)

A CCM is a type of SOAM frame. MEF 30.1 requires MEPs to support CCM messages and processes as defined in IEEE 802.1Q.

The MEPs within an ME exchange CCMs across the network to detect loss of continuity (LOC) and misprovisioning. Each MEP periodically sends a CCM to its peer and expects to periodically receive a CCM from its peer.

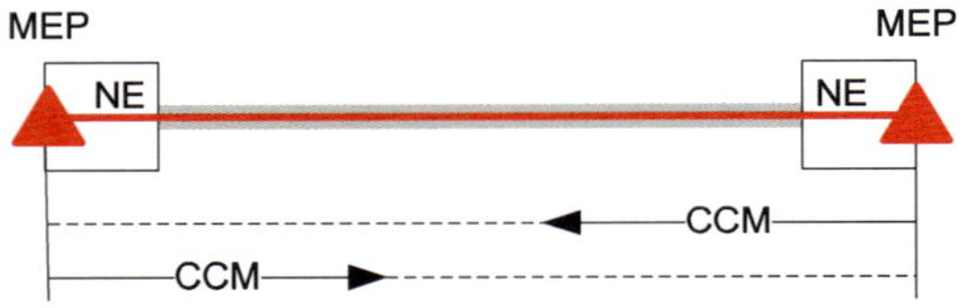

Figure 84: MEPs Exchanging CCMs

Important: *Per MEF 30.1, CCMs can be administratively turned on and off, but compliant NEs must support CCM intervals of 1 second and 10 seconds and should support intervals of 3.33 ms, 10 ms, and 100 ms.*

If a MEP fails to receive a CCM within a provisioned time interval (3 times the CCM interval, by default), the MEP raises an LOC alarm.

27 IEEE 802.1ag-2007 (the original source of IEEE SOAM standards, referenced in older MEF specifications) has been incorporated into and superseded by IEEE 802.1Q-2011.

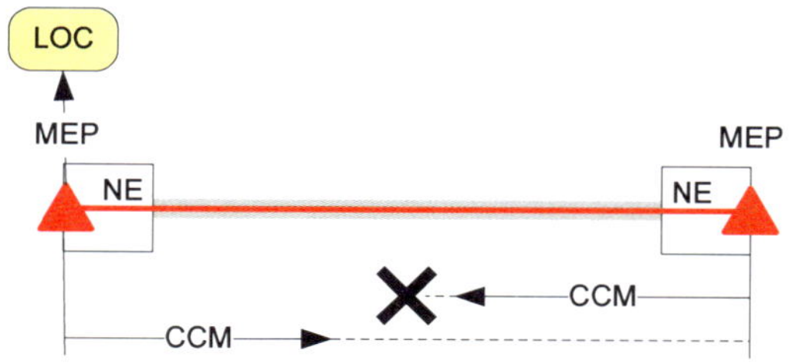

Figure 85: MEP Raising LOC

For availability monitoring, the CCM interval is commonly set to 1 second (or 10 seconds), which results in 3-second (or 30-second) LOC detection.

For protection switching, the CCM interval is commonly set to 10 ms or 3.33 ms.

For sub-50ms protection switching, the CCM interval can be set to 3.33 ms, which provides 10 ms LOC detection.

Note: *CCMs can be used to implement ITU-T G.8031 (Ethernet linear protection switching) or G.8032 (Ethernet ring protection switching), which are covered in another lesson.*

Related Links

Protection Mechanisms and Transport Technologies

9.3.2 Remote Defect Indication (RDI)

MEF 30.1 requires MEPs to support RDI as defined in IEEE 802.1Q[28] (equivalent to ETH-RDI in ITU-T Y.1731 and ITU-T G.8021).

A MEP uses RDI to inform its peer of LOC. When a MEP raises an LOC condition, the MEP encodes RDI in all outgoing CCMs (sets bit 8 of the CCM flags field to 1) until the LOC defect clears.

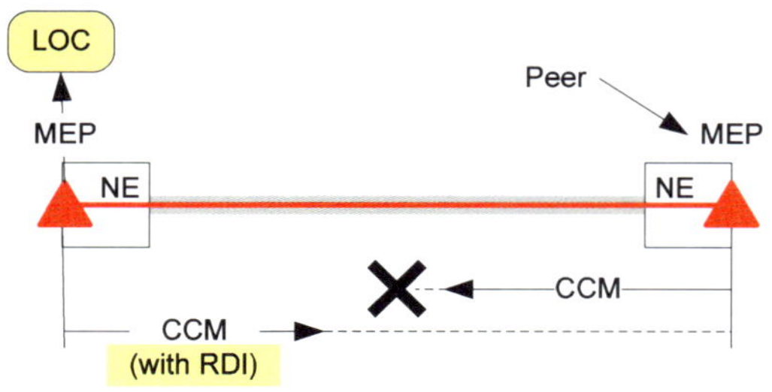

Figure 86: MEP Raising LOC and Signaling RDI

28 IEEE 802.1ag-2007 (the original source of IEEE SOAM standards, referenced in older MEF specifications) has been incorporated into and superseded by IEEE 802.1Q-2011.

9.3.3
Alarm Indication Signal (ETH-AIS)

AIS is a type of SOAM frame that is used to support LOC condition suppression.

Per MEF 30.1, MEPs should support ETH-AIS as defined in ITU-T Y.1731 and ITU-T G.8021.

Note: *Per MEF 30.1, ETH-AIS is not intended for, or recommended for, multipoint services.*

Without LOC condition suppression, a connectivity failure generally results in multiple levels of LOC, one for each MEG spanning the point of failure, as illustrated in the following figure.

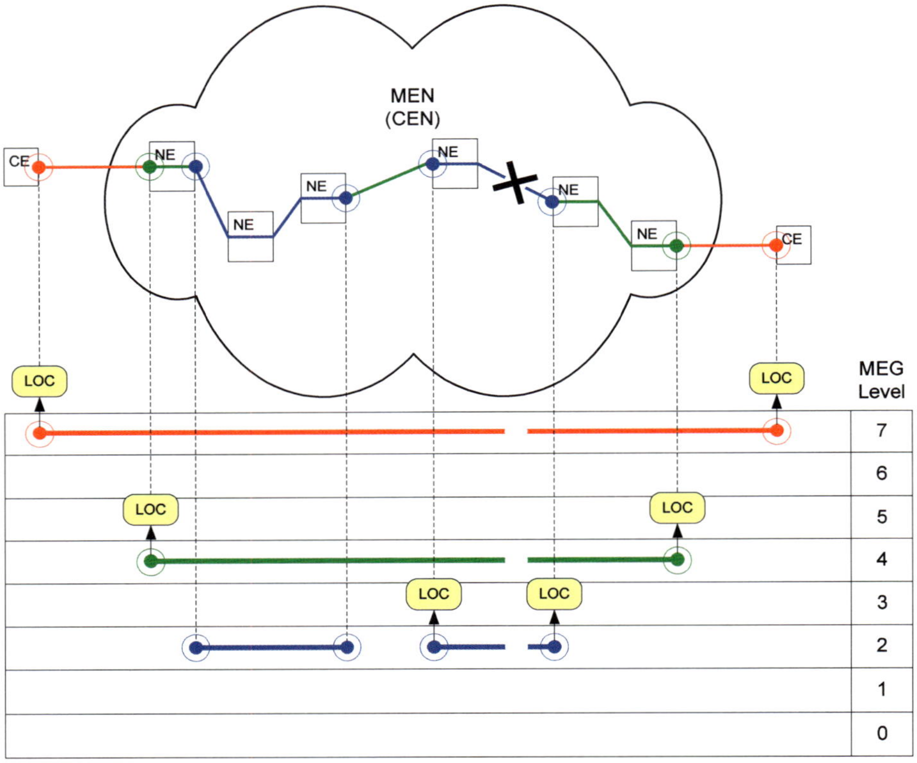

Figure 87: Connectivity Failure without AIS and LOC Condition Suppression

Several administrative domains are alerted to the fault, but each has no immediate way of knowing whether the fault originated at their level, so they can fix it, or at a lower level, which is the responsibility of another organization.

AIS and LOC condition suppression resolves this issue by masking higher-level conditions, leaving only root-level LOC conditions in the network, as shown in the following figure.

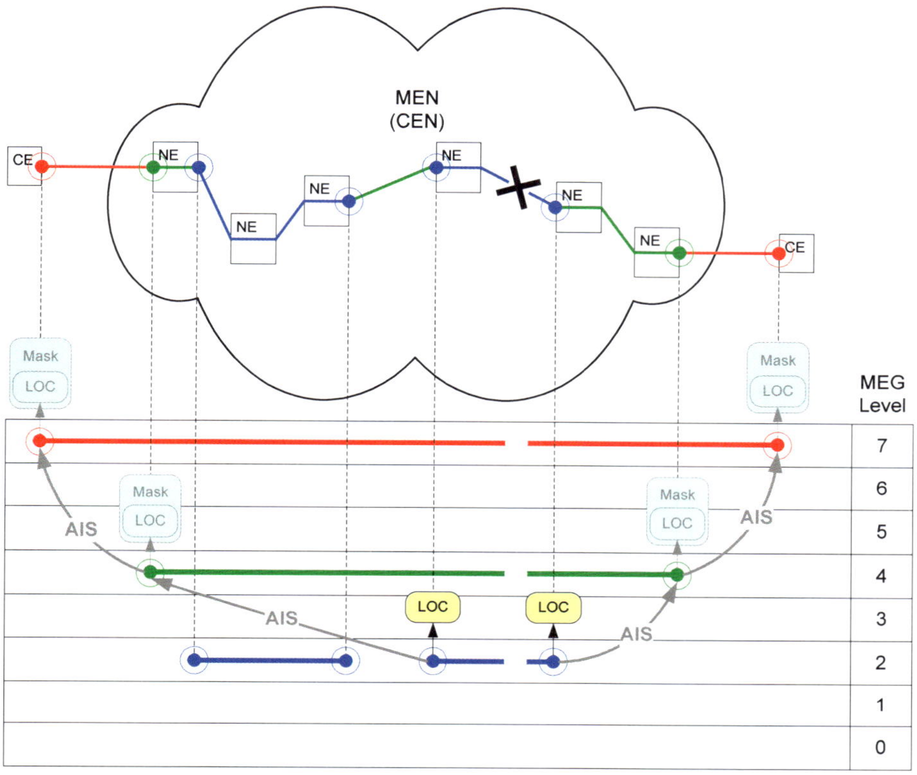

Figure 88: Connectivity Failure with AIS and LOC Condition Suppression

When an LOC condition occurs, MEPs send AIS frames away from their peers, causing higher-level MEPs in remote NEs to suppress LOC conditions.

Now, if an unsuppressed LOC condition occurs, the organization knows that the fault originated at their level and that they need to fix it.

9.3.4
Locked Signal (ETH-LCK)

The ETH-LCK function is an administrative function that does two things when applied to a MEP:

1. Locks the MEP (stops it from forwarding data traffic)

2. Causes the MEP to send LCK signals to higher-level MEGs, indicating that traffic has been interrupted for administrative purposes (rather than by a defect)

Per MEF 30.1, MEPs should support ETH-LCK as defined in ITU-T Y.1731 and ITU-T G.8021.

Note: *Per MEF 30.1, ETH-LCK is not intended for, or recommended for, multipoint services.*

The ETH-LCK function allows an operator to stop traffic flow through a MEP for any purpose, but is generally used to support out-of-service testing (using ETH-Test, for example).

Without LCK signaling, a locked MEP would look like a fault to each higher-level MEG spanning the locked MEP. As shown in the following figure, LOCs would occur because CCMs would not be forwarded through the locked MEP.

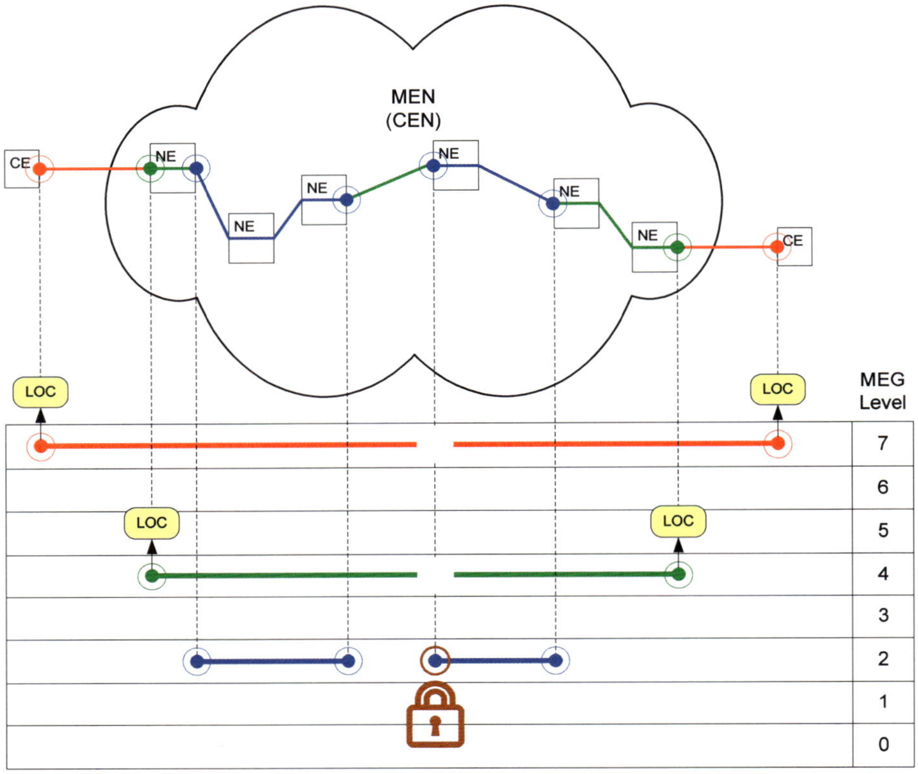

Figure 89: MEP Lock without LCK Signaling and LOC Condition Suppression

However, when the ETH-LCK function is applied to a MEP, the MEP stops forwarding traffic and begins sending LCK frames (a type of SOAM frame) periodically in both directions. These LCK frames support LOC condition suppression as shown in the following figure.

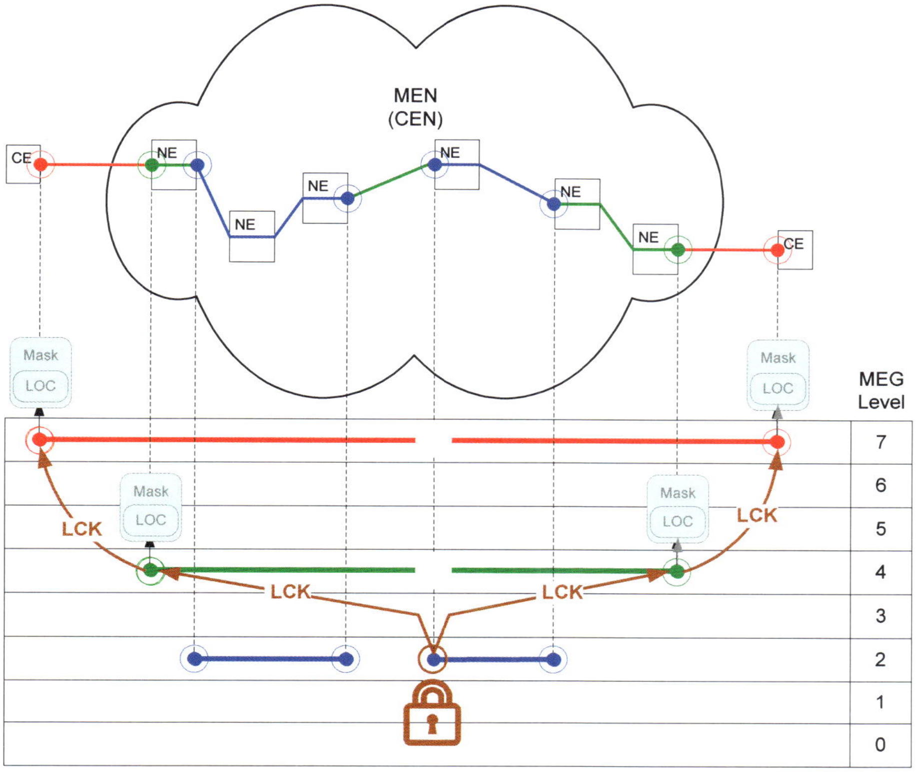

Figure 90: MEP Lock with LCK Signaling and LOC Condition Suppression

Note: *LCK frames, like AIS frames, support LOC condition suppression. However, LCK frames indicate that traffic interruption is due to administrative locking, rather than due to a defect condition.*

9.3.5 Test Signal (ETH-Test)

The ETH-Test function is an administrative function that is applied to a MEP to trigger one-way in-service or out-of-service diagnostic tests between itself and its peer MEP.

The ETH-Test function causes the MEP to send test frames with specified throughput, test patterns, and frame size to its peer MEP. It also causes the peer MEP to monitor received test frames, detect errors, and report errors (such as bit error rates).

Per MEF 30.1, MEPs can support ETH-Test as defined in ITU-T Y.1731 and ITU-T G.8021.

If the MEP is configured for **out-of-service ETH-Test testing**, the MEP is locked (stops forwarding data traffic) and sends LCK frames, as previously described for the ETH-LCK function.

If the MEP is configured for **in-service ETH-Test testing**, the MEP is not locked (continues to forward data traffic). In this case, care must be taken to ensure that bandwidth consumed by test frames is limited (so that services are not unacceptably impacted).

9.3.6
SOAM Loopback

SOAM loopback, defined by IEEE 802.1Q[29], is different from the traditional loopback, which redirects existing traffic flow back toward its source. SOAM loopback is more similar to an Internet Control Message Protocol (ICMP) **ping**, which tests target reachability.

SOAM loopback sends a stream of test frames from a source MEP to a target MEP/MIP with instructions to return the frames to the sender. SOAM loopback also offers options to set frame characteristics, such as size, CoS, and drop-eligibility (green/yellow).

SOAM Loopback uses two types of SOAM frames, loopback messages (LBMs) and loopback replies (LBRs), as shown in the following figure.

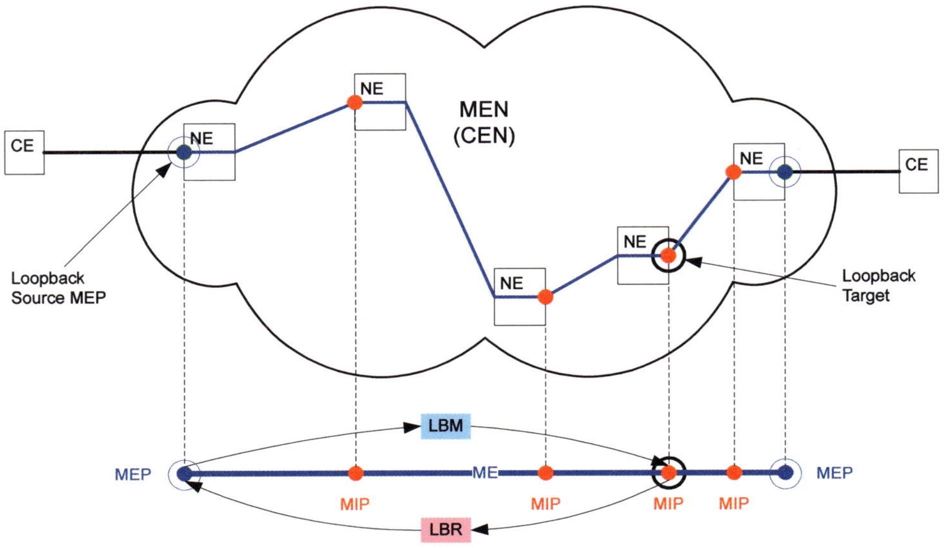

Figure 91: SOAM Loopback

An LBM is sent from a MEP towards its peer targeting a specific MIP or MEP in the same ME. If the target MIP/MEP receives the LBM, it immediately returns an LBR.

LBMs are sent as a loopback session of N (default is 3) consecutive LBMs in a predefined time interval. If an LBR is not received within a given time interval (default is 5 seconds), the MEP declares a reachability fault.

Important: *Unlike LOAM loopback, SOAM Loopback is not service disrupting.*

---

29 IEEE 802.1ag-2007 (the original source of IEEE SOAM standards, referenced in older MEF specifications) has been incorporated into and superseded by IEEE 802.1Q-2011.

9.3.7 SOAM Link Trace

SOAM Link Trace, defined by IEEE 802.1Q[30], is used to test connectivity and to troubleshoot connectivity issues.

Note: *SOAM Link Trace is analogous to Internet Control Message Protocol (ICMP)* ***traceroute***.

SOAM Link Trace uses two types of SOAM frames, link trace messages (LTMs) and link trace replies (LTRs), as shown in the following figure.

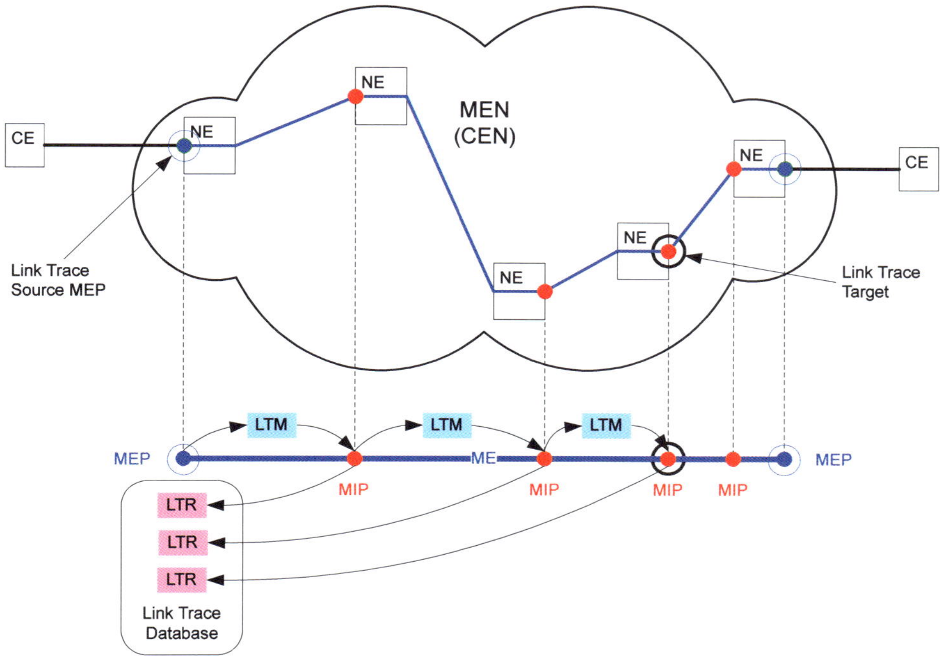

Figure 92: SOAM Link Trace

An LTM is sent from a MEP towards its peer targeting a specific MIP or MEP in the same ME.

The LTM travels to the first MIP/MEP encountered with a matching MEG level (in the same ME). There it triggers the MIP/MEP to send a link trace reply (LTR) back to the source and (if it is not the target MEP/MIP) to forward the LTM onward toward the target MEP/MIP. The process continues in this way until the target is reached or until TTL (time to live) terminates the process. LTR data, collected at the source MEP, can then be analyzed to troubleshoot connectivity issues.

30 IEEE 802.1ag-2007 (the original source of IEEE SOAM standards, referenced in older MEF specifications) has been incorporated into and superseded by IEEE 802.1Q-2011.

9.4
SOAM Performance Management

In this section:

SOAM Performance Management (PM) addresses the problem of measuring EVC/OVC performance, for example, to verify that performance objectives specified in the SLS are met.

From previous lessons, recall that:

- Frame delivery performance for an EVC-based service is specified using the **EVC Performance** service attribute.
- Frame delivery performance for an OVC-based service is specified using the **Service Level Specification** service attribute.

In either case, performance is specified per CoS in the SLS using eight performance objectives (limiting values for eight performance metrics). As shown in the following table, four of the performance metrics are related to frame loss, and the other four are related to frame delay.

Table 22: Performance Metrics

Frame Loss – Related Performance Metrics	Frame Delay – Related Performance Metrics
• Frame Loss Ratio (FLR) • Availability • High Loss Intervals (HLI) • High Loss Intervals (CHLI)	• Frame Delay (FD) • Frame Delay Range (FDR) • Mean Frame Delay (MFD) • Inter Frame Delay Variation (IFDV)

SOAM PM defines techniques for measuring these performance metrics.

Note: *For clarity, you may want to quickly review the **EVC Performance** service attribute (for EVC-based service) and the **Service Level Specification** service attribute (for OVC-based service), which were described earlier in this study guide.*

Related Links

Service Level Specification (OVC Service Attribute)
EVC Performance Service Attributes

9.4.1 SOAM Framework for Measuring Performance Metrics

The MEF framework for SOAM PM, initially defined in MEF 17, has been extended in MEF 35.

Note: *The initial MEF-CECP exam did not include extensions to SOAM PM introduced in MEF 35.*

The extended framework, defined in MEF 35, is based on the concept of a *PM Solution*:

- **PM Solution**–A set of related requirements that when implemented allow a given set of performance metrics to be measured using a given set of PM functions

As shown in the following figure, a PM Solution can involve components at three layers:

1. Network Element (NE) layer
2. Element Management Systems (EMS) layer
3. Network Management System (NMS) layer

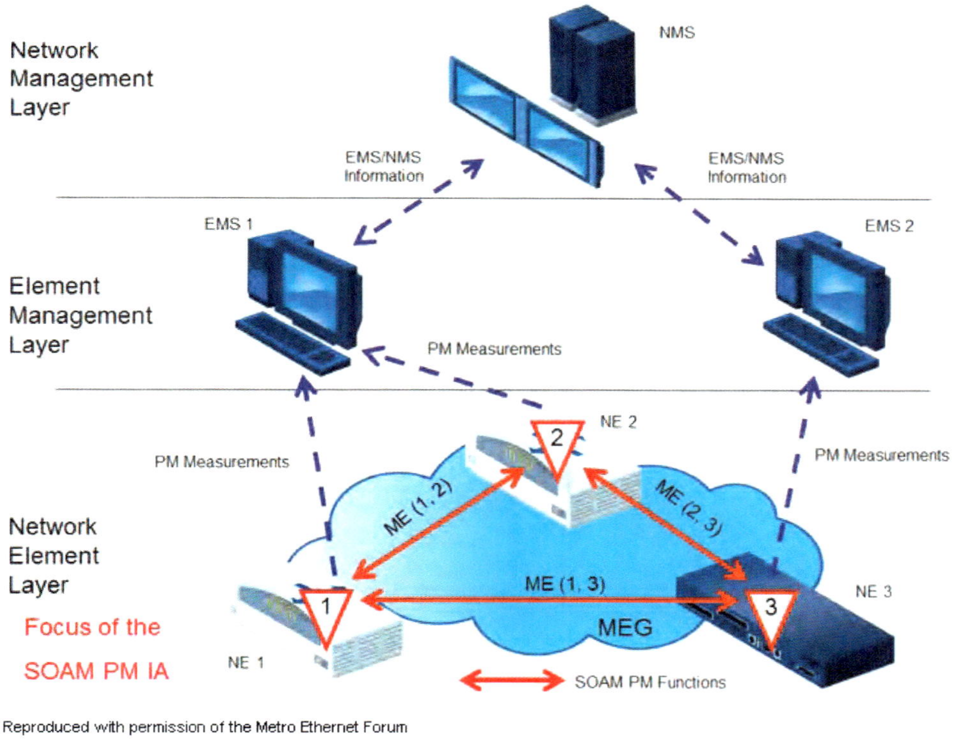

Figure 93: PM Solution Components (Figure 4 in MEF 35)

The NE layer is responsible for conducting performance measurements, while the EMS and NMS layers are responsible for configuring, collecting, and processing performance measurements to determine one or more performance metrics for the MEG.

MEF 35 focuses on the NE layer and defines requirements for three PM solutions (PM-1, PM-2, and PM-3) at the NE layer only. These three PM solutions are summarized in the following figure.

Important: *Support for PM-1 is mandatory, while PM-2 and PM-3 are optional.*

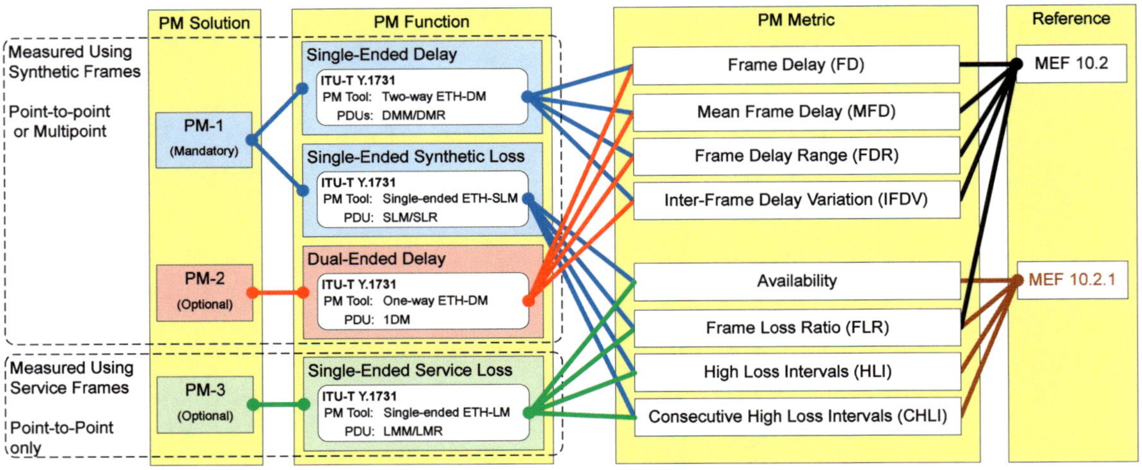

Figure 94: Summary of PM Solutions (Based on MEF 35, Section 8)

As shown in this figure, each PM solution supports the determination of four or eight PM metrics using one or two PM functions.

- **PM Function**—A MEP capability specified for performance monitoring purposes.

MEF 35 defines four PM functions:

- Single-Ended Delay
- Single-Ended Synthetic Loss
- Dual-Ended Delay
- Single-Ended Service Loss.

As shown in the following table, each PM Function is implemented using an ITU-T-defined **PM Tool** that uses one or two ITU-T-defined **PDUs**.

Table 23: PM Functions, Tools, and PDUs (Based on Table 3 in MEF 35)

PM Function	ITU-T Y.1731	
	PM Tool	PDU(s)
Single-Ended Delay	Two-Way ETH-DM	DMM (Delay Measurement Message)
		DMR (Delay Measurement Reply)
Single-Ended Synthetic Loss	Single-Ended ETH-SLM	SLM (Synthetic Loss Message)
		SLR (Synthetic Loss Reply)
Dual-Ended Delay	One-Way ETH-DM	1DM (One-Way Delay Measurement)
Single-Ended Service Loss	Single-Ended ETH-LM	LMM (Loss Measurement Message)
		LMR (Loss Measurement Reply)

9.4.2 Synthetic Frames versus Service Frames

Two out of three PM Solutions (and three out of four PM Functions) use synthetic frames, rather than service frames.

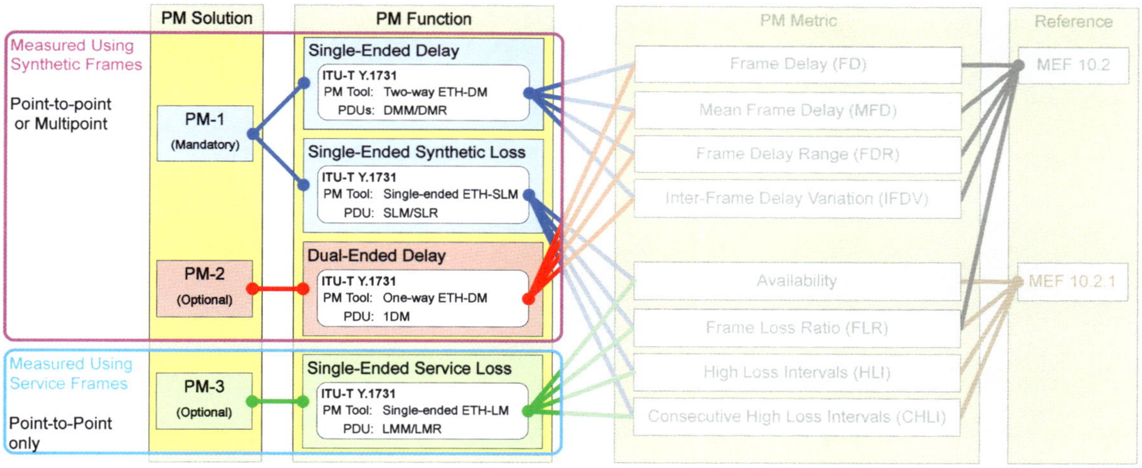

Figure 95: Summary of PM Solutions (Based on MEF 35, Section 8)

Synthetic frames are Ethernet frames that are created to emulate service traffic for performance measurement purposes. They are not service frames, but they are treated in the network like service frames. Synthetic frames are used in place of service frames because they can carry additional information, such as time stamp and/or identification data, which may be needed or useful for calculating frame delay or frame loss.

As shown in the previous figure, PM-3 (which uses service frames) has limited utility compared to PM-1 and PM-2 (which use synthetic frames):

- PM-3 only supports MEGs with two MEPs (point-to-point MEGs), while PM-1 and PM-2 can support MEGs with two or more MEPs (multipoint MEGs).
- PM-3 can only determine frame loss-related metrics (not frame delay-related metrics).

These limitations are due to the limited utility of service frames (compared to synthetic frames) in measuring PM metrics.

9.4.3 Single-Ended versus Dual-Ended PM Functions

Three out of four PM functions are single-ended. One is dual-ended.

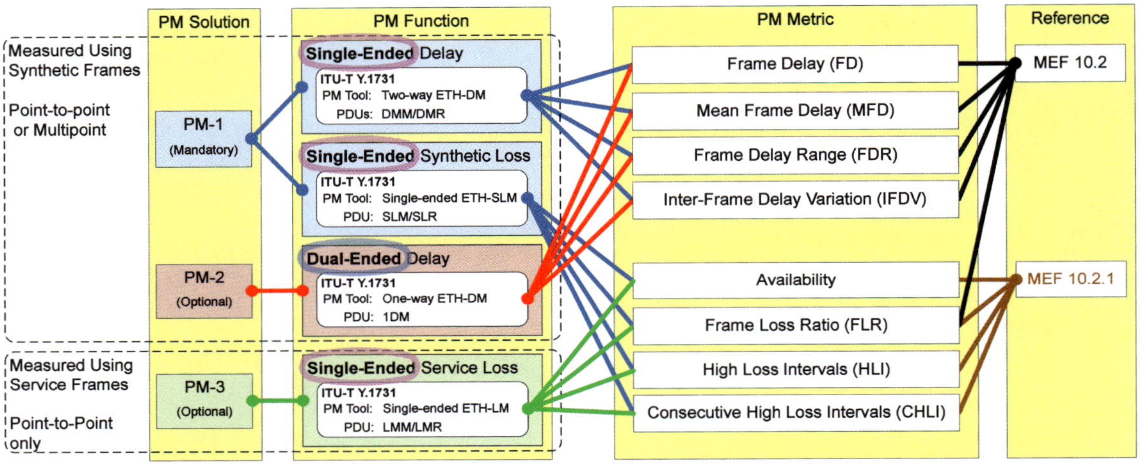

Figure 96: Summary of PM Solutions (Based on MEF 35, Section 8)

Per MEF 35, the following definitions apply:

- **Single-ended** – A type of process where a MEP sends a measurement request and the peer MEP replies with the requested information so the originating MEP can calculate the measurement
- **Dual-ended** – A type of process where a MEP sends measurement information to a peer MEP that will perform the calculations.

Each of the three single-ended PM functions use two SOAM PDUs (a message PDU and a reply PDU). The dual-ended PM function (Dual-ended Delay) uses only one SOAM PDU.

9.4.4
Single-Ended Delay

PM function **Single-Ended Delay** is implemented using PM tool **Two-Way ETH-DM**, as defined by ITU-T Y.1731.

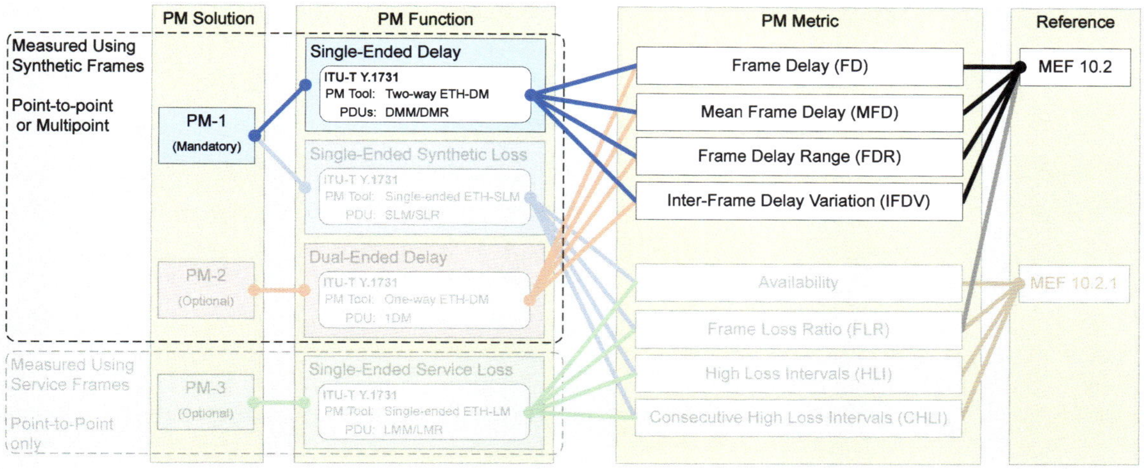

Figure 97: Summary of PM Solutions (Based on MEF 35, Section 8)

As shown in the following figure, two types of SOAM frames are used: the DMM (Delay Measurement Message) and the DMR (Delay Measurement Reply).

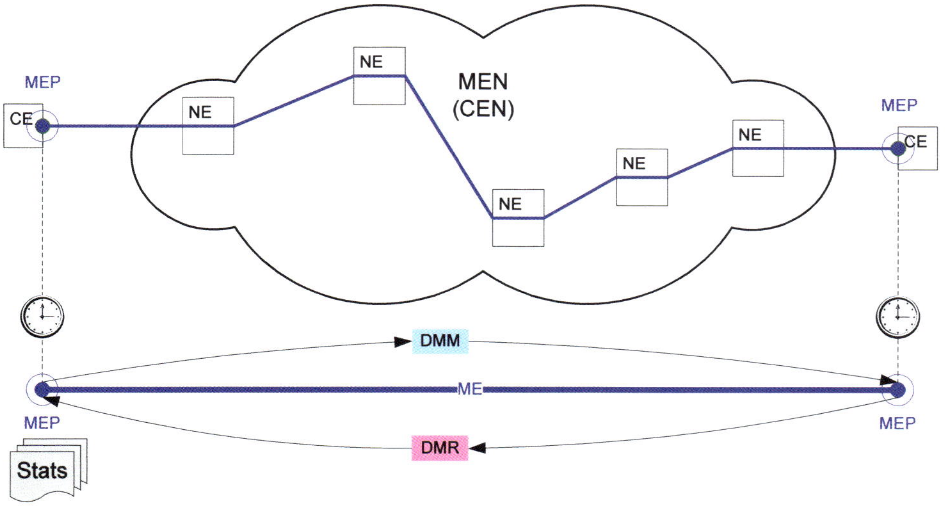

Figure 98: Single-Ended Delay Measurement

In a given ME, one MEP periodically transmits a DMM toward its peer. When the peer MEP receives a DMM, it returns a DMR. Both MEPs encode time stamps into DMMs and DMRs, reflecting time sent and time received.

If the MEPs are time-synchronized, one-way measurements can be taken (one measurement in each direction).

Otherwise, two-way measurements are taken using only the source MEP clock. The peer MEP copies "time sent" from the DMM to the DMR so that elapsed time for two-way travel can be computed when the source MEP receives the DMR.

Single-Ended Delay measurement:

- Measures the delay of synthetic frames (not service frames)
- Provides data for computing frame delay-related PM metrics (FD, MFD, FDR, and IFDV)
- Can be applied to point-to-point MEGs and multipoint MEGs

In multipoint-to-multipoint and rooted-multipoint service applications, multiple MEs may be involved because PM metrics are measured over an agreed subset of ordered UNI pairs, as specified in the SLS.

Separate measurements are made per CoS.

9.4.5 Single-Ended Synthetic Loss

PM function **Single-ended Synthetic Loss** is implemented using PM tool **Single-ended ETH-SLM**, as defined by ITU-T Y.1731.

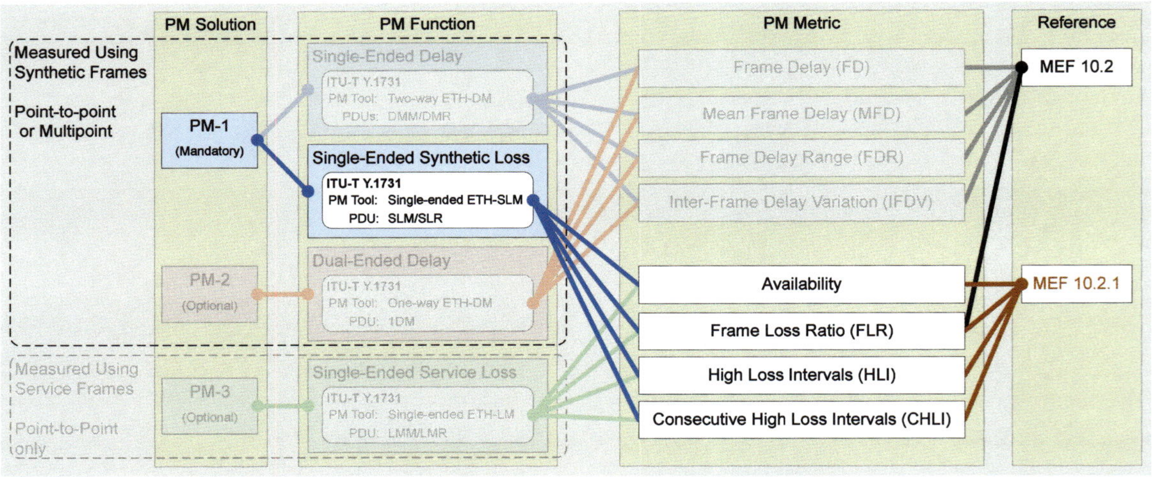

Figure 99: Summary of PM Solutions (Based on MEF 35, Section 8)

As shown in the following figure, two types of SOAM frames are used: the SLM (Synthetic Loss Message) and the SLR (Synthetic Loss Reply).

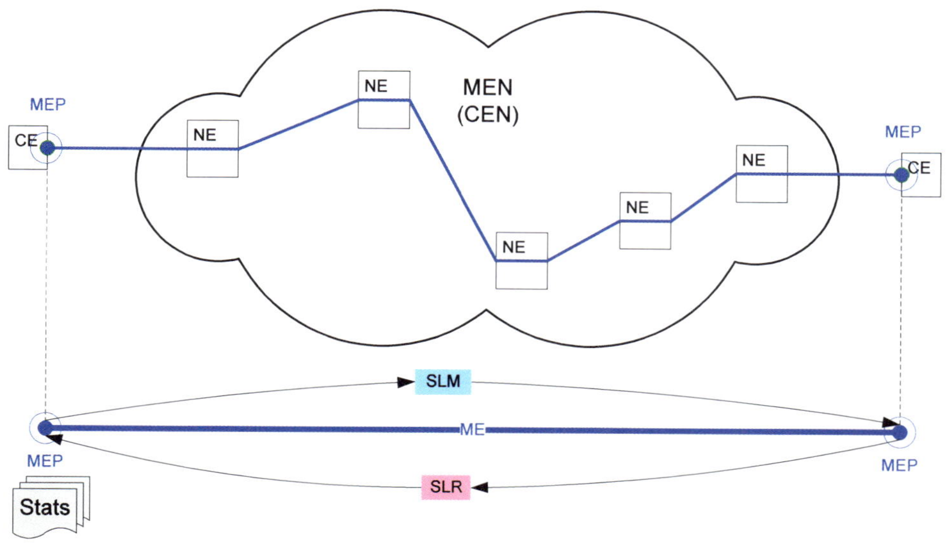

Figure 100: Single-Ended Synthetic Loss Measurement

In a given ME, one MEP periodically transmits an SLM toward its peer. When the peer MEP receives an SLM, it returns an SLR. Both MEPs keep count of synthetic frames sent and received.

Single-Ended Synthetic Loss measurement:

- Measures the loss of synthetic frames (not service frames)
- Provides data for computing frame loss-related PM metrics (Availability, FLR, HLI, and CHLI)
- Can be applied to point-to-point MEGs and multipoint MEGs

In multipoint service applications, multiple MEs may be involved because PM metrics are measured over an agreed subset of ordered UNI pairs, as specified in the SLS.

Separate measurements are made per CoS.

9.4.6 Dual-Ended Delay

PM function **Dual-ended Delay** is implemented using PM tool **One-way ETH-DM**, as defined by ITU-T Y.1731.

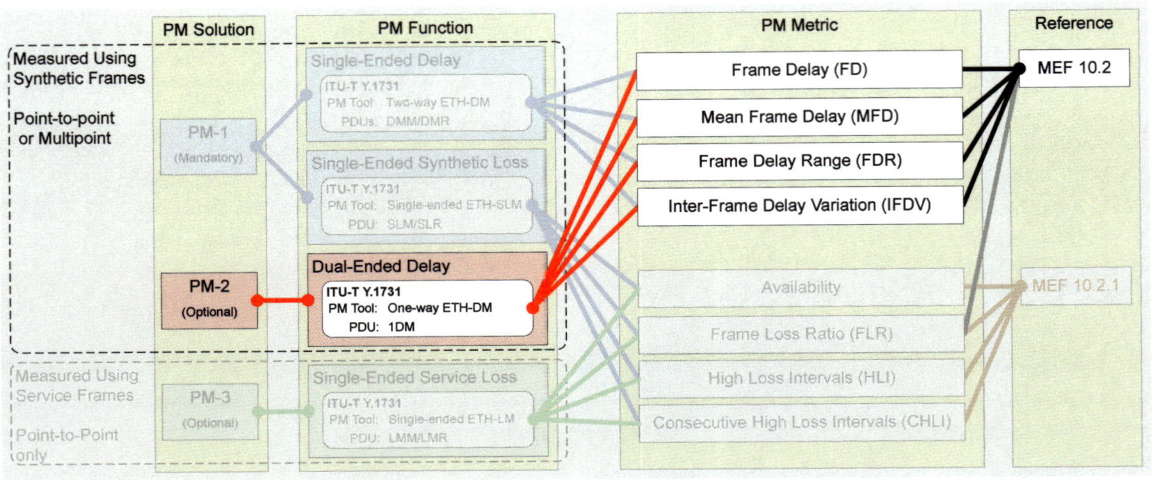

Figure 101: Summary of PM Solutions (Based on MEF 35, Section 8)

As shown in the following figure, one type of SOAM frame is used: the DM1 (One-way Delay Measurement) PDU.

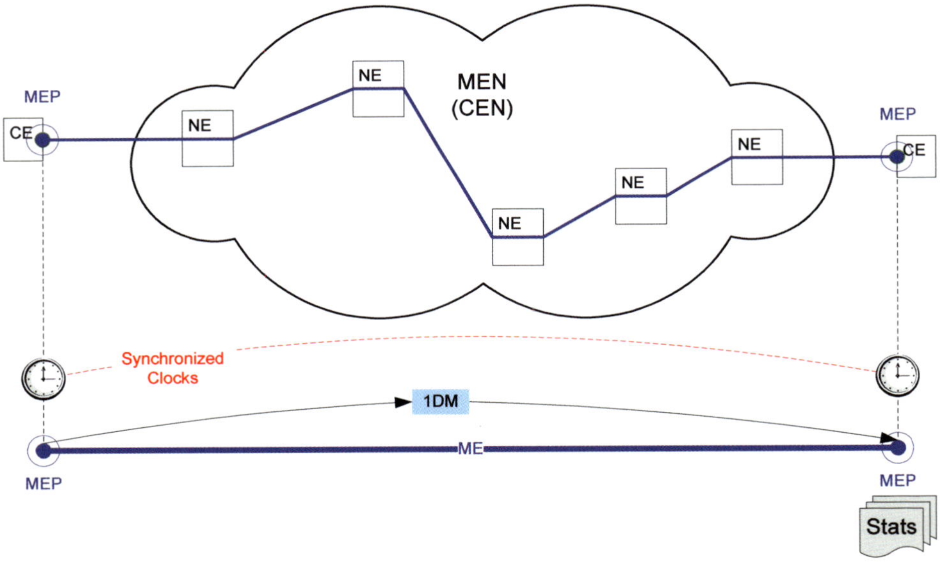

Figure 102: Dual-Ended Delay Measurement

In a given ME, one MEP periodically transmits a time-stamped 1DM toward its peer. When the peer MEP receives the 1DM, it computes delay as the difference between the time stamp value and the current time (the two MEPs are time-synchronized).

If delay measurement is required for both directions, a second instance of Dual-ended Delay measurement is applied in the opposite direction.

Dual-Ended Delay measurement:

- Measures the delay of synthetic frames (not service frames)
- Provides data for computing frame delay-related PM metrics (FD, MFD, FDR, and IFDV)
- Can be applied to point-to-point MEGs and multipoint MEGs

In multipoint service applications, multiple MEs may be involved because PM metrics are measured over an agreed subset of ordered UNI pairs, as specified in the SLS.

Separate measurements are made per CoS.

9.4.7 Single-Ended Service Loss

PM function **Single-ended Service Loss** is implemented using PM tool **Single-ended ETH-LM**, as defined by ITU-T Y.1731.

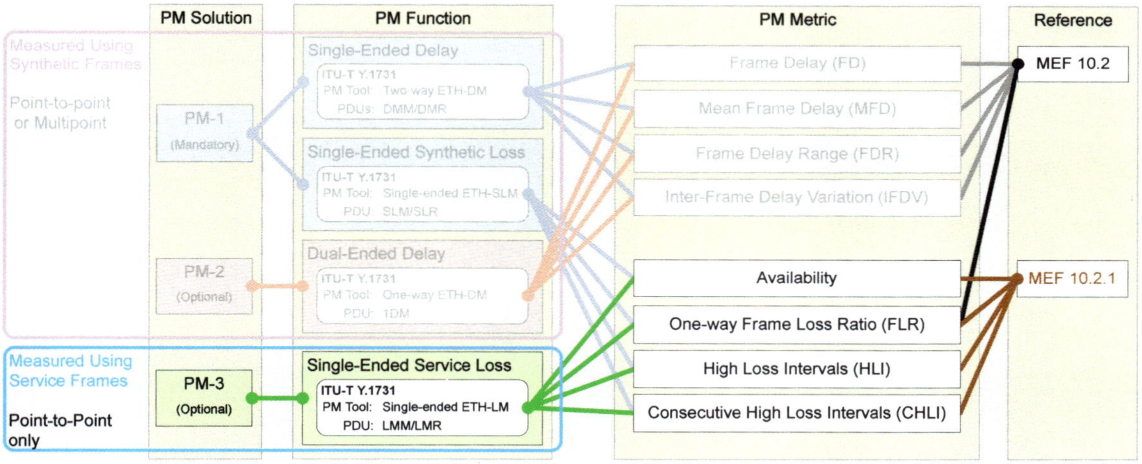

Figure 103: Summary of PM Solutions (Based on MEF 35, Section 8)

As shown in the following figure, two types of SOAM frames are used: the LMM (Loss Measurement Message) and the LMR (Loss Measurement Reply).

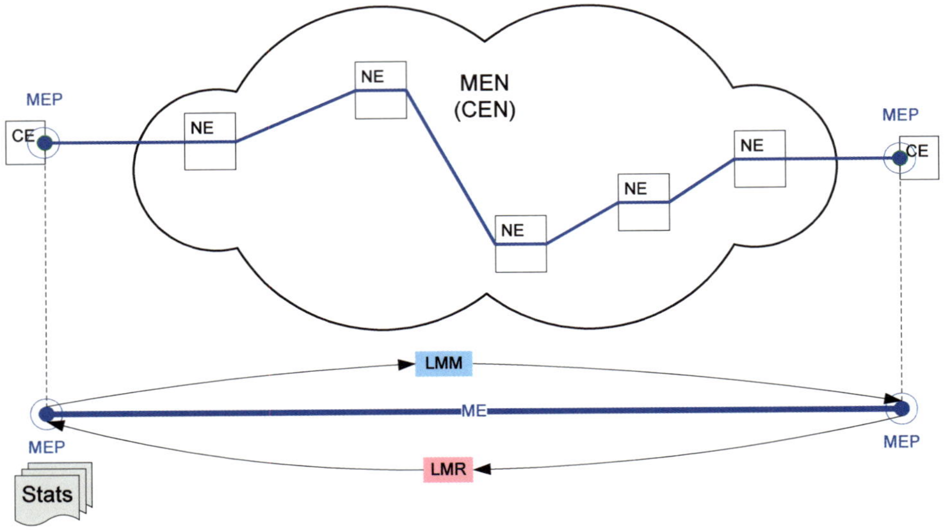

Figure 104: Single-Ended Service Loss Measurement

In a given ME, one MEP periodically transmits an LMM toward its peer. When the peer MEP receives an LMM, it returns an LMR. Both MEPs keep count of service frames sent and received.

Single-Ended Service Loss measurement:

- Measures the loss of service frames (not synthetic frames)
- Provides data for computing frame loss-related PM metrics (Availability, FLR, HLI, and CHLI)
- Is applicable to MEGs with two MEPs (point-to-point MEGs) only

Separate measurements are made per CoS.

9.4.8 The Performance Monitoring Process

Per MEF 35, the performance monitoring process is made up of a number of performance monitoring instances, known as PM Sessions.

- **PM Session**—an instance of a PM Solution between a given pair of MEPs using a given CoS frame set over a given (possibly indefinite) period of time

A PM Session is initiated on a MEP to take performance measurements for a given CoS frame set in conjunction with its peer MEP. A PM Session can be classified as either a loss measurement session or a delay measurement session, depending on the PM function used.

Various parameters define a PM session, including start time, end time, and frequency of SOAM frame transmission. Time within a PM session is divided into discrete, non-overlapping periods of time called measurement intervals.

- **Measurement Interval**–A period of time within a PM session in which PM session measurements are performed and results are gathered. Measurements initiated during one measurement interval are kept separate from measurements taken during other measurement intervals.

Note: *A measurement interval is different from T (the time interval over which a performance objective is defined).*

Results for loss measurement sessions are stored directly, per measurement interval, as shown in the following example. This example is for a MEP running a Single-Ended Synthetic Loss PM function that measures loss separately for each direction.

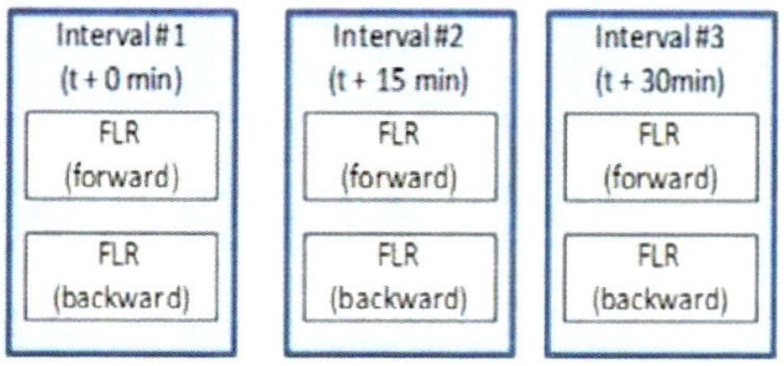

Figure 105: Example of FLR Measurements (Figure 8 in MEF 35)

Results for delay measurement sessions are stored, per measurement interval, using measurement bins.

- **Measurement Bin**–a counter that stores the number of delay measurements falling within a specified range during a measurement interval

The following figure illustrates the relationship between measurement intervals and measurement bins.

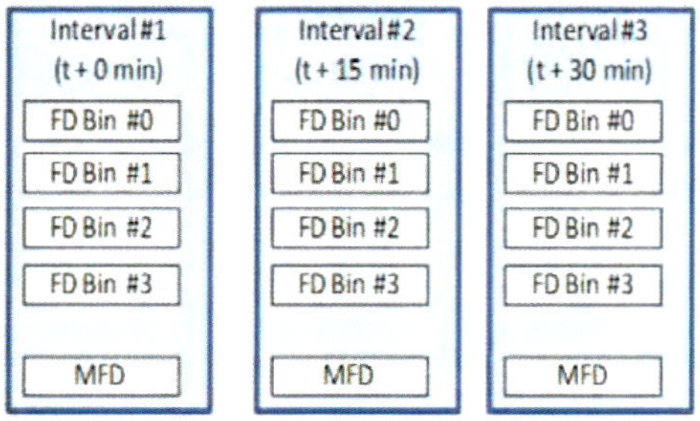

Figure 106: Example of Measurement Intervals and Bins (Figure 7 in MEF 35)

Each bin is defined by a configurable threshold value, as illustrated in the following example.

Bin	Threshold	Range
bin 0	0 µs	0 µs ≤ measurement < 5,000 µs
bin 1	5,000 µs	5,000 µs ≤ measurement < 10,000 µs
bin 2	10,000 µs	10,000 µs ≤ measurement < 15,000 µs
bin 3	15,000 µs	15,000 µs ≤ measurement < ∞

Figure 107: Example Measurement Bin Configuration (Table 5 in MEF 35)

Measurement intervals and measurement bins reduce that amount of memory that an NE needs for storing PM data:

- Rather than store the time at which the measurement is made (individually for each measurement), each measurement is simply associated to a predefined measurement interval.
- Rather than store values individually for each delay measurement, each delay measurement simply increments the counter for a predefined measurement bin.

Each PM Session uses a PM Function. Each PM Function uses a specific ITU-T PM Tool which in turn uses specific ITU-T PDU(s).

Multiple PM Sessions can be run simultaneously between the MEPs, allowing for multiple classes of service to be tested.

PM Sessions define how PM data is measured and stored at the NE layer. As previously explained, the NE layer is responsible for conducting performance measurements, while the EMS and NMS layers are responsible for configuring, collecting, and processing performance measurements to determine performance metrics.

Related Links

SOAM Framework for Measuring Performance Metrics

9.5
Review

1. Which one of the following statements describes how frame loss ratio (FLR) is measured for an EPL service?

    a. For each EVC CoS, count all green and yellow frames sent and received over an agreed time interval.

    b. For each EVC CoS, count all green frames sent and received over an agreed time interval.

    c. For an EVC, count all frames sent and received (regardless of CoS) over an agreed time interval.

    d. For an EVC, count all green and yellow frames sent and received over an agreed time interval.

2. NE A and NE B are connected by an Ethernet link. Brian, an operator, has access only to NE B (NE B is in Brian's domain of operations, but NE A is not). Which two options below can be available (with correct provisioning) to Brian? (Choose two.)

    a. Use SOAM loopback to redirect all traffic at NE A back toward NE B.

    b. Use SOAM loopback to verify link connectivity.

    c. Use LOAM loopback to redirect all traffic at NE B back toward NE A.

    d. Use LOAM loopback to redirect all traffic at NE A back toward NE B.

    e. Use RDI to initiate LOAM link discovery.

3. Which of the following is <u>not</u> an SOAM frame?

    a. Connectivity Check Message (CCM)

    b. Alarm Indication Signal (AIS)

    c. Remote defect indication (RDI)

    d. Loss Measurement Message (LMM)

4. Which two of the following statements are correct ? (Choose two.)

    a. LOAM support is required for Type 2.1 UNIs and at ENNIs.

    b. SOAM CCMs can support 50 ms protection switching.

    c. SOAM domains can overlap.

    d. An ME can include more than two MEPs.

    e. An ME can include any number of MIPs.

5. If a MEP sends a Loss Measurement Message (LMM), where does it go?
    a. To all MEPs in every ME
    b. To peer MIPs in the same ME
    c. To all MIPs in every ME
    d. To all MEPs in the MEG
    e. To the peer MEP in the same ME
6. Per MEF 30.1, to whom does the EVC MEG belong?
    a. Subscriber
    b. Service Provider
    c. Operator
    d. Subscriber and Operator
    e. Operator and Service Provider
7. MEP clock synchronization is required to measure which two of the following? (Choose two.)
    a. One-way delay with the Single-Ended Delay function
    b. Two-way delay with the Single-Ended Delay function
    c. One-way delay with the Dual-Ended Delay function
    d. Two-way delay with the Dual-Ended Delay function
8. To isolate a fault within a point-to-point OVC, an operator can:
    a. Inspect all continuity check messages (CCMs) for the appropriate ME.
    b. Initiate SOAM Loopback over the appropriate ME.
    c. Send a frame loss measurement message (LMM) to the appropriate peer MEP.
    d. Process the most recent remote defect indicator (RDI) per IEEE 802.1Q.
    e. Initiate SOAM Link Trace over the appropriate ME.

9. Which Maintenance Entity (ME) should the service provider use to monitor the performance of an Ethernet service that spans two Operator MENs?

    a. Subscriber ME

    b. EVC ME

    c. Operator ME

    d. ENNI ME

    e. UNI ME

10. Which two SOAM frames are used for performance monitoring? (Choose two.)

    a. CCM (Continuity Check Message)

    b. LMM (Loss Measurement Message)

    c. LBM (Loopback Message)

    d. LTM (Link Trace Message)

    e. DMM (Delay Measurement Message)

11. One-way frame delay for a service frame in an EVC-based service... (Choose two.)

    a. Measures time of travel, CE-to-CE, for the first bit in the service frame.

    b. Includes the time of transmission between the UNI-C and the UNI-N at both UNIs.

    c. Includes the time that the service frame waits in queue at the CE to be scheduled for transmission.

    d. Depends on the length of the service frame.

12. Which four of the following items are connectivity fault management (CFM) functions? (Choose four.)

    a. Remote defect indication (RDI)

    b. Link Trace

    c. Frame Delay Measurement

    d. Alarm Indication Signal (AIS)

    e. Loopback

    f. Frame Loss Measurement

13. How many ordered UNI pairs are in an EVC containing three UNIs?

- **a.** Two
- **b.** Three
- **c.** Six
- **d.** Nine

14. Which two statements about SOAM frames are true? (Choose two.)

- **a.** AIS frames are sent away from the direction of failure.
- **b.** AIS frames are sent toward the direction of failure.
- **c.** CCM frames containing RDI are sent away from the direction of failure.
- **d.** CCM frames containing RDI are sent toward the direction of failure.

Related Links

Answers: Ethernet OAM

10
Access Technologies

In this chapter:

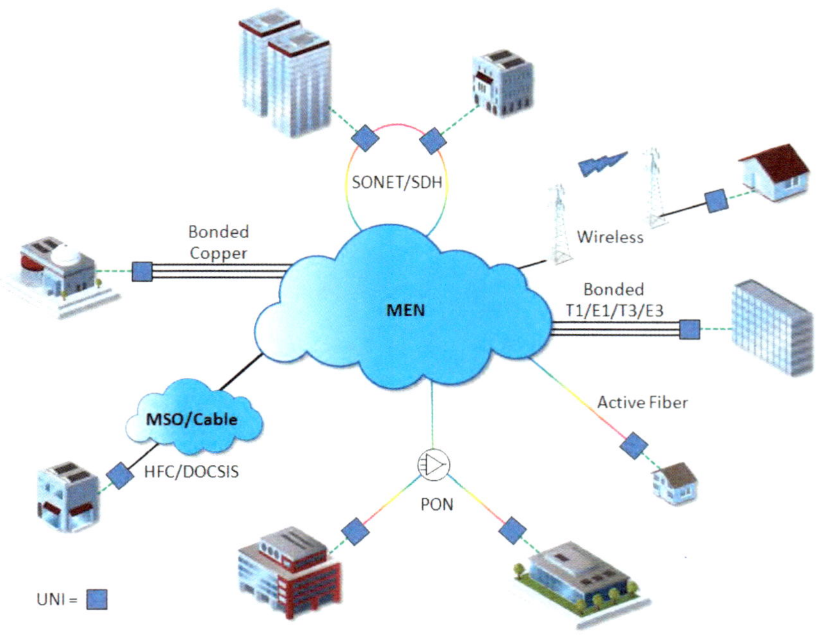

Service providers need to support subscribers through a wide variety of access technologies, which are also called "first-mile" or "last-mile" technologies.

The Access Technologies

The following table, taken from the MEF white paper *Delivering Ubiquitous Ethernet Services using an Array of Access Technologies*, provides a summary of Carrier Ethernet access technologies.

Hint: *In the MEF White Papers area of the MEF Web site (MetroEthernetForum.org), the link to this white paper is titled Carrier Ethernet – the Technology of Choice for Access Networks.*

Summary of Carrier Ethernet Access Technologies			
Carrier Ethernet Access Method	Technology Alternatives	Deployment Scenarios (When to use the technology)	Advantages
Ethernet over Fiber	- Active Ethernet - Ethernet over SONET/SDH - Passive Optical Network	- On-net buildings - Greenfield - Dense Metro area - 1Gbit/s or greater bandwidth requirements	- Highest bandwidth - Noise immunity - Security - Long reach - SONET/SDH leverage existing - Growth potential via xWDM
Ethernet over PDH	- Bonded T1/E1 - DS3/E3 and bonded DS3/E3	- Remote branch offices - Off-net customer locations (out of region, type 2) - SMB	- Leverage existing transport - Universally deployable - Lower CAPEX - No reach limitations - Well understood provisioning - Resiliency through bonding
Ethernet over Copper	- 2BASE-TL - 10PASS-TS	- Remote branch offices - On-net or off-net - SMB - Campus settings - Traffic monitoring	- Ubiquitous copper availability - Rapid deployment - Low cost unbundled local loop - Resiliency through bonding
Wireless Ethernet	- Terrestrial microwave - WiMAX - Broadband wireless - Free space optics - WiFi	- Remote branch office - Campus setting - No fiber or copper available - Mobility required	- Installation requires no trenching - Rapid deployment - Some alternatives offer mobility
Hybrid Fiber Coax	DOCSIS 2.x/3.x	- Work at home - SOHO/SMB - Remote branch office	- Extensive coverage - High performance options - Deep penetration into residential and suburban geographies

For the MEF-CECP 2.0 exam, it is important to know how each technology operates and their merits. Given an application scenario, you should also be able to identify which access technologies meet stated requirements.

Hint: *Consider jumping ahead to preview Figure 111 (a study aid that appears in the summary to this lesson). This study aid may not make much sense at this time, but it may help you to visualize what you are expected to take away from this lesson.*

Key Areas for Comparing the Access Technologies

The access technologies differ in the following key areas:

- **Transmission Medium**—Through what substance is traffic supported? For example, is traffic carried over optical fiber, copper wire, or coaxial cable, or is it transmitted through the air?
- **Infrastructure Availability**—Does the technology use infrastructure that is already in place (commonly available), or does it require new infrastructure to be deployed?
- **Bandwidth (Speed and Symmetry)**—What speeds of Carrier Ethernet service can the technology support (for example, 10 Mb/s, 1 Gb/s, or 40 Gb/s)? Is the technology neutral with respect to upstream versus downstream bandwidth support, or is it inherently asymmetric?
- **Distance**—Over what distances can the technology operate?
- **Reliability**—Does the technology support protection/resiliency mechanisms?
- **Cost**

Attention: *By necessity, this lesson makes generalizations about relative cost and performance characteristics, such as maximum bandwidth and distance capabilities. These generalizations align with the MEF-CECP 2.0 exam, which tests for a general understanding of the relative strengths of each technology at the time the initial MEF-CECP exam was developed (circa 2010). These generalizations are not meant to capture performance records or to be perfectly current.*

10.1
Ethernet over Optical Fiber

In this section:

Summary of Carrier Ethernet Access Technologies

Carrier Ethernet Access Method	Technology Alternatives	Deployment Scenarios (When to use the technology)	Advantages
Ethernet over Fiber	- Active Ethernet - Ethernet over SONET/SDH - Passive Optical Network	- On-net buildings - Greenfield - Dense Metro area - 1Gbit/s or greater bandwidth requirements	- Highest bandwidth - Noise immunity - Security - Long reach - SONET/SDH leverage existing - Growth potential via xWDM
Ethernet over PDH	- Bonded T1/E1 - DS3/E3 and bonded DS3/E3	- Remote branch offices - Off-net customer locations (out of region, type 2) - SMB	- Leverage existing transport - Universally deployable - Lower CAPEX - No reach limitations - Well understood provisioning - Resiliency through bonding
Ethernet over Copper	- 2BASE-TL - 10PASS-TS	- Remote branch offices - On-net or off-net - SMB - Campus settings - Traffic monitoring	- Ubiquitous copper availability - Rapid deployment - Low cost unbundled local loop - Resiliency through bonding
Wireless Ethernet	- Terrestrial microwave - WiMAX - Broadband wireless - Free space optics - WiFi	- Remote branch office - Campus setting - No fiber or copper available - Mobility required	- Installation requires no trenching - Rapid deployment - Some alternatives offer mobility
Hybrid Fiber Coax	DOCSIS 2.x/3.x	- Work at home - SOHO/SMB - Remote branch office	- Extensive coverage - High performance options - Deep penetration into residential and suburban geographies

10.1.1
Ethernet over Active Fiber

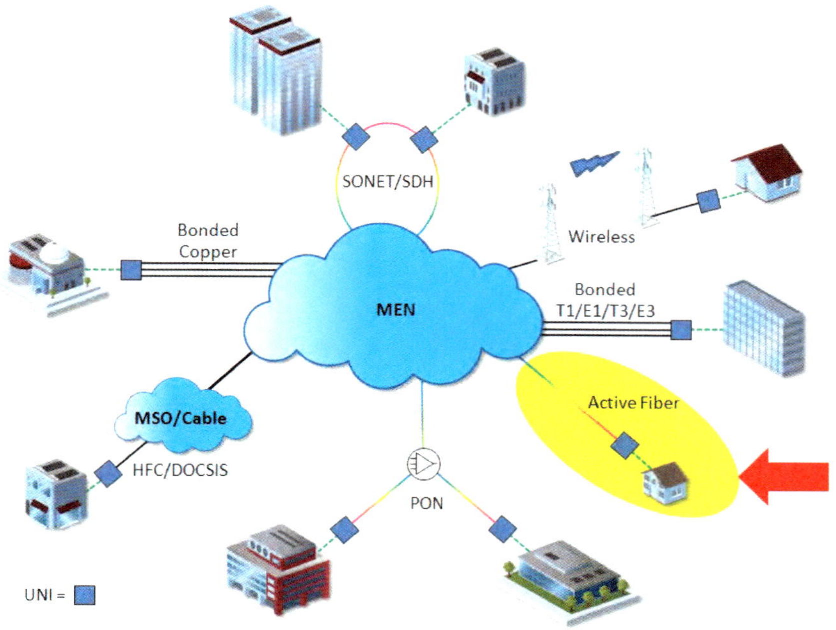

Ethernet over Active Fiber is a generic term that refers to the delivery of Ethernet traffic over optical fiber, without reference to particular protocols or standards, but limited by the performance of current technology. Consider these attributes:

- **Speed**—Up to 100 Gb/s is supported.[31]
- **Distance**—Up to 150 km can be easily achieved using standard optics. Greater distances are possible with amplification/regeneration.

These attributes demonstrate how fiber optic technology is virtually "future-proof" with respect to bandwidth and distance requirements.

31 As of March 2012, a single optical fiber can support 100 Gb/s without wavelength division multiplexing (WDM). With WDM, a single fiber-optic cable can support up to 88 DWDM wavelengths (at 100 Gb/s each) or 8.8 Tb/s.

10.1.2
Ethernet over SONET/SDH

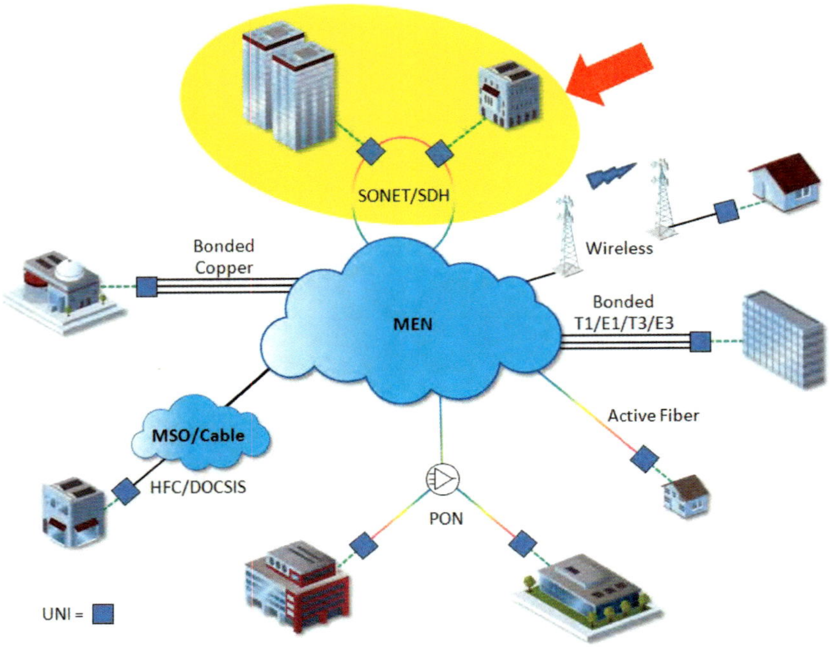

Ethernet over SONET/SDH (EoS) is a point-to-point technology that uses fixed-size TDM time slots to transport Ethernet frames over fiber-optic cables. The time slots can be concatenated together to match bandwidth requirements.

Advantages of EoS include:

- Widely deployed and available worldwide
- Highly reliable and traditionally regarded as the "gold standard" for resiliency (sub-50 ms protection/restoration, Layer 1 security)
- Fully standardized OAM tools for performance monitoring and fault detection
- Capable of simultaneously supporting both TDM and Ethernet
- Capable of supporting a wide range of bandwidths from 1.5 Mb/s (using subrate virtual tributary signals) up to 40 Gb/s
- No distance limitation

Note: *SONET/SDH is described in more detail in the Transport Technology lesson.*

Related Links

SONET/SDH (Synchronous Optical Network / Synchronous Digital Hierarchy)

10.1.3
Ethernet over PON

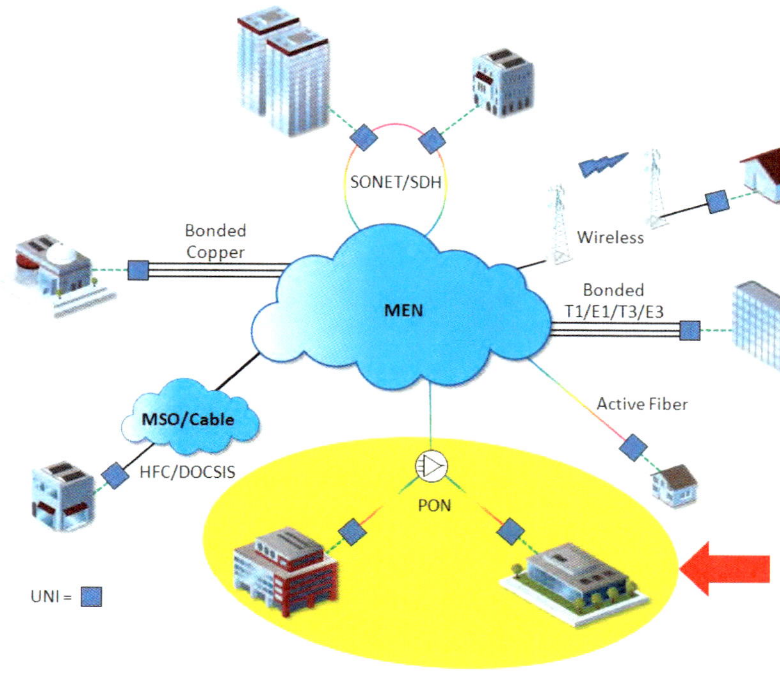

A **passive optical network (PON)** is a point-to-multipoint optical fiber technology that uses unpowered optical splitters to enable a single optical fiber to serve multiple UNIs.

There are two PON standards:

- **EPON** (Ethernet Passive Optical Network)–Developed by the IEEE, this standard supports 1 Gb/s symmetrical communication over distances up to at least 20 km (per the IEEE standard) or up to about 40 km in commercial application[32].

- **GPON** (Gigabit Passive Optical Network)–Developed by ITU-T, this standard supports 1.25 Gb/s upstream and 2.5 Gb/s downstream over distances up to at least 20 km (per the ITU-T standard) or up to about 40 km in commercial application.

Additionally, 10 Gb/s versions of both standards (10G-EPON and 10G-PON) are now available.

A PON consists of an optical line terminal (OLT) at the service provider's central office and a number of optical network terminals (ONTs) at customer premises. A PON reduces the amount of fiber and central office equipment required compared with point-to-point architectures.

32 Per the MEF access technologies white paper *Delivering Ubiquitous Ethernet Services using an Array of Access Technologies*, in the MEF White Papers area of the MEF Web site (MetroEthernetForum.org), under the link titled *Carrier Ethernet – the Technology of Choice for Access Networks*.

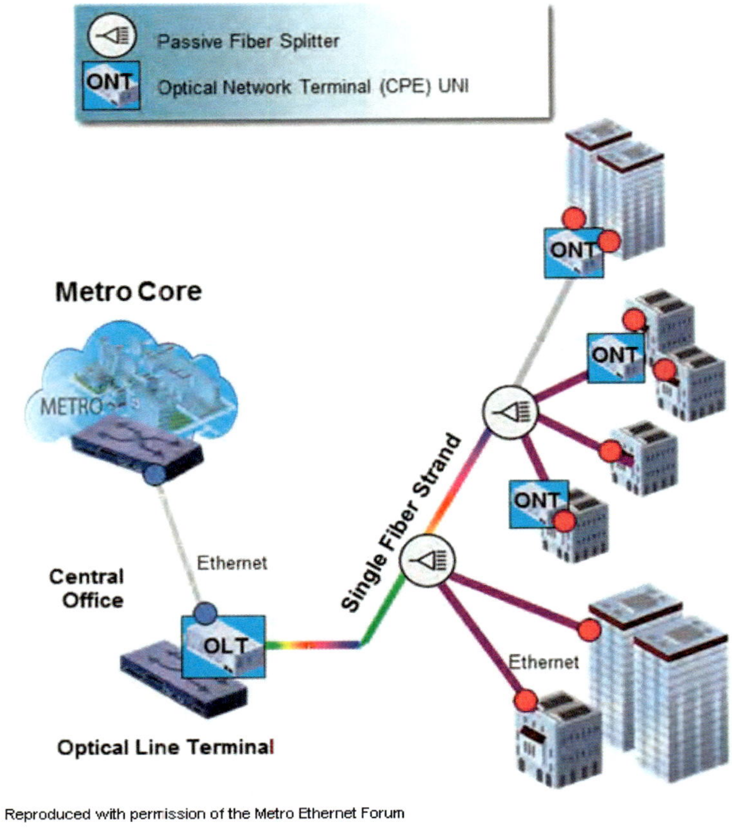

Figure 108: Ethernet over PON

Downstream signals are broadcast to all ONTs sharing the same OLT fiber. Each ONT then filters the downstream signal appropriately (encryption can prevent eavesdropping), as shown in the following figure.

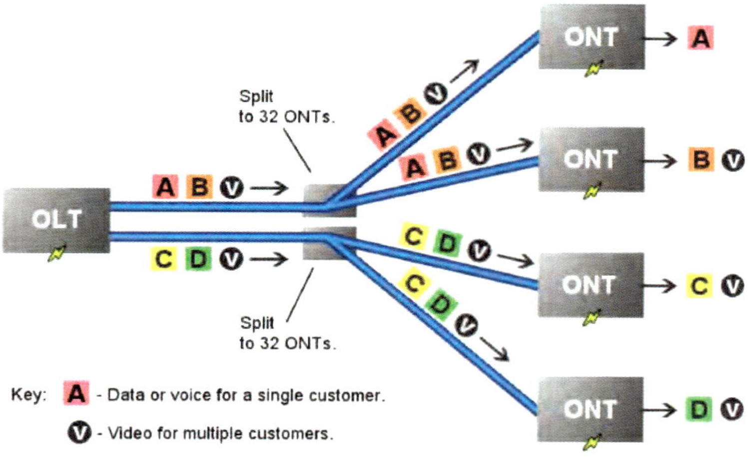

Figure 109: PON Downstream Broadcast

This figure also illustrates how PON naturally supports multicasting: the video is multicast to three of the four subscribers.

Upstream signals are combined using a multiple access protocol, usually TDMA (time division multiple access).

Advantages of PON include:

- High bandwidth capability compared to legacy copper technologies
- Lower deployment and operational costs compared to active optical networks
- Reduced fiber requirements due to point-to-multipoint topology
- More efficient multicasting

10.2
Ethernet over PDH

Summary of Carrier Ethernet Access Technologies			
Carrier Ethernet Access Method	Technology Alternatives	Deployment Scenarios (When to use the technology)	Advantages
Ethernet over Fiber	- Active Ethernet - Ethernet over SONET/SDH - Passive Optical Network	- On-net buildings - Greenfield - Dense Metro area - 1Gbit/s or greater bandwidth requirements	- Highest bandwidth - Noise immunity - Security - Long reach - SONET/SDH leverage existing - Growth potential via xWDM
Ethernet over PDH	- Bonded T1/E1 - DS3/E3 and bonded DS3/E3	- Remote branch offices - Off-net customer locations (out of region, type 2) - SMB	- Leverage existing transport - Universally deployable - Lower CAPEX - No reach limitations - Well understood provisioning - Resiliency through bonding
Ethernet over Copper	- 2BASE-TL - 10PASS-TS	- Remote branch offices - On-net or off-net - SMB - Campus settings - Traffic monitoring	- Ubiquitous copper availability - Rapid deployment - Low cost unbundled local loop - Resiliency through bonding
Wireless Ethernet	- Terrestrial microwave - WiMAX - Broadband wireless - Free space optics - WiFi	- Remote branch office - Campus setting - No fiber or copper available - Mobility required	- Installation requires no trenching - Rapid deployment - Some alternatives offer mobility
Hybrid Fiber Coax	DOCSIS 2.x/3.x	- Work at home - SOHO/SMB - Remote branch office	- Extensive coverage - High performance options - Deep penetration into residential and suburban geographies

Reproduced with permission of the Metro Ethernet Forum

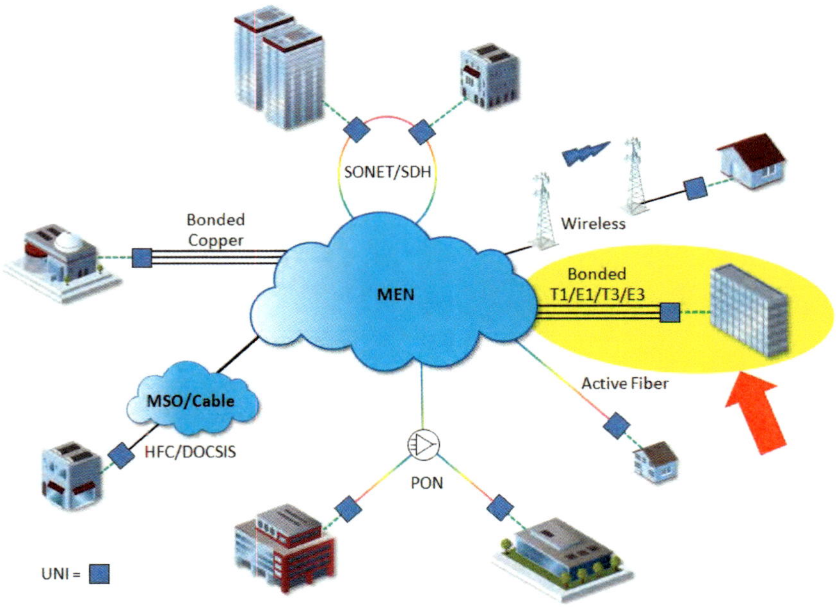

PDH (Plesiochronous Digital Hierarchy) refers to a telecommunications access technology that has, until recently, dominated business voice and data telecommunications for decades.

Ethernet over PDH leverages the telecommunications infrastructure that supports the following carrier signals:

- **T1 and T3 lines**–Used in the US and Canada

  Note: *DS designations (DS1 and DS3) are used in connection with T1 and T3. Strictly speaking, a DS1/DS3 is the data carried on a T1/T3 circuit, but in practice the terms are used interchangeably.*

- **E1 and E3 lines**–Used in the rest of the world

Ethernet over PDH (EoPDH) can support a wide variety of Carrier Ethernet service speeds (up to approximately 100 Mb/s) and with no distance limitations using bonded signals, as shown in the following table.

Carrier Type	Number of Bonded Signals	Maximum Bandwidth	Distance Limitation
T1 (E1)	1	1.5 Mb/s (2.0 Mb/s)	None
	2	3.0 Mb/s (4.0 Mb/s)	
	3	4.5 Mb/s (6.0 Mb/s)	
	4	6.0 Mb/s (8.0 Mb/s)	
	5	7.5 Mb/s (10.0 Mb/s)	
	6	9.0 Mb/s (12.0 Mb/s)	
	7	10.5 Mb/s (14.0 Mb/s)	
	8	12.0 Mb/s (16.0 Mb/s)	
T3 (E3)	1	45 Mb/s (34 Mb/s)	
	2	90 Mb/s (68 Mb/s)	

Note: *Bonding is the aggregation of multiple lines/circuits to provide additional bandwidth and resiliency through load sharing. If an individual line/circuit fails in a bonded group, the service remains active, but with less bandwidth. EoPDH is deployed predominately with bonded T1/E1 signals. It is rarely deployed with bonded T3/E3 signals.*

Advantages of PDH include:

- Widely deployed and available worldwide
- Reliable and well understood technology
- Supports rapid turn-up
- Supports access over any distance
- Supports resiliency through bonding

10.3
Ethernet over Copper

In this section:

Summary of Carrier Ethernet Access Technologies			
Carrier Ethernet Access Method	Technology Alternatives	Deployment Scenarios (When to use the technology)	Advantages
Ethernet over Fiber	- Active Ethernet - Ethernet over SONET/SDH - Passive Optical Network	- On-net buildings - Greenfield - Dense Metro area - 1Gbit/s or greater bandwidth requirements	- Highest bandwidth - Noise immunity - Security - Long reach - SONET/SDH leverage existing - Growth potential via xWDM
Ethernet over PDH	- Bonded T1/E1 - DS3/E3 and bonded DS3/E3	- Remote branch offices - Off-net customer locations (out of region, type 2) - SMB	- Leverage existing transport - Universally deployable - Lower CAPEX - No reach limitations - Well understood provisioning - Resiliency through bonding
Ethernet over Copper	- 2BASE-TL - 10PASS-TS	- Remote branch offices - On-net or off-net - SMB - Campus settings - Traffic monitoring	- Ubiquitous copper availability - Rapid deployment - Low cost unbundled local loop - Resiliency through bonding
Wireless Ethernet	- Terrestrial microwave - WiMAX - Broadband wireless - Free space optics - WiFi	- Remote branch office - Campus setting - No fiber or copper available - Mobility required	- Installation requires no trenching - Rapid deployment - Some alternatives offer mobility
Hybrid Fiber Coax	DOCSIS 2.x/3.x	- Work at home - SOHO/SMB - Remote branch office	- Extensive coverage - High performance options - Deep penetration into residential and suburban geographies

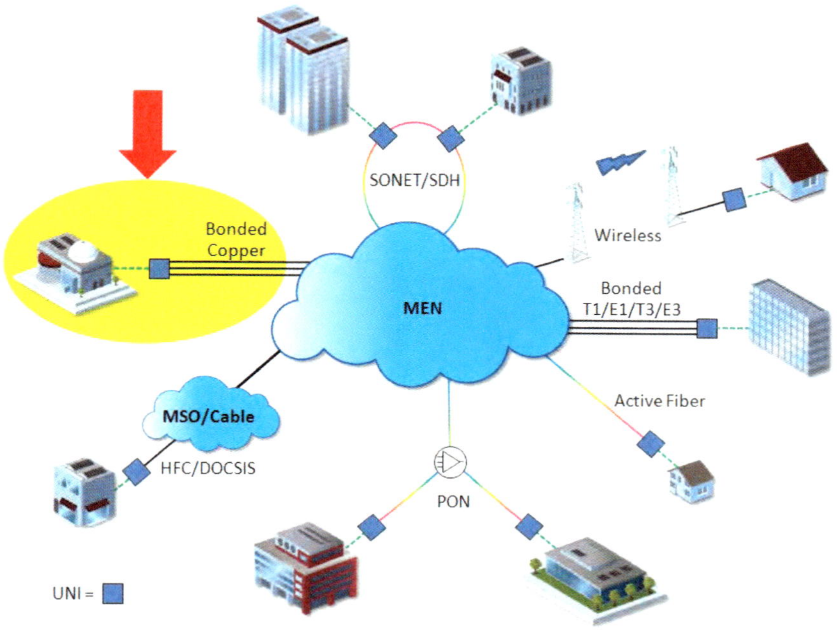

Ethernet over Copper (EoCu) allows fast deployment of resilient Ethernet access links over existing voice-grade copper infrastructure, providing a very economical alternative to fiber. There are two standardized alternatives: 2BASE-TL and 10PASS-TS.

10.3.1 2BASE-TL (G.SHDSL)

2BASE-TL for Ethernet over Copper combines two technologies:

- **2BASE-TL** (2 Mb/s, baseband, twisted pair, long reach) technology based on IEEE 802.3 (physical layer Ethernet)
- **SHDSL** (symmetric high-speed digital subscriber line) technology based on ITU-T G.991.2 (commonly known by its draft name: ITU-T **G.SHDSL**)

Note: *G.SHDSL is a type of digital subscriber line (DSL) technology.*[33]

Hint: *In the MEF-CECP 2.0 exam, either term, **2-BASE-TL** or **G.SHDSL**, may refer to this combination of technologies.*

33 *G.SHDSL technology differs from ADSL (asymmetric digital subscriber line) technology, which is used in residential applications:*

- *ADSL supports asymmetric data communication over phone lines in parallel with analog, plain old telephone service (POTS), voice communication using a frequency splitter (DSL filter) to separate ADSL data frequencies from POTS frequencies.*
- *G.SHDSL supports higher bandwidth symmetrical data communication over phone lines using all frequencies, including POTS frequencies. Unlike ADSL, G.SHDSL cannot be used together with POTS on the same copper pair. G.SHDSL is a popular choice for business applications.*

2Base-TL (G.SHDSL) can support either of two bandwidths over a single copper pair:

- 2 Mb/s (symmetric bandwidth) over distances up to 6 km (3.7 miles)
- 5.69 Mb/s (symmetric bandwidth) over distances up to 3 km (1.9 miles)

10.3.2 10PASS-TS (VDSL)

10PASS-TS for Ethernet over Copper, combines two technologies:

- **10PASS-TS** (10 Mb/s, twisted pair, short reach) technology based on IEEE 802.3 (physical layer Ethernet)
- **VDSL** (very-high-bit-rate digital subscriber line) technology based on ITU-T G.993.1

Hint: *In the MEF-CECP 2.0 exam, either term,* ***10PASS-TS*** *or* ***VDSL****, may refer to this combination of technologies.*

10PASS-TS (VDSL) supports high bandwidth asymmetrical data communication over short distances of voice grade phone line. For distances up to 0.75 km (about 0.5 miles), 10PASS-TS over a single copper pair supports:

- 10 Mb/s downstream
- 1-2 Mb/s upstream

10.3.3 Bonded Copper Pairs

Both EoCu technologies (2BASE-TL and 10PASS-TS) support bonding of multiple copper pairs to provide higher bandwidth and resiliency. Bonded 10PASS-TS can provide bandwidth in excess of 100 Mb/s (downstream).

Note: *Bonding is the aggregation of multiple lines/circuits to provide additional bandwidth and resiliency through load sharing. If an individual line/circuit fails in a bonded group, the service remains active, but with less bandwidth.*

10.3.4 EoCu versus EoPDH

Ethernet over Copper and **Ethernet over PDH** use the same transport medium: voice-grade copper wire.

EoPDH is a TDM (time division multiplexing) technology. EoCu is not.

EoCu is generally cheaper than EoPDH, but has significant distance limitations (EoPDH has no distance limitations).

10.4
Ethernet over Wireless Network

Summary of Carrier Ethernet Access Technologies			
Carrier Ethernet Access Method	Technology Alternatives	Deployment Scenarios (When to use the technology)	Advantages
Ethernet over Fiber	- Active Ethernet - Ethernet over SONET/SDH - Passive Optical Network	- On-net buildings - Greenfield - Dense Metro area - 1Gbit/s or greater bandwidth requirements	- Highest bandwidth - Noise immunity - Security - Long reach - SONET/SDH leverage existing - Growth potential via xWDM
Ethernet over PDH	- Bonded T1/E1 - DS3/E3 and bonded DS3/E3	- Remote branch offices - Off-net customer locations (out of region, type 2) - SMB	- Leverage existing transport - Universally deployable - Lower CAPEX - No reach limitations - Well understood provisioning - Resiliency through bonding
Ethernet over Copper	- 2BASE-TL - 10PASS-TS	- Remote branch offices - On-net or off-net - SMB - Campus settings - Traffic monitoring	- Ubiquitous copper availability - Rapid deployment - Low cost unbundled local loop - Resiliency through bonding
Wireless Ethernet	- Terrestrial microwave - WiMAX - Broadband wireless - Free space optics - WiFi	- Remote branch office - Campus setting - No fiber or copper available - Mobility required	- Installation requires no trenching - Rapid deployment - Some alternatives offer mobility
Hybrid Fiber Coax	DOCSIS 2.x/3.x	- Work at home - SOHO/SMB - Remote branch office	- Extensive coverage - High performance options - Deep penetration into residential and suburban geographies

Reproduced with permission of the Metro Ethernet Forum

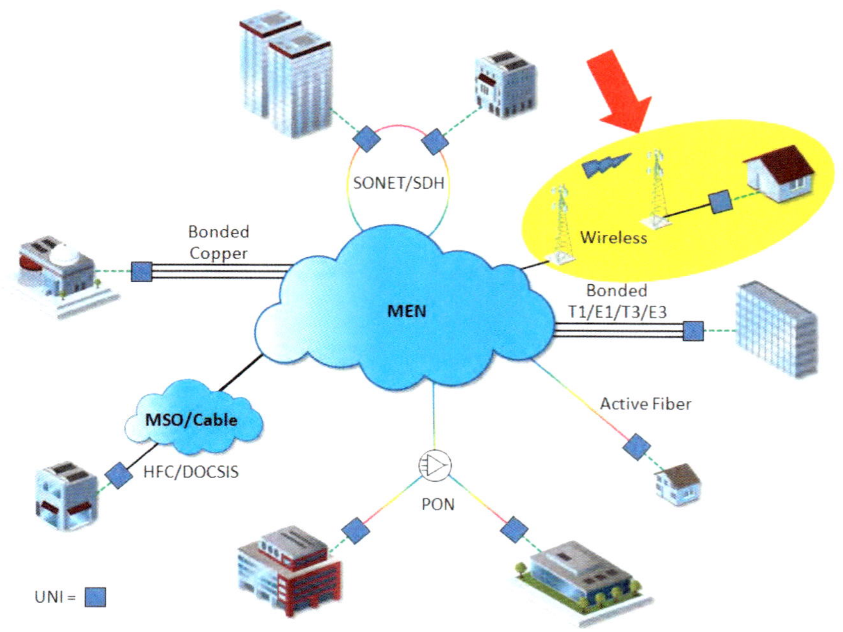

Wireless access networks are used when wireline services are not available or practical, including:

- Across rugged terrain
- Where mobility is required, such as to support a laptop computer with a broadband (WiFi) capability
- To establish a quick connection (without wire/cable deployment)

A variety of wireless technologies are available. The following table lists representative examples.

Table 24: Ethernet over Wireless Access Technologies

Technology	Transmission Medium	Description	Bandwidth over Distance
Free Space Optical	Light	Line of sight, point-to-point communication using light in place of electromagnetic signals	1.25 Gb/s over 11 km (7 miles) is supported.
Terrestrial Microwave	High frequency electromagnetic radiation (Microwave*)	Line of sight, point-to-point communication using either licensed spectrum (allocated and regulated by government agencies) or low-power unlicensed spectrum (within unlicensed regulatory limits)	100 Mb/s can be achieved reliably up to 19 km (12 miles).
WiMAX	High frequency electromagnetic radiation	IEEE 802.16 standards for point-to-multipoint wireless broadband communication	1 Gb/s is shared between users over a 50 km (30 mile) signal radius.
WiFi	High frequency electromagnetic radiation	IEEE 802.11 standards for point-to-multipoint wireless local area connection (WLAN) communication	IEEE 802.11n supports four streams of 150 Mb/s over a 250 m (820 ft) signal radius outdoors.

*Microwave communication uses high-frequency (> 1 GHz) electromagnetic signals.

Note: *Numbers in this table are generalizations selected to align with the MEF-CECP 2.0 exam, which tests for a general understanding of the relative strengths of each technology at the time the initial MEF-CECP exam was developed (circa 2010). The actual capabilities of each wireless technology advance over time and depend on many factors.*

Advantages of wireless Ethernet include:

- Deployable where no fiber or copper infrastructure is available
- Easily installed (no wires or cables)
- Rapidly deployed

10.5
Ethernet over HFC

Summary of Carrier Ethernet Access Technologies			
Carrier Ethernet Access Method	Technology Alternatives	Deployment Scenarios (When to use the technology)	Advantages
Ethernet over Fiber	- Active Ethernet - Ethernet over SONET/SDH - Passive Optical Network	- On-net buildings - Greenfield - Dense Metro area - 1Gbit/s or greater bandwidth requirements	- Highest bandwidth - Noise immunity - Security - Long reach - SONET/SDH leverage existing - Growth potential via xWDM
Ethernet over PDH	- Bonded T1/E1 - DS3/E3 and bonded DS3/E3	- Remote branch offices - Off-net customer locations (out of region, type 2) - SMB	- Leverage existing transport - Universally deployable - Lower CAPEX - No reach limitations - Well understood provisioning - Resiliency through bonding
Ethernet over Copper	- 2BASE-TL - 10PASS-TS	- Remote branch offices - On-net or off-net - SMB - Campus settings - Traffic monitoring	- Ubiquitous copper availability - Rapid deployment - Low cost unbundled local loop - Resiliency through bonding
Wireless Ethernet	- Terrestrial microwave - WiMAX - Broadband wireless - Free space optics - WiFi	- Remote branch office - Campus setting - No fiber or copper available - Mobility required	- Installation requires no trenching - Rapid deployment - Some alternatives offer mobility
Hybrid Fiber Coax	DOCSIS 2.x/3.x	- Work at home - SOHO/SMB - Remote branch office	- Extensive coverage - High performance options - Deep penetration into residential and suburban geographies

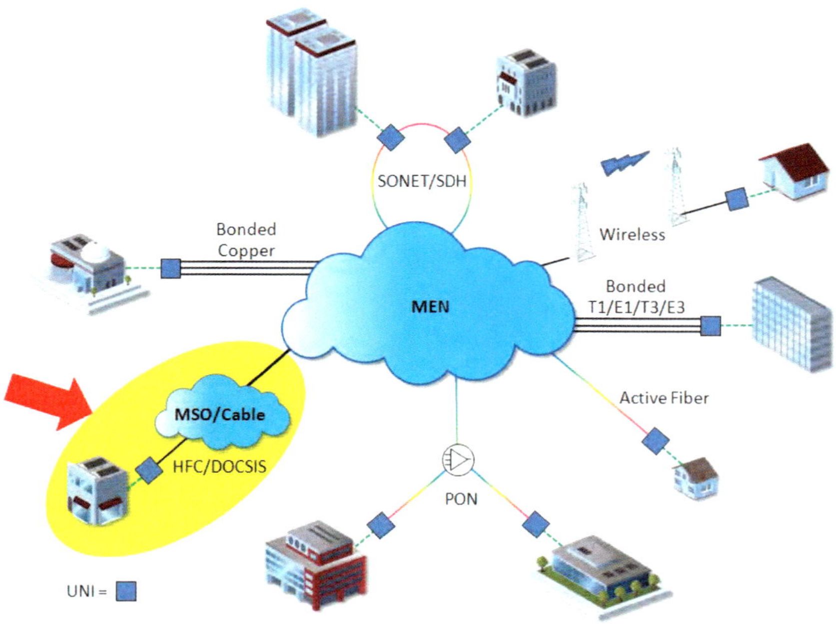

Hybrid Fiber Coax (HFC) is a broadband network that combines optical fiber and coaxial cable and has been commonly deployed by MSO/Cable TV operators since the early 1990s.

The fiber-optic network extends from the cable operators' regional head-end to a neighborhood's node, which serves anywhere from 25 to 2000 homes through coaxial cables. A master head-end will usually have satellite dishes, antennas, and/or fiber-optic connections to other video sources for reception of distant video signals. Some master head-ends also house telephony equipment for providing telecommunications services.

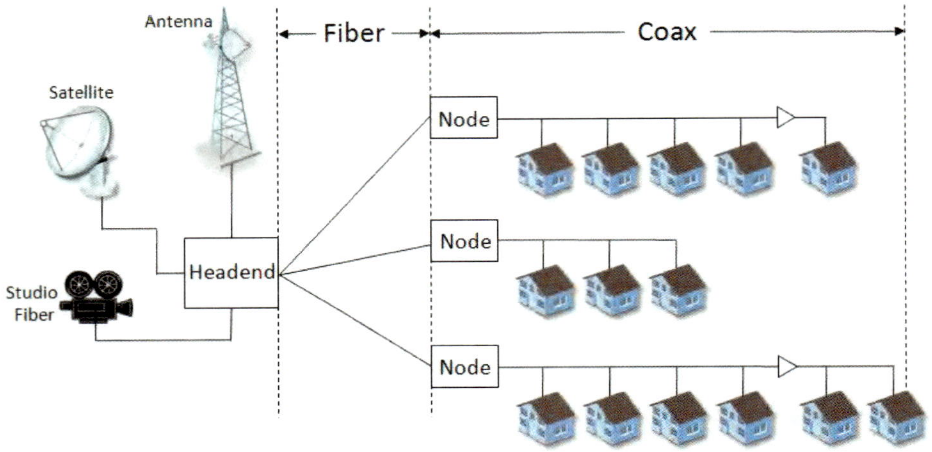

Figure 110: HFC Network

Using frequency division multiplexing, an HFC network may carry a variety of services simultaneously, including analog TV, digital TV (standard definition and HDTV), telephony, and high-speed data services.

Cable modems connect to the HFC plant to provide always-on, high-speed access to the Internet. The DOCSIS (Data Over Cable Service Interface Specifications) standard for cable modems have enabled low-cost, interoperable cable modems. DOCSIS 3.0 provides a peak downstream rate of approximately 350 Mb/s and a peak upstream rate of approximately 125 Mb/s.

Ethernet over HFC/DOCSIS can support MEF services. However, most HFC deployments today do not support Ethernet over DOCSIS, but only IP services over DOCSIS.

Advantages of HFC include:

- Supports bandwidth up to 350 Mb/s downstream per user (subject to lower performance if many users are on the network at the same time)
- Widely available in many major markets

10.6
Summary

The MEF-CECP 2.0 exam includes questions to test your ability to choose an access technology for a given set of requirements. These questions usually include speed and distance requirements.

The following figure places the access technologies on a graph showing approximate maximum distance and speed capabilities.

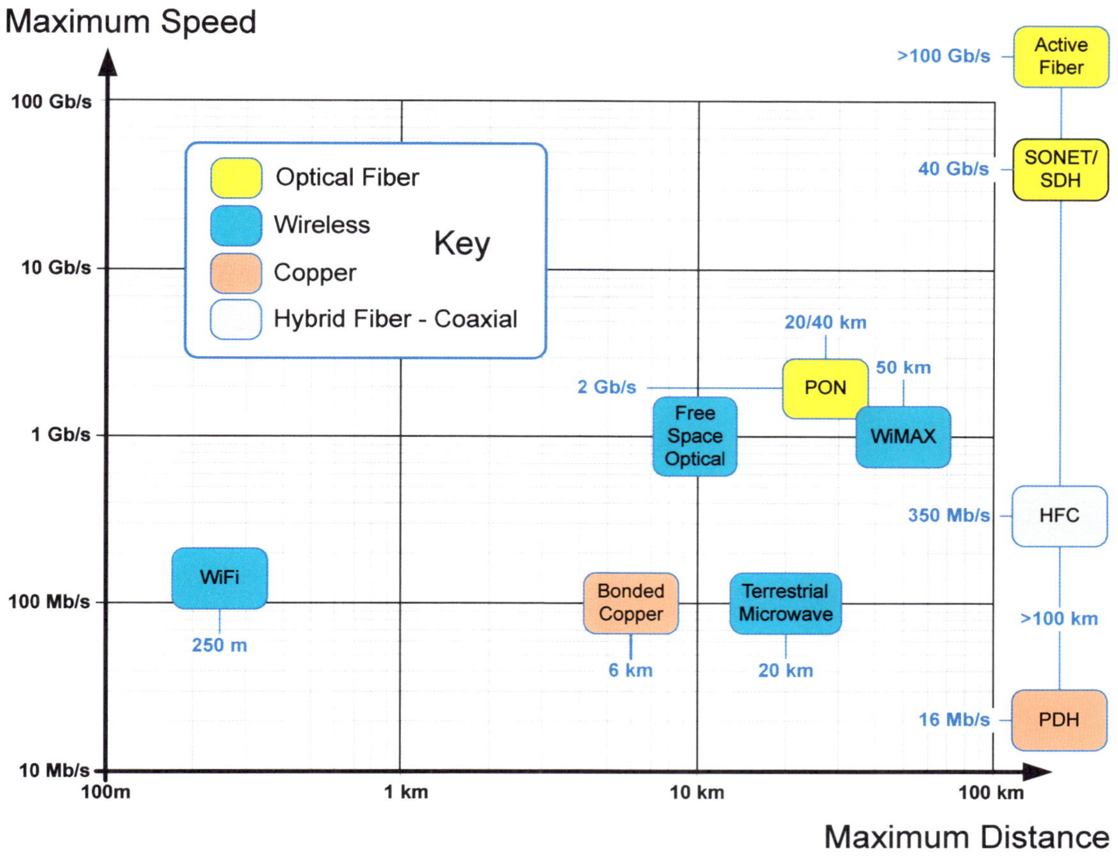

Figure 111: Approximate Maximum Distance and Speed Capabilities of Carrier Ethernet Access Technologies

Note: *Values shown in the preceding figure are approximations and are selected to align with the MEF-CECP 2.0 exam, which tests for a general understanding of the relative strengths of each technology at the time the initial MEF-CECP exam was developed (circa 2010). The values shown are based principally on MEF access technologies white paper.*[34] *The actual capabilities of each technology advance over time and depend on many factors.*

34 *Delivering Ubiquitous Ethernet Services using an Array of Access Technologies, in the* MEF White Papers *area of the MEF Web site (* MetroEthernetForum.org*), under the link titled Carrier Ethernet – the Technology of Choice for Access Networks.*

Many other factors can also influence the choice of access technology, including those listed in the following table.

Table 25: Highlights of Carrier Ethernet Access Technologies.

Access Technology		Transmission Medium	Other Characteristics
Ethernet over Optical Fiber	Active fiber	Fiber-optic cable	High performance
	SONET/SDH	Fiber-optic cable	Widely deployed Excellent reliability
	PON	Fiber-optic cable	Low cost Efficient multicast Asymmetric bandwidth
Ethernet over PDH		Voice-grade copper wire	Widely deployed Resiliency through bonding Bandwidth may be asymmetric
Ethernet over Copper		Voice-grade copper wire	Low cost Widely deployed Resiliency through bonding Bandwidth may be asymmetric
Wireless Ethernet	Free space optics	Light	Rapid deployment No wire/fiber infrastructure WiMAX and WiFi are point-to-multipoint technologies.
	Terrestrial microwave	Microwave	
	WiMAX	Microwave	
	WiFi	Microwave	
Ethernet over HFC		Hybrid fiber - coaxial cable	Asymmetric bandwidth Widely deployed in some markets

10.7
Review

1. What are three advantages offered by *Ethernet over Bonded Copper*? (Choose three.)

    a. Easy to deploy in rugged terrain

    b. Low cost

    c. No distance limitation

    d. Supports rates up to 1 Gb/s through bonding of additional copper pairs

    e. Resiliency through bonding copper pairs

    f. Widely available copper infrastructure

2. An office, located 3 miles (4.8 km) from the service provider network, requires 2 Mb/s of symmetric connectivity to an EVP-LAN service. A single copper pair is available. Which of the following access technologies is most appropriate?

    a. Ethernet over DOCSIS

    b. G.SHDSL

    c. Ethernet over PDH

    d. 10PASS-TS

3. What is the advantage of having Ethernet service over WiMAX?

    a. Bandwidth split among multiple users

    b. No cabling

    c. Resilience

    d. No reach limitation

4. Which access technology can support Ethernet services in parallel with TV services as commonly deployed by MSO/Cable TV operators?

    a. Ethernet over GPON

    b. Ethernet over Bonded Copper

    c. Ethernet over HFC

d. Ethernet over PDH

e. Ethernet over Terrestrial Microwave

5. A mobile backhaul operator needs to reach a cell tower located in an undeveloped rural area. Which two of the following technologies could be most quickly deployed? (Choose two.)

   a. Ethernet over PDH

   b. Ethernet over Terrestrial Microwave

   c. Ethernet over Free Space Optics

   d. Ethernet over SONET/SDH

   e. Ethernet over bonded copper pairs

   f. Ethernet over Fiber

6. What are two advantages of *Ethernet over SONET/SDH*? (Choose two.)

   a. Supports multiple classes of service

   b. Highly reliable

   c. Supports multipoint services (EP-LAN, EP-Tree)

   d. Provides dedicated bandwidth that is not impacted by other user traffic

   e. Supports circuit emulation

7. An incumbent provider (ILEC/PTT) wants to provide an Ethernet service at 6 Mb/s over its existing copper network infrastructure. The UNI will be 20 km (12.4 miles) away from the nearest switching office. Which technology do you recommend for this application?

   a. Ethernet over bonded copper pairs (2BASE-TL)

   b. Ethernet over HFC/DOCSIS

   c. Ethernet over bonded T1/E1 PDH circuits

   d. Ethernet over bonded T3/E3 PDH circuits

8. What is an advantage of *Ethernet over PDH* compared to *Ethernet over Copper*?

   a. Uses widely deployed infrastructure

   b. Easy, low-cost, and rapid to deploy

   c. Supports eight classes of service

  d. Supports resiliency through bonding

  e. No distance limitation

9. Which two of the following access technologies support rates over 1 Gb/s? (Choose two.)

  a. Ethernet over SONET/SDH

  b. Ethernet over PDH

  c. Ethernet over HFC/DOCSIS

  d. Ethernet over Active Fiber

10. Which wireless access technology uses licensed frequency spectrum in high-power applications?

  a. Ethernet over DOCSIS

  b. Ethernet over terrestrial microwave

  c. Ethernet over free space optical

  d. Ethernet over WiFi

11. Which of the following access technologies has the shortest reach?

  a. Ethernet over Terrestrial Microwave

  b. Ethernet over Bonded Copper

  c. Ethernet over PDH

  d. Ethernet over HFC/DOCSIS

  e. Ethernet over Active Fiber

12. Which of the following is a disadvantage of *Ethernet over Active Fiber*?

  a. Distance limited to under 40 km

  b. Lack of widespread availability for fiber infrastructure

  c. Speed limited to 100 Mb/s

d. Dated technology

e. Poor bandwidth granularity

13. What are two advantages of using *Ethernet over GPON* versus *Ethernet over copper*? (Choose two.)

a. The physical media is widely deployed.

b. Higher bandwidth

c. Supports efficient multicasting

d. Resiliency through bonding

Related Links

Answers: Access Technologies

11

Transport Technologies

In this chapter:

Transport technologies are technologies used within a Carrier Ethernet network (MEN/CEN or Operator MEN/CEN) to implement Ethernet services. MEF standards focus on service requirements at UNI and ENNI interfaces and requirements for Ethernet frame delivery between UNIs and/or ENNIs. They do not address how the service provider (or operator) achieves these requirements.

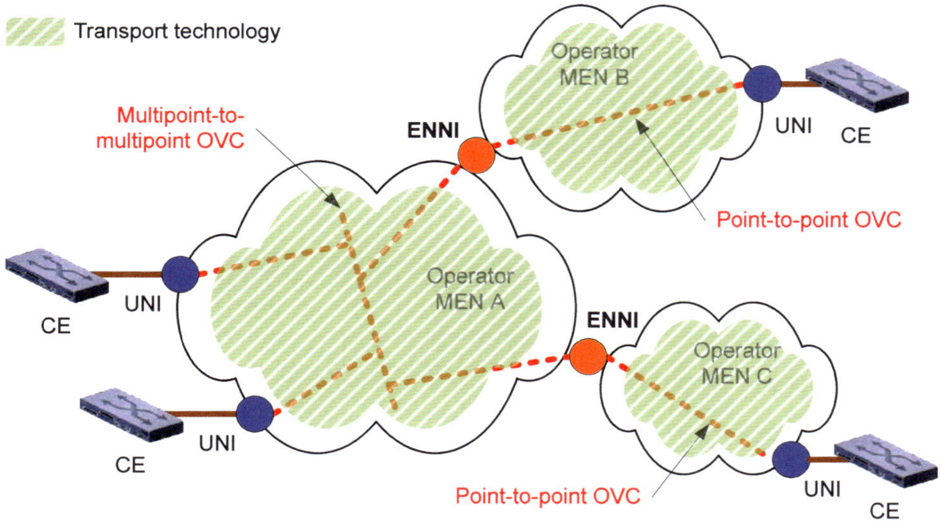

Figure 112: Transport Technologies are Internal to the MEN or Operator MEN

Note: *The preceding figure is only meant to help visualize where transport technologies are used (for service implementation within a MEN/CEN or Operator MEN/CEN).*

The Transport Technologies

Several technologies can be used to transport MEF services. Some are simple Layer 1 technologies, some are Layer 2 technologies, and some are Layer "2.5" technologies (a "shim" between Layers 2 and 3). The following transport technologies are reviewed in this lesson:

Table 26: Transport Technologies for Carrier Ethernet Services

OSI Model Layer	Transport Technology
Layer 1	• Synchronous Optical Network (SONET)/Synchronous Digital Hierarchy (SDH) • Optical Transport Network (OTN)—ITU-T G.709 • Wavelength Division Multiplexing (WDM)
Layer 2	• Bridging—IEEE 802.1D and 802.1Q–2005 • Provider Bridging (PB)—IEEE 802.1ad • Provider Backbone Bridging (PBB)—IEEE 802.1ah • Provider Backbone Bridging with Traffic Engineering extensions (PBB-TE)—IEEE 802.1Qay • Ethernet Tag Switching—Fujitsu implementation of connection-oriented Ethernet (COE) leveraging the IEEE 802.1ad frame format
Layer "2.5"	• Multiple Protocol Label Switching Virtual Private Wire Service (MPLS VPWS) • Multiple Protocol Label Switching Virtual Private LAN Service (MPLS VPLS) • Multiple Protocol Label Switching Transport Profile (MPLS-TP)

All of these technologies can operate over any physical network topology: mesh, partial mesh, tree, or a set of rings.

For the MEF-CECP 2.0 exam, it is important to have a basic understanding of how each technology operates and a clear understanding of the support capabilities, advantages, and disadvantages of each technology.

Hint: *Consider jumping ahead to preview Figure 129 (a study aid that appears in the summary to this lesson). This study aid will not make much sense at this time, but it may help you to visualize what you are expected to take away from this lesson.*

Key Areas for Comparing the Transport Technologies

The transport technologies differ in their ability to support Carrier Ethernet services in the following key areas:

- **Multipoint (MAC Learning)**–MAC learning is required to support EP-LAN, EVP-LAN, EP-Tree, and EVP-Tree services. Some technologies are designed to support only point-to-point services (EPL, EVPL, Access EPL, and Access EVPL).
- **CE-VLAN awareness**–CE-VLAN awareness is required to support EVPL, EVP-LAN, EVP-Tree, and Access EVPL services.
- **S-VLAN awareness**–S-VLAN awareness is required to support ENNIs and OVC-based services, including Access EPL and Access EVPL services.
- **CoS support**–Some technologies support CoS, while others do not. Technologies that do not support CoS treat all service frames identically, providing the same quality of service (QoS) to all service frames.
- **Bandwidth**–Technologies differ in terms of maximum bandwidth capability and in their ability to support service bandwidths of arbitrary size (granularity).
- **Scalability**–Layer 2 and Layer "2.5" technologies differ in the number of Ethernet services they allow a single network to support (this is scalability). For some technologies, scalability is limitless. For other technologies, various factors limit scalability, such as the number of unique service identifiers and/or vulnerability to MAC table overflow.
- **Service protection mechanisms**–Transport technologies differ in fault detection and protection switching mechanisms. Sub-50 ms protection switching/restoration is an important benchmark for real-time communications.

Note: *A brief overview of basic Ethernet service protection technologies appears near the end of this lesson.*

Related Links

Ethernet Service Protection Technologies

11.1 Layer 1 Transport Technologies

In this section:

Three Layer 1 transport technologies provide sufficient bandwidth to support Carrier Ethernet services:

- SONET/SDH—Synchronous Optical Network/Synchronous Digital Hierarchy
- OTN—Optical Transport Network
- WDM—Wavelength Division Multiplexing

These fiber-optic technologies can be used alone or in combination to provision highly reliable, fixed-bandwidth circuits for the transparent transport of Ethernet service frames in point-to-point applications.

Important: *When a Layer 1 transport technology supports an Ethernet service, a fixed amount of network bandwidth is consumed by the service, regardless of actual service traffic. Bandwidth in a Layer 1 network is shared using time division multiplexing (TDM). Each service is allocated fixed bandwidth (dedicated time slots), end-to-end, through the network. That bandwidth is consumed by the service, regardless of actual data throughput (another service cannot use that bandwidth).*

Layer 1 technologies are not VLAN aware and do not support frame-based routing/switching, CoS, or multicasting. They only support port-based point-to-point EVC-based services (EPL services).

When used together, they are used in one of the following combinations:

- SONET/SDH within OTN
- OTN within WDM
- SONET within OTN, then both within WDM

SONET/SDH and OTN include OAM functions and protection mechanisms. WDM does not include OAM functions or protection mechanisms because it is typically used in combination with SONET/SDH and/or OTN, which already provide that functionality.

11.1.1 SONET/SDH (Synchronous Optical Network / Synchronous Digital Hierarchy)

SONET and SDH are similar technologies, but developed with different standards. Both are widely deployed today. SONET is used in the U.S. and Canada, and SDH is used in the rest of the world.

SONET/SDH technology was originally designed to support circuit-switched real-time voice communication, but has evolved to support a wide variety of payload types, including Ethernet frames.

The following figure shows an EPL service implemented using Ethernet over SONET/SDH (EoS).

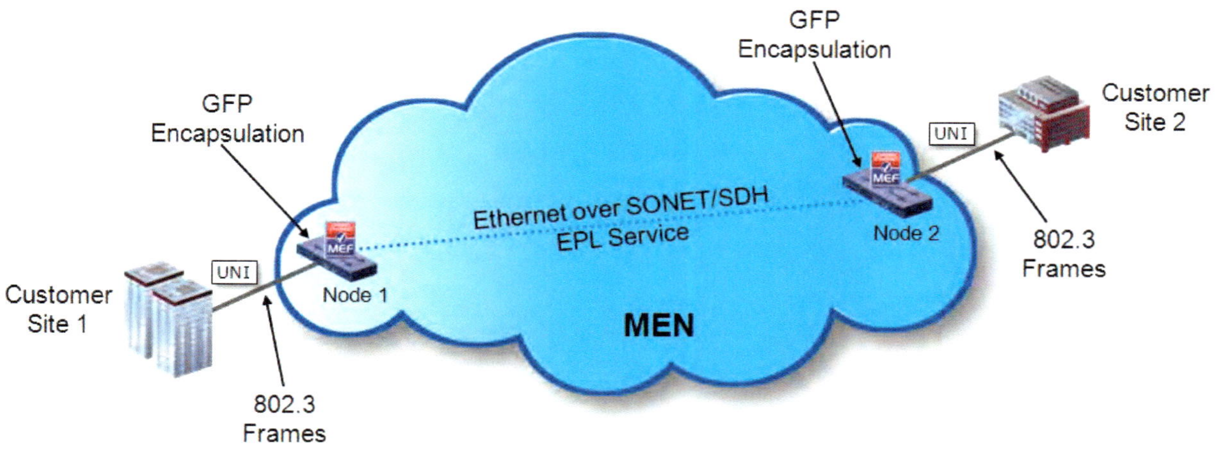

Figure 113: Ethernet over SONET/SDH

In this application, Ethernet frames are transported as follows:

1. Customer Site 1 sends 802.3-compliant Ethernet frames to the service provider's Node 1.
2. Node 1 encapsulates the Ethernet frames in GFP (generic framing procedure) frames.
3. GFP frames are cross connected into a SONET payload.
4. The SONET network sends the encapsulated Ethernet frame to Node 2.
5. Node 2 de-encapsulates the Ethernet frame (removes the Ethernet frame from the GFP frame) and sends it out on the UNI to Customer Site 2.

Note: *Transport in the opposite direction is similar.*

Bandwidth

Bandwidth in a SONET/SDH network is shared through time division multiplexing (TDM). Each service is allocated fixed bandwidth (dedicated time slots), end-to-end, through the network, and that bandwidth is consumed by the service, regardless of usage.

As shown in the following table, standard SONET/SDH signals do not provide much flexibility with respect to the size of payload bandwidth. For example, a service requiring 100 Mb/s would use an OC-3 SONET signal, providing 150 Mb/s, 50% more bandwidth than needed.

Table 27: SONET/SDH Bandwidth

SONET Optical Carrier Level	SONET Frame Format	SDH Frame Format	Payload (kb/s)	Line Rate (kb/s)	Standard Ethernet Speed Supported
OC-1	STS-1	STM-0	50,112	51,814	10 Mb/s
OC-3	STS-3	STM-1	150,336	155,520	100 Mb/s
OC-12	STS-12	STM-4	601,344	622,080	n/a
OC-24	STS-24	n/a	1,202,688	1,244,160	1 Gb/s
OC-48	STS-48	STM-16	2,405,376	2,488,320	n/a
OC-192	STS-192	STM-64	9,621,504	9,953,280	10 Gb/s*
OC-768	STS-768	STM-256	38,486,016	39,813,120	40 Gb/s*
* Cannot be transported at full rate transparently					

Bandwidth flexibility can be increased using **virtual concatenation (VCAT)**, which permits bandwidth to be custom sized in any multiple of the basic rate (STS-1, or 50 Mb/s, for SONET) or in any multiple of a smaller VT (virtual tributary) subrate (VT1.5, or 1.5 Mb/s, for SONET low-order VCAT). This greatly increases the range of bandwidth sizes that the service provider can offer. SDH provides support for virtual concatenation that is similar to SONET's support for virtual concatenation, but using different bandwidth container sizes.

Protection and OAM

SONET/SDH includes OAM performance monitoring and fault detection and supports 1+1, UPSR (unidirectional path-switched ring) protection, and BLSR (bidirectional line-switched ring) protection switching, each with sub-50 ms protection switching/restoration.

SONET/SDH Summary

Advantages	Limitations / Issues
Rapid service turn-up: • Leverages widely deployed equipment and optical fiber • Ubiquitous availability world wide Highly resilient and secure: • Sub-50ms protection/restoration • Layer 1 security Flexible bandwidth options: • STS-3/STM-1 (155 Mb/s) up to STS-768/STM-256 (39.8 Gb/s) • Sub-rate and NxSTS/STM available with bonding using VCAT circuits Comprehensive OAM: • Performance monitoring • Fault detection and isolation	Larger, less granular, fixed bandwidth increments may not correlate well with Ethernet bandwidth requirements. The maximum rate of any SONET/SDH interface is 40 Gb/s. Only EPL service is supported (no VLAN awareness). Only single-MEN/CEN applications are supported (no ENNI or OVC support). Because only Layer 1 functionality is supported, SONET/SDH: • Is not frame aware • Does not support classes of service (supports one CoS) • Does not support multicasting or any other Layer 2 functions

Related Links

Ethernet over SONET/SDH

11.1.2 OTN (Optical Transport Network)

The ITU-T developed OTN to provide SONET/SDH-like functionality for narrowband optical channels within WDM (wavelength division multiplexing) technology. OTN, also called G.709 digital-wrapper technology, has expanded beyond WDM into a generic, independent technology for switching, multiplexing, and aggregating any traffic type (not just within WDM applications).

Bandwidth

OTN supports higher bandwidths compared to SONET/SDH as shown in the following table.

Table 28: OTN Bandwidth

OTN Level	OTU Rate (Gb/s)	ODU Rate (Gb/s)	OPU Rate (Gb/s)	Payload Rate (Gb/s)	Standard Ethernet Speed Supported
OTU1	2.66	2.4999	2.489	2.488	2 x 1 Gb/s
OTU2	10.709	10.037	10.001	9.995	n/a
OTU2e	11.096	10.400	10.361	10.356	10 Gb/s
OTU3	43.018	40.319	10.171	40.151	40 Gb/s
OTU4	111.809	104.794	104.410	104.356	100 Gb/s

Also observe that OTU signals are sized to align well with standard Ethernet speeds.

Similar to Ethernet over SONET/SDH, Ethernet service frames are encapsulated with GFP, prior to OTN encapsulation, as shown in the following figure.

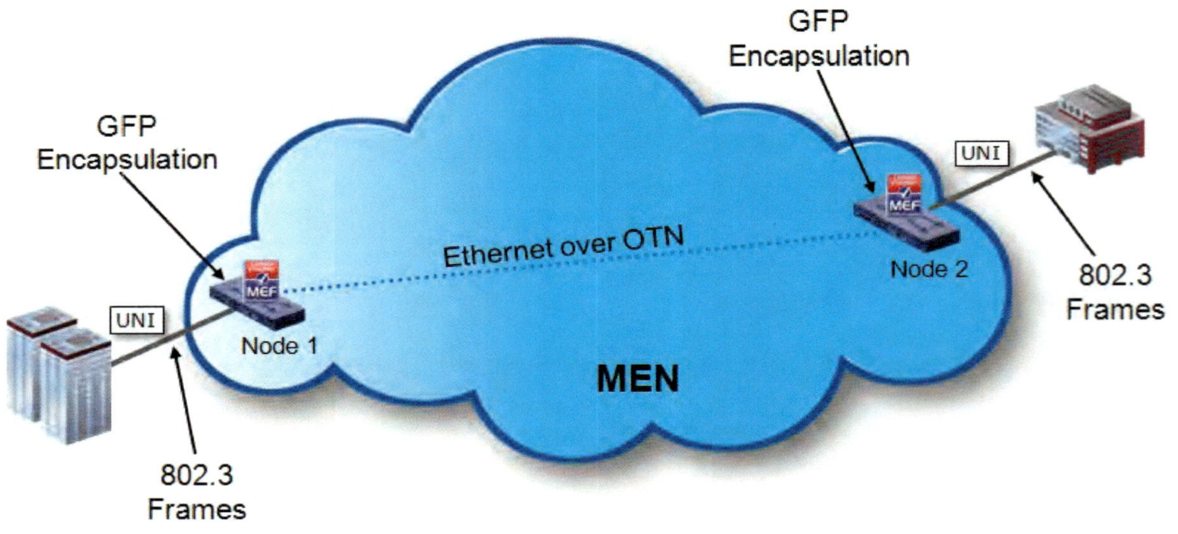

Figure 114: Ethernet over OTN

Like SONET/SDH, OTN provides channel bonding, similar to SONET/SDH virtual concatenation, using a technology called ODUflex that allows concatenation of ODUs to create different bandwidth options. Additionally, SONET/SDH entities (including subrate/bonded VCAT entities) can be encapsulated and transported within the OTN entity to support more granular service bandwidth.

Protection and OAM

OTN includes OAM performance monitoring and fault detection (superior to SONET/SDH OAM) and supports 1+1, 1:1, and 1:N optical protection switching with sub-50 ms protection switching/restoration.

OTN supports forward error correction (FEC), which increases OTN's tolerance to physical-layer transmission errors, allowing longer transmission distances compared to all other Carrier Ethernet transport options.

OTN Summary

Advantages	Limitations / Issues
Highly resilient and secure: • Sub-50ms protection/restoration • Layer 1 security Higher bandwidth (up to 100 Gb/s) compared to SONET Superior OAM: • Stronger forward error correction than other Carrier Ethernet transport technologies • Supports up to 6 TCM (Tandem Connection Monitoring) levels that provide connectivity fault management (similar to SOAM) • Enhanced performance monitoring at multiple layers	Larger, less granular, fixed bandwidth increments may not correlate well with Ethernet bandwidth requirements. Only EPL service is supported (no VLAN awareness). Only single-MEN/CEN applications are supported (no ENNI or OVC support). Because only Layer 1 functionality is supported, OTN: • Is not frame aware • Does not support classes of service (supports one CoS) • Does not support multicasting or any other Layer 2 functions

11.1.3 WDM (Wavelength Division Multiplexing)

WDM (wavelength division multiplexing) technology allows a single fiber-optic cable to support multiple data streams, with each data stream assigned a different wavelength (or "color") of laser light.

Each wavelength in a WDM link has capacity equivalent to an individual fiber-optic cable (using a single standard wavelength) or to an individual OTN signal. Thus, WDM greatly improves fiber utilization and reduces the need for new fiber installation.

Within a WDM network, wavelengths can be individually added, dropped, or rerouted at NEs using optical splitting, filtering, and blocking. These mechanisms allow narrowband wavelength signals to follow independent paths through the WDM network.

WDM wavelengths use the G.709 OTN frame.

The ITU-T defines two types of WDM technologies, CWDM (coarse wavelength division multiplexing) and DWDM (dense wavelength division multiplexing), which are compared in the following table.

Table 29: CWDM versus DWDM

Attribute	CWDM	DWDM
Cost	Less expensive	More expensive
Transmission distance	Short range	Long-range
Spacing between wavelengths	Wide (20 nm) providing 1 GHz frequency spacing	Narrow (0.4 nm) providing 50 GHz frequency spacing
Number of channels	16 or fewer (typical)	Up to 88 (typical)
Laser precision	Less precise	More precise
Laser rate	Up to 10 Gb/s (typical)	Up to 100 Gb/s (typical)
Maximum capacity per fiber-optic cable	160 Gb/s (10 Gb/s x 16 wavelengths per fiber)	8.8 Tb/s (100 Gb/s x 88 wavelengths per fiber)

Bandwidth

WDM provides the highest transport capacity of all Carrier Ethernet transport options: Up to 88 DWDM wavelengths (at 100 Gb/s each) per fiber-optic cable (as of March 2012).

If more granular service bandwidth is required, it can be supported using OTN and/or SONET/SDH subsignals within the WDM wavelength signal, as shown in the following example:

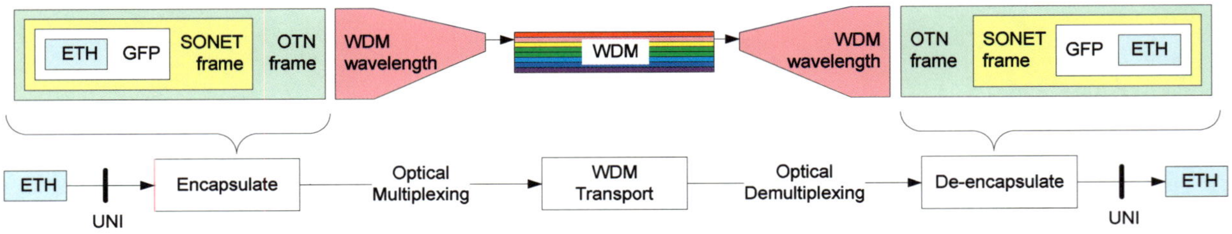

Figure 115: Example of Ethernet over WDM Transport

In this example:

1. The Ethernet service frame is encapsulated, as required. GFP and OTN are mandatory. SONET/SDH is optional.
2. The OTN frame is converted to a narrowband WDM wavelength and optically multiplexed into the WDM signal.
3. Data is transported across the WDM network.
4. The WDM wavelength is optically demultiplexed and converted to an OTN frame.
5. The ETH service frame is de-encapsulated from OTN, SONET/SDH, and GFP frames, as required.

Protection and OAM

WDM wavelengths use the G.709 OTN frame for optical-layer OAM functions and protection. WDM networks provide the same OAM features and protection functions as OTN.

WDM Summary

Table 30: WDM Summary

Advantages	Limitations / Issues
Most efficient fiber-optic cable utilization: • Enables multiple wavelengths per fiber-optic cable • Economical use of fiber-optic cable • Provides highest capacity transport—Up to 88 DWDM wavelengths (at 100 Gb/s each) per fiber optic cable (as of March 2012). • Minimizes new fiber-optic cable installs Cost effective: • Small form-factor pluggable (SFP) transceivers, multiplexers, and media converters enable WDM wavelengths with existing infrastructure equipment Flexibility: • Wavelengths can support different network protocols, for example, a mix of Ethernet and TDM services over the same fiber-optic cable Superior OAM and protection (identical to OTN)	Only EPL service is supported (no VLAN awareness). Only single-MEN/CEN applications are supported (no ENNI or OVC support). Because only Layer 1 functionality is supported, WDM: • Is not frame aware • Does not support classes of service • Does not support multicasting or any other Layer 2 functions

11.2
Layer 2 Transport Technologies

In this section:

Layer 2 transport technologies are based on IEEE 802.1 standards:

- **Bridging**—IEEE 802.1Q-2005
- **PB** (Provider Bridging)—IEEE 802.1ad
- **PBB** (Provider Backbone Bridging)—IEEE 802.1ah
- **PBB-TE** (Provider Backbone Bridging with Traffic Engineering extensions)—IEEE 802.1Qay
- **Ethernet Tag Switching**—IEEE 802.1ad

Important: *The IEEE standards shown in the preceding list are now obsolete, superseded by IEEE 802.1Q-2011. IEEE 802.1Q-2011 updated IEEE 802.1Q-2005 and incorporated many amendments to IEEE 802.1Q-2005, including IEEE 802.1ad (Provider Bridging), IEEE 802.1ah (Provider Backbone Bridging), and IEEE 802.1Qay (Provider Backbone Bridging with Traffic Engineering Extensions). IEEE 802.1Q-2011 is now the proper current reference for all of these technologies. This lesson references the older, now obsolete, IEEE standards for purposes of continuity.*

Note: *Ethernet tag switching is a Fujitsu implementation of connection-oriented Ethernet (COE) that leverages the IEEE 802.1ad frame format to forward/route Ethernet frames.*

Layer 2 transport technologies fall naturally into two groups:

1. **Bridging**, **PB**, and **PBB** technologies support MAC learning, which is needed for multipoint services.
2. **PBB-TE** and **Ethernet Tag Switching** do not use MAC learning (they use provisioned, traffic-engineered paths to forward frames). Both are connection-oriented Ethernet (COE) technologies designed to support point-to-point services.

11.2.1
Bridging

Note: *This material assumes that you understand MAC addressing, MAC bridging, MAC learning, and Spanning Tree Protocol.*

Bridging supports multipoint-to-multipoint and rooted-multipoint Ethernet services based on two standards:

- **IEEE 802.1D bridging**—MAC addressing, MAC bridging, MAC learning, and Spanning Tree Protocol (STP)
- **IEEE 802.1Q Customer VLAN tagging**—C-Tag or CE-VLAN tag

Recall that all Ethernet service frames are either untagged or include an IEEE 802.1Q customer tag (CE-VLAN tag) in the Ethernet header.

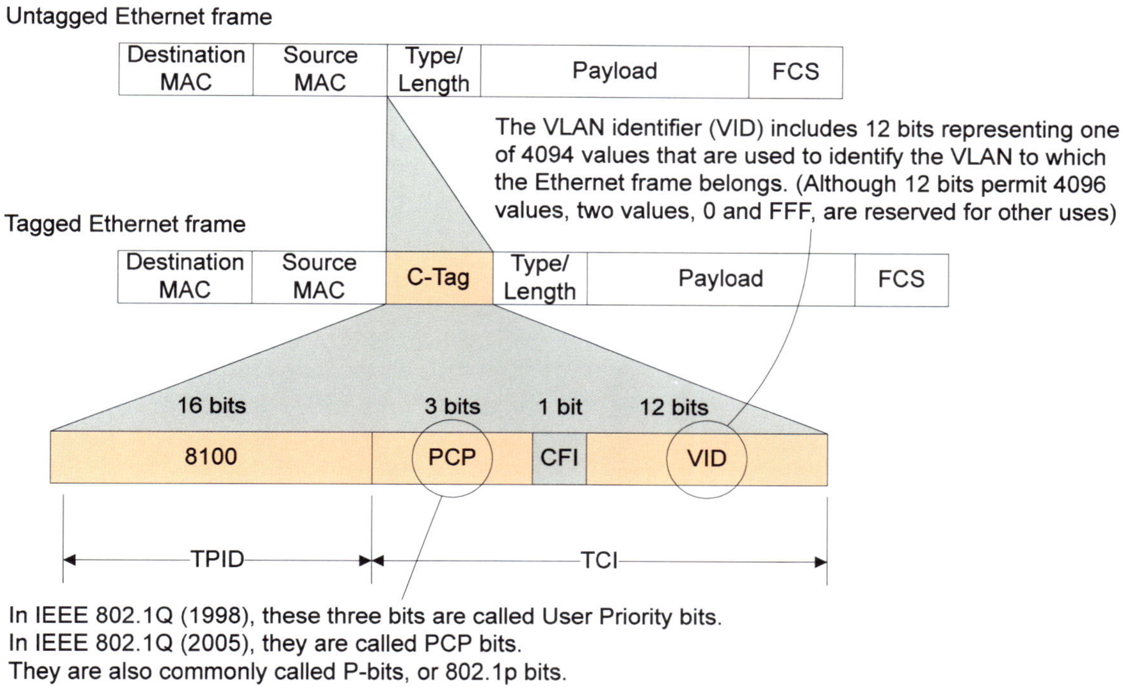

Figure 116: IEEE 802.1Q–2005 Ethernet Frame

Bridging technology is widely available and familiar but is difficult to apply in Carrier Ethernet networks because the service provider must repurpose the 802.1Q customer VLAN tag (used by subscribers for CE-VLAN tagging at UNIs) to support service identification, CoS identification, and drop-eligibility identification within the MEN (CEN). In addition, L2CP frame tunneling becomes difficult to support, and ENNIs are not supported because S-Tags are unsupported.

If bridging is used for transport within the MEN (CEN), every Ethernet service frame must include an 802.1Q customer VLAN tag, and that tag is used within the MEN (CEN) as follows:

- The VID is used to identify the service (service identifier).
- The PCP bits are used to identify both CoS and drop-eligibility (green/yellow).

This usage limits the service provider's ability to support Ethernet services, as defined by the MEF, in several ways.

First, the subscriber cannot use the CE-VLAN tag at the UNI in any way that conflicts with the service provider's system of usage.

Table 31: Conflicts and Implications of Subscriber CE-VLAN Tag Usage

Service type	Conflict	Implications
Port-based (EPL, EP-LAN, and EP-Tree)	The service provider uses the VLAN-ID to identify the service in the MEN (CEN), so the subscriber must ensure one of the following: 1. All service frames are untagged. 2. All service frames share the same CE-VLAN ID.	All-to-one bundling is not supported. The service can support one VLAN, at most, in the subscriber domain, making it equivalent to an unbundled VLAN-based service.
VLAN-based (EVPL, EVP-LAN, EVP-Tree)	The service provider uses the VLAN-ID to identify the service in the MEN (CEN), so the subscriber can associate only one CE-VLAN ID with a service.	Bundling is not supported.
	The subscriber cannot use a CE-VLAN ID that conflicts with the VLAN-IDs used by other services in the MEN (CEN).	The subscriber and service provider must coordinate the value of CE-VLAN ID in some way, such as one of the following: • The service provider assigns the CE-VLAN ID value. • The subscriber chooses a CE-VLAN ID value from a limited set of currently unused VLAN ID values. • The service provider translates the CE-VLAN ID value at MEN (CEN) ingress and restores it at MEN (CEN) egress.
All	The subscriber cannot use PCP bits in the CE-VLAN tag in any way that conflicts with the service provider's system of usage.	Subscribers conforms to a compromise, such as one of the following: • The subscriber does not use PCP bits in CE-VLAN tags. • The subscriber conforms to the service provider's system for PCP bit usage. • The subscriber allows the service provider to overwrite the PCP bits at MEN (CEN) ingress (this implies that the CE-VLAN CoS preservation service attribute must be No).

Additionally, the service provider's usage of VLAN tags makes it difficult to support tunneling of L2CP frames. To tunnel L2CP frames within a service, the service provider must push a C-Tag (802.1Q customer VLAN tag) onto each L2CP frame at MEN (CEN) ingress and pop the C-Tag at MEN (CEN) egress. However, it becomes difficult to distinguish L2CP frames from other service frames at MEN (CEN) egress because they all have C-Tags.

Scalability

Bridging transport technology is not very scalable. The number of services a MEN (CEN) can support is limited to approximately 4K, which is the number of unique VLAN IDs supported by the 802.1Q–2005 Ethernet frame.

MAC table growth is another factor limiting scalability. At each network bridge, MAC tables grow and may overflow as services are added and MAC addresses multiply.

Note: *One way service providers deal with MAC address scalability is to place a router, rather than a bridge, at the CE. The router "hides" all of the MAC addresses in the local subscriber network from the service provider network so that the service provider network sees only router interface MAC addresses.*

Protection

Bridging supports protection with xSTP (any variant of Spanning Tree Protocol). However, restoration time is limited by the speed of spanning tree convergence (sub-50 ms protection switching can be difficult to achieve, depending on network size).

Bridging Summary

Considering all of its limitations, bridging transport technology is only suitable for small-scale MEN (CEN) applications.

Advantages	Limitations / Issues

• Supports all types of MEF services, excluding Access EPL and Access EVPL (with limitations)
• Widely available and familiar technology
	• MEF service support is limited (and implementation is complicated) because the service provider must repurpose CE-VLAN tags to support transport within the MEN (CEN).
	• There is no support for CE-VLAN bundling or all-to-one bundling.
	• Increased coordination (between subscriber and service provider) is needed for CE-VLAN tagging.
	• There is no support for ENNIs due to the lack of S-Tag support. This technology cannot support OVCs, Access EPLs, Access EVPLs, or multi-Operator MEN/CEN service applications.
	• Scalability is very limited because:
	    • The maximum number of services supported per MEN (CEN) is approximately 4K.
	    • MAC tables grow with service addition, leading to possible overflow (unless routers are used at CEs).
	• Sub-50 ms protection switching can be difficult to achieve with xSTP (depending on network size).

11.2.2
PB (Provider Bridging)

Provider Bridging, also called "Q-in-Q" or VLAN tag stacking, was developed to address the limitations of Bridging. PB uses the same IEEE 802.1D bridging standard (for MAC addressing, MAC bridging, MAC learning, and STP) used by Bridging, but it replaces the VLAN tagging standard (IEEE 802.1Q-2005) with IEEE 802.1ad, which adds a second VLAN tag to the Ethernet frame, as follows:

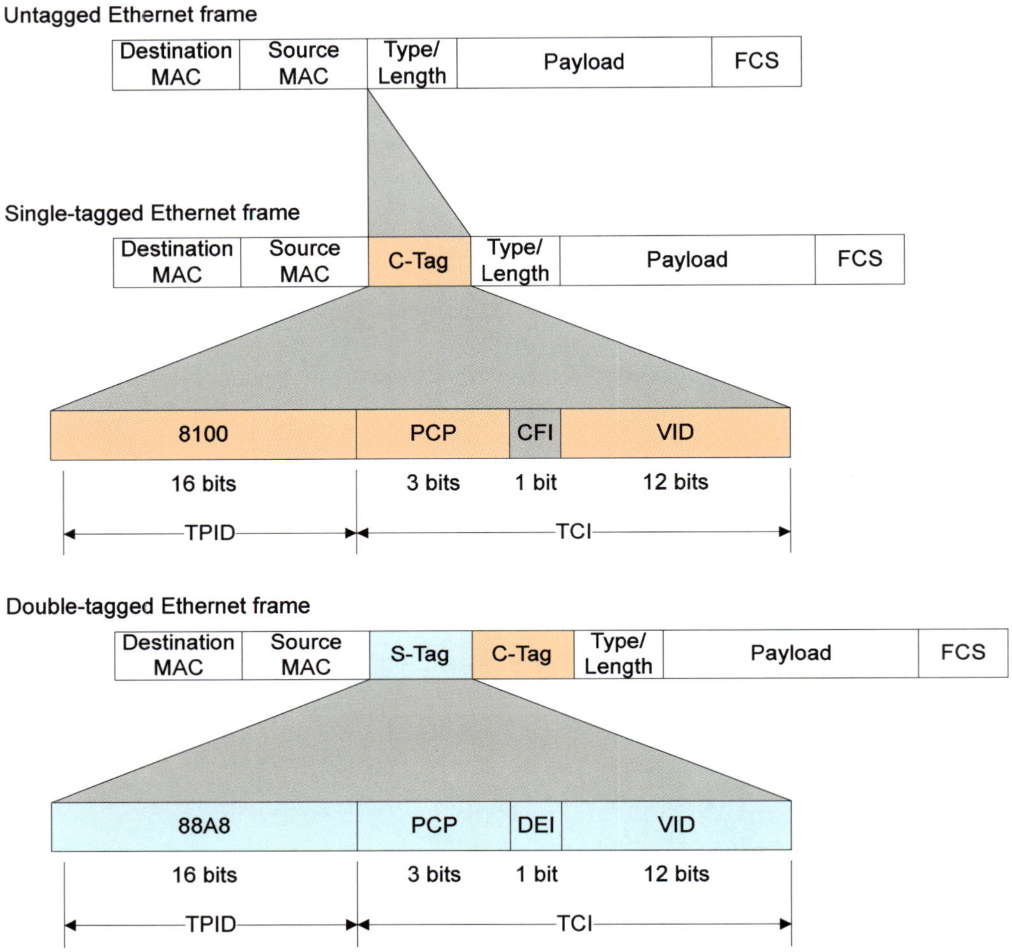

Figure 117: IEEE 802.1ad Ethernet Frame

By now, this frame format should be familiar. The C-Tag corresponds to the CE-VLAN tag used at UNIs. The S-Tag is the additional tag required at ENNIs to capture CoS, Color, and OVC end-point mapping information.

Used as a transport technology within a MEN/CEN (or Operator MEN/CEN), PB overcomes the inherent limitations of Bridging. The service provider no longer needs to use the C-Tag to support service identification, CoS identification, and drop-eligibility within the MEN (CEN). All of this information is instead captured in the S-Tag, which is added (pushed) to each service frame at MEN (CEN) ingress and removed (popped) at MEN (CEN) egress. The C-Tag now more truly belongs to the subscriber[35]. The service provider is able to:

- Transport and preserve CE-VLAN tags
- Support all types of MEF service (including bundling) without limitations
- Support ENNIs (multi-Operator MEN/CEN applications)

Additionally, L2CP frames in port-based services (EPL, EP-LAN, and EP-Tree services) can be tunneled with the service (using the S-Tag).

Scalability

Provider Bridging is more scalable than Bridging, but is still not highly scalable. If services are identified within the MEN (CEN) based on S-VLAN ID alone, PB supports approximately 4K services (the number of unique S-VLAN IDs supported by 802.1ad). Although this same number is supported by Bridging, S-VLAN IDs (unlike C-VLAN IDs) do not have to be coordinated with subscribers.

MAC table growth is another factor limiting scalability. At each network bridge, MAC tables grow and may overflow as services are added and MAC addresses multiply.

Note: *One way service providers deal with MAC address scalability is to place a router, rather than a bridge, at the CE. The router "hides" all of MAC addresses in the local subscriber network from the service provider network so that the service provider network sees only router-interface MAC addresses.*

Protection

PB supports two protection options:

- **xSTP** (some variant of Spanning Tree Protocol)—Can be applied over any network topology. Restoration time is limited by the speed of spanning tree convergence, which can make sub-50 ms protection switching difficult to achieve (depending on network size).
- **G.8032** (ITU-T Ethernet Ring Protection Switching)—Can be applied over ring topologies only. Provides sub-50 ms protection switching.

35 Of course, even using PB, the service provider may use the C-Tag at MEN (CEN) ingress to identify which service frames belong to a VLAN-based service and/or to identify CoS.

Provider Bridging Summary

Advantages	Limitations / Issues

• Supports all types of MEF services
• Widely available and familiar technology
• Supports bundling
• S-Tag supports CoS and drop-eligibility with no impact on subscriber CE-VLAN tag
• Supports ENNIs
• Supports sub-50 ms protection switching with G. 8032 ERPS
	• Scalability is limited:
	  • A maximum of approximately 4K services per MEN (CEN) is supported.
	  • MAC tables grow with service addition, leading to possible overflow (unless routers are used at CEs).
	• Sub-50 ms protection switching can be difficult to achieve using only xSTP (depending on network size).

11.2.3
PBB (Provider Backbone Bridging)

Provider Backbone Bridging (IEEE 802.1ah), also called "MAC-in-MAC," was developed to address scalability limitations of Provider Bridging (802.1ad), including the following:

- **Service instance scalability**—PB can only support about 4K services (limited by the number of unique S-VLAN ID values).
- **MAC address scalability**—MAC tables in a PB network grow and may overflow as services are added and MAC addresses multiply.

To address these problems, PBB defines the 802.1ah Ethernet frame, which adds a second MAC header to the 802.1ad Ethernet frame:

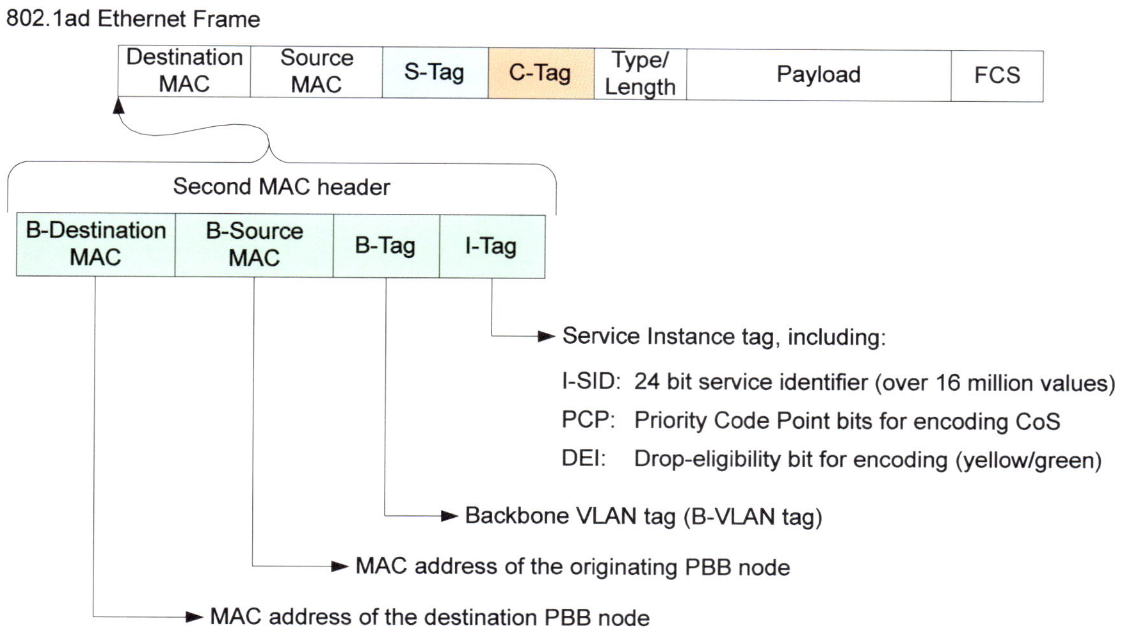

Figure 118: IEEE 802.1ah Ethernet Frame

The second MAC header is used as follows:

1. At Ingress to the PBB network (MEN/CEN or Operator MEN/CEN), a Provider Edge Bridge encapsulates each incoming Ethernet frame (802.1Q–2005 or 802.1ad), adding a second MAC header.

2. Within the PBB network, Provider Core Bridges forward 802.1ah Ethernet frames based solely on the second MAC header.

3. On egress from the PBB network, the original Ethernet frame is de-encapsulated from the 802.1ah Ethernet frame.

The I-Tag is used for service identification.

The B-Tag (which is identical to an S-Tag in format and TPID) supports B-VLAN functionality within the PBB network. B-VLANs are used to confine services to a virtual subnetwork within the PBB network as illustrated by the following example. It is not used for service identification.

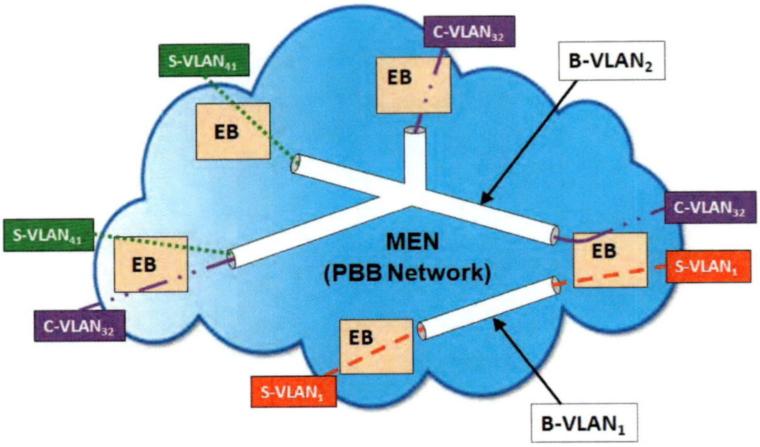

Figure 119: B-VLANs

CoS and drop-eligibility information can be encoded in the I-Tag.

Important: *The second MAC header isolates subscriber BPDUs from service provider BPDUs, improving security.*

Scalability

The 24-bit service identifier (I-SID, within the I-Tag) makes service instance scalability virtually unlimited. The I-SID provides over 16 million unique identifiers, each of which can be associated with over 4K unique S-VLAN IDs (and over 4K unique C-VLAN IDs).

Provider Core Bridges within the PBB network are completely unaware of subscriber-level MAC addresses. Switching and learning (except at the network edge) are based on the B-VLAN tag and B-MAC addresses only. Isolation from subscriber-level MAC addresses protects the PBB network from MAC table overflow, making it scalable with respect to MAC addressing.

Example of Subscriber-Level MAC Learning

This example explains how the PBB network learns to switch subscriber traffic using subscriber-level MAC learning at network-edge bridges only.

The PBB network shown in the following figures has been operating for some time, PBB-level xSTP has converged, and B-MAC learning is complete. An EP-LAN service, connecting four subscriber locations (Denver, Atlanta, Boston, and Chicago), is in operation.

Host 1 in Denver, now sends an Ethernet frame to Host 3 (in Boston). The PBB network has not previously seen traffic from Host 1 or Host 3. The following steps explain how the PBB network supports the unknown unicast (Host 1 to Host 3) and the reply (Host 3 to Host 1) and learns to connect the two hosts without broadcasting. The first figure illustrates Steps 1 through 4. The second figure illustrates Steps 5 through 9.

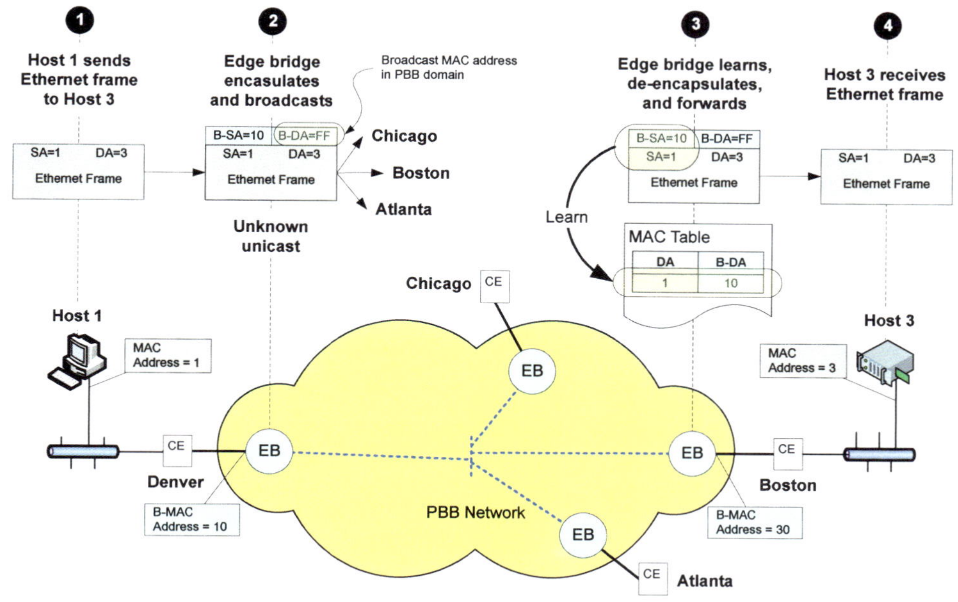

Figure 120: PBB Network Example, Part 1: Unknown Unicast

1. Host 1 initiates an unknown unicast to host 3 addressed with SA=1 and DA=3.

2. The PBB edge bridge (EB) in Denver receives this frame through port 10 (the UNI), but has no entry for Host 3 in its MAC tables. So it treats the frame as an unknown unicast:

   a. It applies a second MAC header to the frame with:

      - B-Tag and I-Tag set appropriately for the service
      - B-SA=10 (the B-MAC address of the originating port, port 10)
      - B-DA=FF (FF represents *FFFF,FFFF,FFFF*, the broadcast MAC address)

   b. It forwards the frame out of all ports associated with the service, flooding the service domain.

3. Edge bridges in Atlanta, Boston, and Chicago each receive the frame. Each of these bridges then:

   a. Learns,for future reference, that DA=1 traffic should be addressed to B-DA=10

   b. De-encapsulates the subscriber frame

   c. Forwards the subscriber frame out the appropriate subscriber port (UNI)

4. Host 3 in Boston receives the frame.

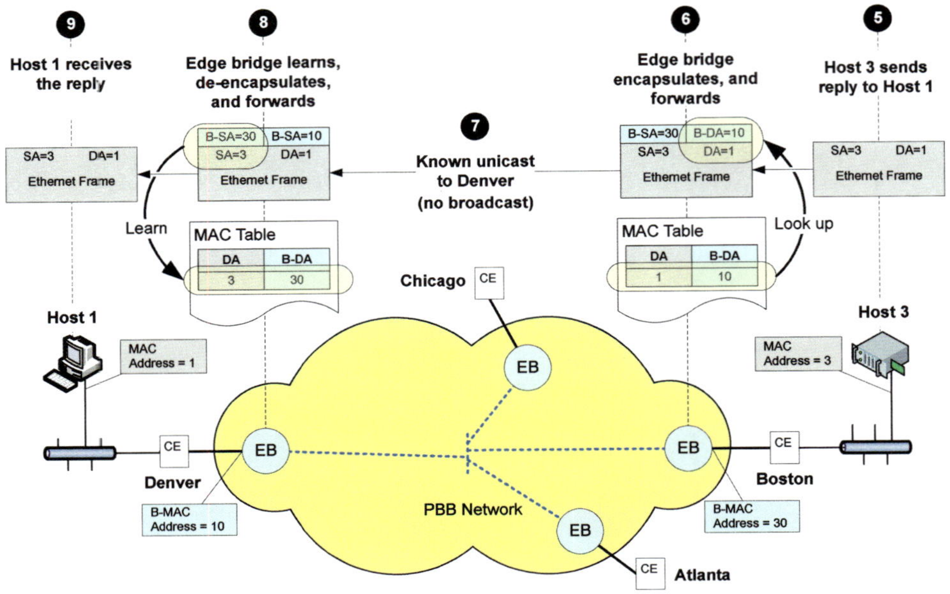

Figure 121: PBB Network Example, Part 2: Reply

5. Host 3 sends a reply with SA=3 and DA=1.

6. The edge bridge in Boston receives the reply frame through the UNI and knows where to send it based on previous learning (Step 3a), so it treats the frame as a known unicast:

   a. It applies a second MAC header to the frame with:

      - B-Tag and I-Tag set appropriately for the service
      - B-SA=30 (the B-MAC address of the originating port)
      - B-DA=10 (the B-MAC address of the target port, learned in Step 3a)

   b. It forwards the frame out the port associated with B-DA=10.

7. The frame travels through the PBB network directly to Denver because it is a known unicast frame. Based on previous learning, all bridges receiving the frame (with B-DA=10) know which port to use for forwarding.

8. The edge bridge in Denver receives the frame and then:

   a. Learns, for future reference, that DA=3 traffic should be addressed to B-DA=30

   b. De-encapsulates the subscriber frame

   c. Forwards the subscriber frame out the appropriate subscriber port (UNI)

9. Host 1 in Denver receives the reply.

The PBB network can now support unicast traffic between Host 1 and Host 3 without broadcasting. Subscriber-level MAC addresses were learned only at the network edge.

Protection

PBB supports two protection options:

- **xSTP** (some variant of Spanning Tree Protocol)—Can be applied over any network topology. Restoration time is limited by the speed of spanning tree convergence, which can make sub-50 ms protection switching difficult to achieve (depending on network size). However, fast protection switching and restoration may be easier to achieve in a PBB network (compared to a PB or Bridging network) because the spanning tree is isolated from subscriber networks and changes less frequently.
- **G.8032** (ITU-T Ethernet Ring Protection Switching)—Can be applied over ring topologies only. Provides sub-50 ms protection switching.

PBB Summary

Advantages	Limitations / Issues
• Supports all types of MEF services • Unlimited scalability • Can coexist in the same network with PBB-TE • Subscriber L2CP frames are isolated from service-provider L2CP frames, improving security. • Supports sub-50 ms protection switching with G.8032 ERPS • PBB simplifies network operations and troubleshooting by isolating/protecting the service provider's network from the subscribers' networks.	• PBB is more complex than PB or Bridging • Sub-50 ms protection switching may be difficult to achieve, using only xSTP (depending on network size)

11.2.4 PBB-TE (Provider Backbone Bridge Traffic Engineering)

PBB-TE (defined by IEEE 802.1Qay) is based on PBB but is designed to support high-performance, traffic-engineered, connection-oriented Ethernet (COE) services.

PBB-TE can be deployed with PBB in the same network (using a separate PBB-TE domain) and uses the same IEEE 802.1ah Ethernet frame structure as PBB.

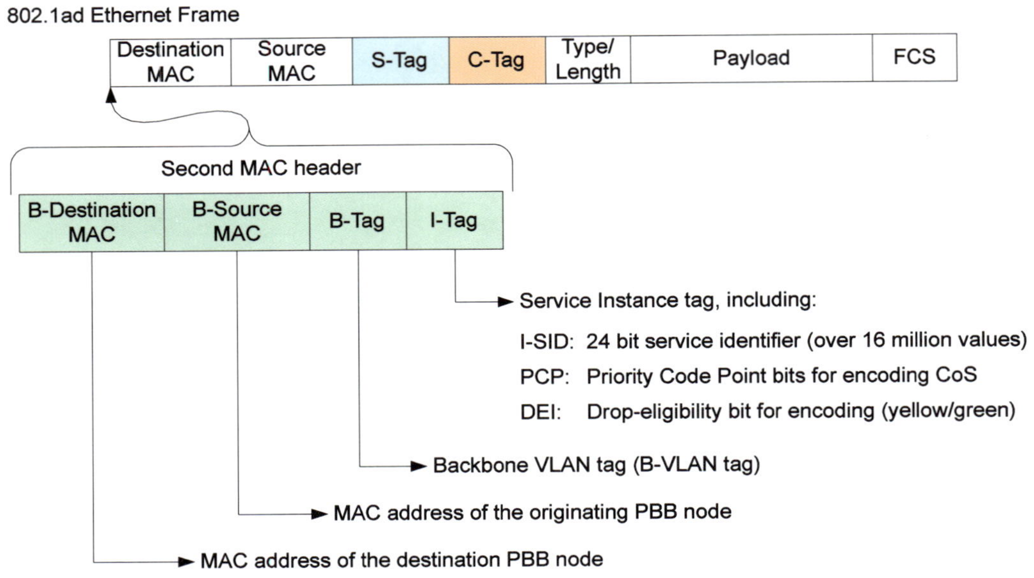

Figure 122: IEEE 802.1ah Ethernet Frame

Unlike PBB, PBB-TE only supports point-to-point services (EPL, EVPL, Access EPL, and Access EVPL services).

PBB-TE supports provisioning that explicitly selects traffic engineered paths within 802.1ah networks, disabling bridging behavior such as flooding (unknown unicast), MAC address learning, and xSTP.

Compared to bridging technologies (Bridging, PB, or PBB), PBB-TE behaves more predictably, and its behavior can be more easily controlled by provisioning connections at each bridge along the forwarding path. This creates a point-to-point tunnel, or Ethernet Switched Path (ESP), that is accessed using the B-VLAN ID and the source and destination B-MAC addresses.

Scalability

The second MAC header gives PBB-TE unlimited scalability, just like PBB.

Protection and OAM

PBB-TE supports:

- Resource reservation across the ESP (Ethernet Switched Path) to guarantee EVC bandwidth
- OAM performance monitoring and fault detection
- G.8031 Ethernet protection switching with sub-50 ms protection switching/restoration

PBB-TE Summary

Advantages	Limitations / Issues
• Supports sub-50 ms protection switching and restoration with G.8031 EPS • Unlimited scalability • Can co-exist in the same network with PBB • Subscriber L2CP frames are isolated from service provider L2CP frames, improving security.	• Supports only point-to-point services (EPL or EVPL)

11.2.5 ETS (Ethernet Tag Switching)

Important: *ETS is not included in the MEF-CECP exam. However, ETS is interesting, not difficult to understand, and highlights the fact that service implementation is not limited to the transport technologies covered in the MEF-CECP exam (any transport technology that can support MEF-compliant services can be used).*

Ethernet Tag Switching (ETS) is Fujitsu's implementation of connection-oriented Ethernet (COE). It supports high-performance, traffic-engineered, point-to-point, MEF-compliant services (EPL and EVPL services) similar to PBB-TE, but using simpler mechanisms. ETS uses 802.1ad frame format (used by PB), rather than the more complex 802.1ah frame format used by PBB-TE.

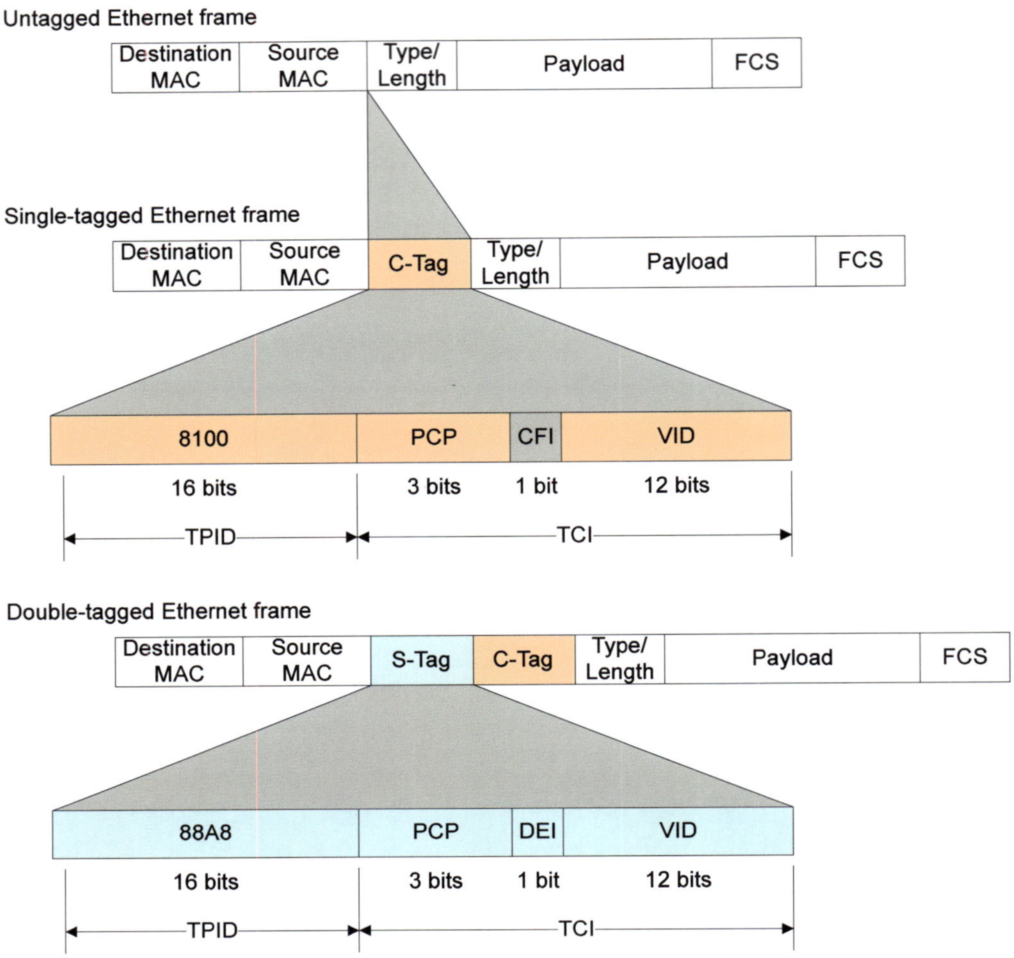

Figure 123: IEEE 802.1ad Ethernet Frame

As with PB, an S-Tag is pushed onto service frames at MEN (CEN) ingress and is used to capture transport information:

- The VID is used for service identification (but, unlike PB, only locally).
- PCP bits encode CoS.
- DEI bit encodes drop-eligibility (green/yellow).

As with PBB-TE, bridging behavior, such as flooding (unknown unicast), MAC address learning, and xSTP Frames, are all disabled.

Frames are forwarded based on S-VLAN IDs and/or C-VLAN IDs, rather than the MAC address. VLAN "cross-connects" are explicitly provisioned through the network with the aid of an element management system (EMS), and frames are forwarded (switched) based on these VLAN "cross-connects."

Scalability

Scalability is unlimited because VLAN tags have only local significance and may be swapped along the path.

Protection and OAM

Similar to PBB-TE, ETS supports:

- Resource reservation through the network to guarantee EVC bandwidth
- Service-level OAM performance monitoring and fault detection
- G.8031 protection switching with sub-50 ms protection switching/restoration

Ethernet Tag Switching Summary

Advantages	Limitations / Issues
• Supports sub-50 ms protection switching and restoration (G.8031) • Provides unlimited scalability • Provides COE functionality similar to PBB-TE using only the 802.1ad frame format and simpler mechanisms	• Supports only point-to-point services (EPL, EVPL, Access EPL, and Access EVPL)

11.3 Layer 2.5 Technologies (Multiprotocol Label Switching)

In this section:

Three transport technologies are based on MPLS (Multiprotocol Label Switching) technology:

- **MPLS VPWS**—MPLS Virtual Private Wire Service
- **MPLS VPLS**—MPLS Virtual Private LAN Service
- **MPLS-TP**—MPLS Transport Profile

MPLS operates at a layer that is generally considered to lie between Layers 2 and 3 and thus is often called a Layer "2.5" protocol. It was designed to provide a unified data-carrying service for both circuit-based clients and packet-switching clients and supports many different kinds of traffic, including IP, ATM, SONET, and Ethernet frames.

MPLS uses an MPLS label (or a stack of MPLS labels) inserted between Layer 2 and Layer 3 of the PDU. The following figure shows MPLS labels inserted into an Ethernet frame.

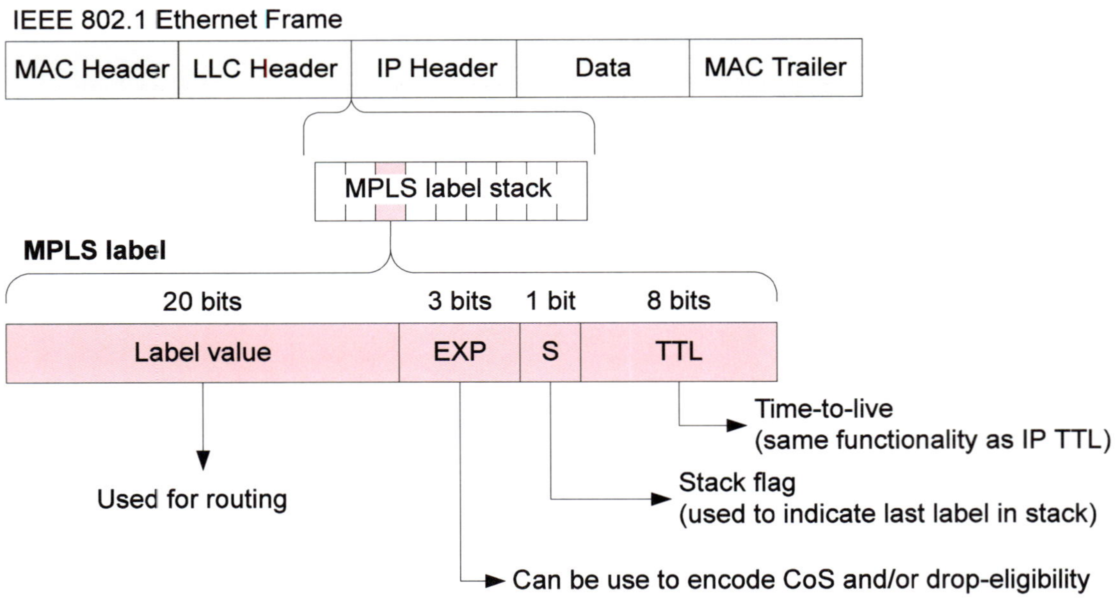

Figure 124: MPLS Label Stack Inserted in IEEE 802.1 Ethernet Frame

Attention: *MPLS can use one or more labels. Stacks of labels are used to support hierarchical routing and are not considered going forward in this study guide.*

The MPLS label value governs packet routing but is only locally significant. At each router, the MPLS label value determines where the packet is sent next, but each router swaps the MPLS label value before forwarding the packet. The following figure shows an example of MPLS packet routing (the figure also introduces some MPLS terminology).

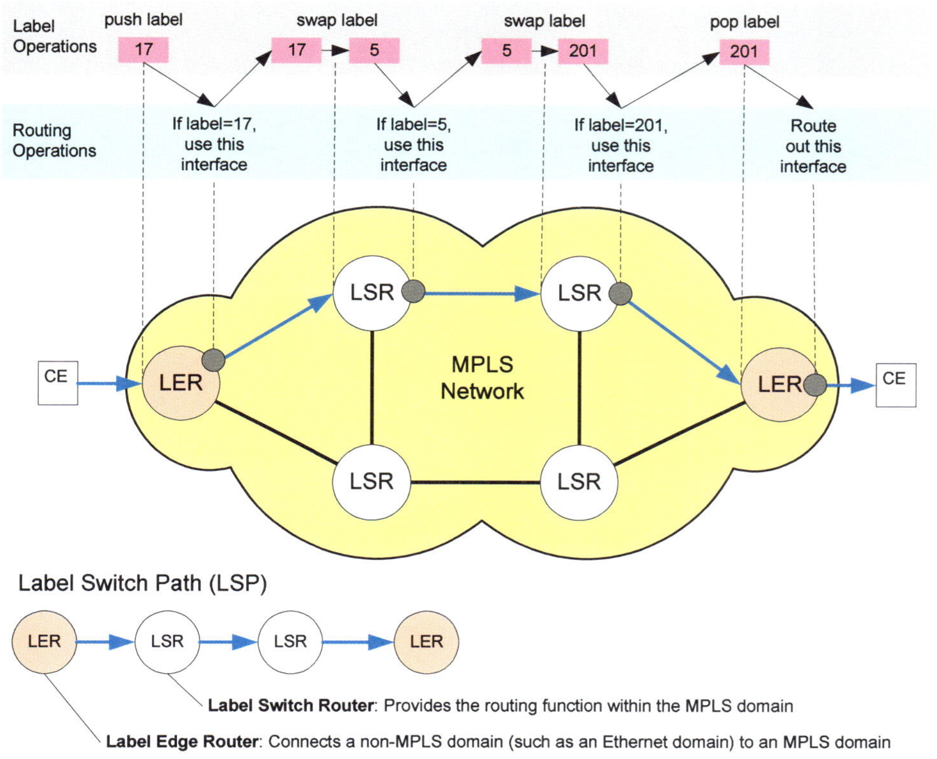

Figure 125: Example of MPLS Packet Routing

In this example, the PDU (Ethernet frame, for example) enters the MPLS network from the left. At ingress, the Label Edge Router (LER) pushes label 17 onto the PDU, and this binds the PDU to a specific Label-Switched Path (LSP). Along the LSP, each router uses the incoming label to make a routing decision and then swaps the label before forwarding the PDU. Eventually the PDU reaches the egress LER, which pops the label and routes the PDU out the appropriate interface.

Key points of MPLS operation include:

- Prior to use, the MPLS network is provisioned to support many Label-Switched Paths (LSPs), like the one shown in the figure.
- When a PDU enters the network, it is assigned an MPLS label value (Label Value=17, in the example), and this value binds the PDU to a specific LSP.

- Because there are more than 1 million unique MPLS label values (20 bits), a Label Edge Router (LER) can provide gateway to more than 1 million LSPs.
- Service instance scalability is limitless because:
  - The MPLS network can include many LERs.
  - Each LER can terminate more than 1 million LSPs.
  - Each LSP can support more than 16 million EVCs/OVCs (based on the number of unique combinations of C-VLAN IDs and S-VLAN IDs) or more than 4,000 EVCs/OVCs (based on unique S-VLAN IDs only).

This explains how MPLS works after LSPs are established. But how are LSPs established?

Establishment of LSPs is a complex topic that is mostly beyond the scope of this study guide. An LSP can be set up (planned) manually or automatically using an algorithm such as OSPF (Open Shortest Path First). LDP (Label Distribution Protocol) allows MPLS routers to exchange label mapping information and to maintain forwarding databases. For the MEF-CECP 2.0 exam, it is enough to understand that there are practical methods for planning and creating LSPs in the network, both manually and automatically.

All three types of MPLS transport technologies (MPLW VPWS, MPLS VPLS, and MPLS-TP) route PDUs in this way.

11.3.1 MPLS VPWS (MPLS Virtual Private Wire Service)

MPLS VPWS (Virtual Pseudowire Service) is used to deliver EPL and EVPL services, providing point-to-point EVCs between two UNIs. In VPWS applications, a pseudowire emulates a simplex Ethernet link (wire) between two UNIs. Therefore, two pseudowires are configured to provide duplex operation, one in the "forward" direction and one in the "reverse" direction, as shown in the following figure.

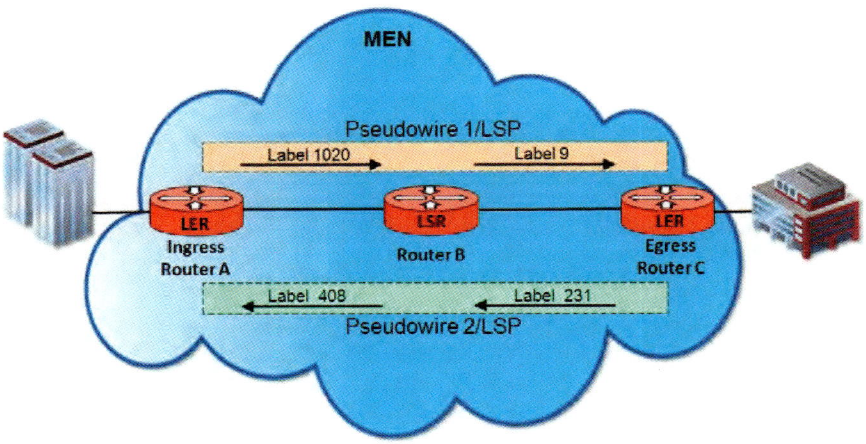

Figure 126: MPLS VPWS Implementation of an EPL or EVPL Service

Two LSPs are required to establish the VPWS, one for each direction. Because LSPs can support multiple pseudowires (as well as other MPLS traffic), suitable LSPs may already exist in the network. Otherwise, they have to be created.

The unidirectional pseudowires are associated to LSPs by provisioning Label Edge Routers (LERs) appropriately:

- At ingress, the LER is provisioned to recognize service frames and to push the appropriate MPLS label onto them, binding them to the appropriate LSP.
- At egress, the LER is provisioned to pop MPLS labels and forward service frames out the appropriate interface.

Service frames are routed through the network using the previously described MPLS routing mechanism.

Scalability

MPLS VPWS has unlimited service instance scalability, like all MPLS transport technologies.

Protection

Protection can be provided using MPLS Fast Reroute (FRR).

Note: *MPLS FRR provides segment-based protection. Each LSP passing through a facility is protected by a backup path that originates at the node immediately upstream to that facility.*

MPLS VPWS Summary

Advantages	Limitations / Issues
• Unlimited scalability • Can coexist in the same network with MPLS VPLS	• Only point-to-point services (EPL, EVPL, Access EPL, and Access EVPL) are supported. • Service-level OAM performance monitoring and fault detection are limited compared to Ethernet-centric OAM, such as Y.1731 and 802.1Q (SOAM).

11.3.2 MPLS VPLS (MPLS Virtual Private LAN Service)

MPLS Virtual Private LAN Service (VPLS) is used to deliver all MEF services (E-Line, E-LAN, and E-Tree), providing multipoint-to-multipoint EVCs between two or more UNIs and MAC learning. VPLS is "private" in that customer equipment (CE) devices that belong to different VPLSs cannot interact. VPLS is "virtual" in that multiple VPLSs can be offered over a common packet-switched network.

Because pseudowires and LSPs are point-to-point and unidirectional, a full mesh is required to create a multipoint-to-multipoint service, as shown by the example in the following figure.

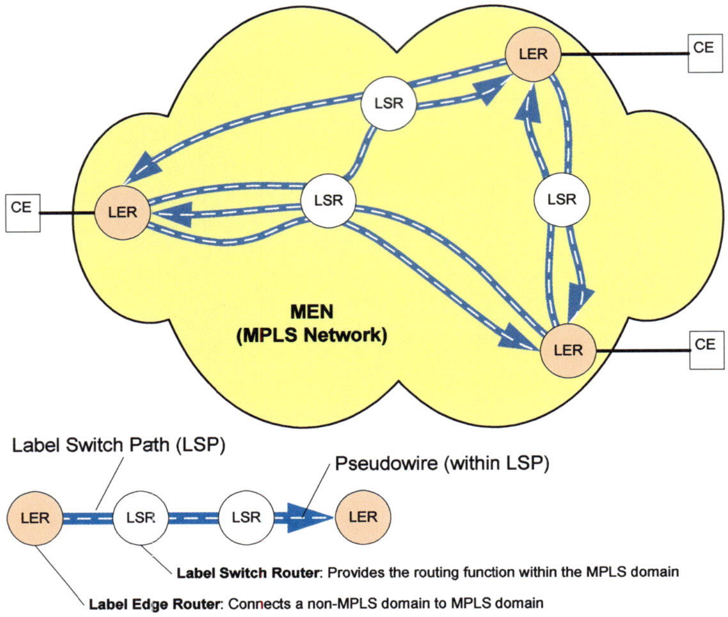

Figure 127: VPLS Connecting Three Subscriber Sites

Six pseudowires/LSPs are needed to create a full mesh between three UNIs.

To support bridging, LERs keep MAC tables and support MAC learning as follows:

1. If an unknown unicast is received from the CE domain, the LER broadcasts the frame out all pseudowires/ LSPs associated with the service.
2. Each LER receiving the broadcast:
    a. Forwards it out the appropriate UNI
    b. Adds a MAC table entry associating the SA of the unknown unicast with the appropriate pseudowire/LSP (not the one it arrived from, but its partner going in the opposite direction)
3. When the target (host) replies, the ingress LER recognizes the DA of the reply unicast frame because of the MAC learning described previously, so it will not broadcast it. The reply is assigned to the correct pseudowire/LSP and goes directly to where it is supposed to go.
4. The LER receiving the unicast reply:
    a. Forwards it out the appropriate UNI
    b. Adds a MAC table entry associating the SA of the unknown unicast reply with the appropriate pseudowire/LSP (not the one it arrived from, but its partner going in the opposite direction)

A connection has now been established between the two hosts.

Observe the similarity between this description of MAC learning and the explanation of MAC learning for PBB (Figure 120 and Figure 121).

Scalability

MPLS-VPLS has unlimited service instance scalability, like all MPLS transport technologies.

Similar to PBB, MAC learning is confined to the network edge (LERs), improving MAC address scalability and reducing its vulnerability to MAC table overflow.

Protection

Protection can be provided using MPLS Fast Reroute (FRR).

Note: *MPLS FRR provides segment-based protection. Each LSP passing through a facility is protected by a backup path that originates at the node immediately upstream to that facility.*

MPLS VPLS Summary

Advantages	Limitations / Issues
• Supports all types of MEF services • Unlimited scalability • Can coexist in the same network with MPLS VPWS	• Service-level OAM performance monitoring and fault detection are limited compared to Ethernet-centric OAM, such as Y.1731 and 802.1Q (SOAM).

11.3.3 MPLS-TP (MPLS Transport Profile)

MPLS-TP (Transport Profile) is a traffic-engineered transport that was developed to simplify MPLS and make it more like a Layer 1 transport protocol. MPLS-TP technology does not typically use control plane protocols to establish connections. Instead, it uses the management plane (an external Network Management System) to establish LSPs. Paths are traffic engineered, resulting in a highly predictable, connection-oriented services.

Like MPLS VPWS, MPLS-TP supports only EVP and EVPL services. However, unlike VPWS, MPLS-TP supports bidirectional pseudowires and LSPs (in addition to unidirectional pseudowires and LSPs).

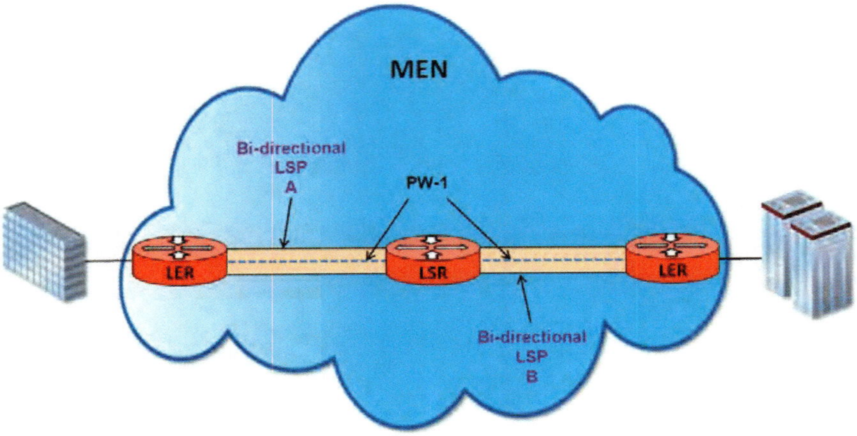

Figure 128: MPLS-TP Implementation of an EPL or EVPL Service

MPLS-TP services are also engineered (they are not automatically created using a protocol such as OSPF) and can provide higher performance.

Scalability

MPLS-TP has unlimited service instance scalability, like all MPLS transport technologies.

Protection and OAM

MPLS-TP supports COE functionality similar to PBB-TE (or ETS, Fujitsu's COE implementation), including:

- Resource reservation through the network to guarantee EVC bandwidth
- Service-level OAM performance monitoring and fault detection (under development)
- Linear protection per RFC 6378 and ring protection (under development)

MPLS-TP

Advantages	Limitations / Issues
• Supports sub-50 ms protection switching and restoration • Unlimited scalability • Support for traffic engineered paths • Simplifies MPLS to make it more like a Layer 1 transport protocol	• Only point-to-point services (EPL, EVPL, Access EPL, and Access EVPL) are supported. • Service-level OAM performance monitoring and fault detection are limited compared to Ethernet-centric OAM, such as Y.1731 and 802.1Q (SOAM).

11.4
Ethernet Service Protection Technologies

In this section:

Networks offer protection through redundancy by providing alternative resources when working resources fail.

11.4.1
1+1 Linear Protection

1+1 linear protection supports point-to-point connectivity only.

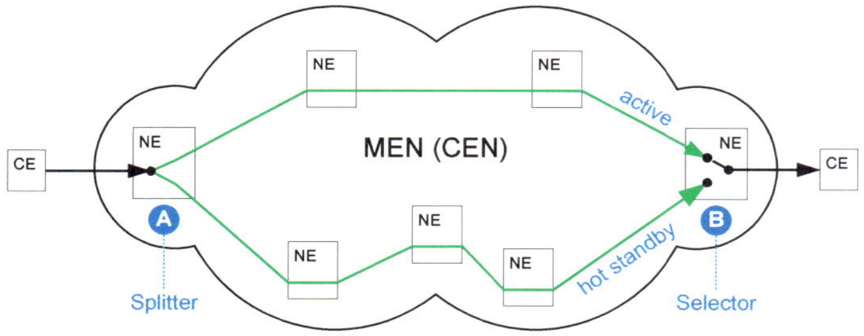

Unidirectional traffic from A to B is protected as follows:

1. Ingress traffic at A is duplicated and sent to B along two distinct network paths.
2. Two copies of the traffic are received at B.
3. At B, one of the two copies is selected for forwarding.
4. Failure in the active signal is detected at B and triggers switching.

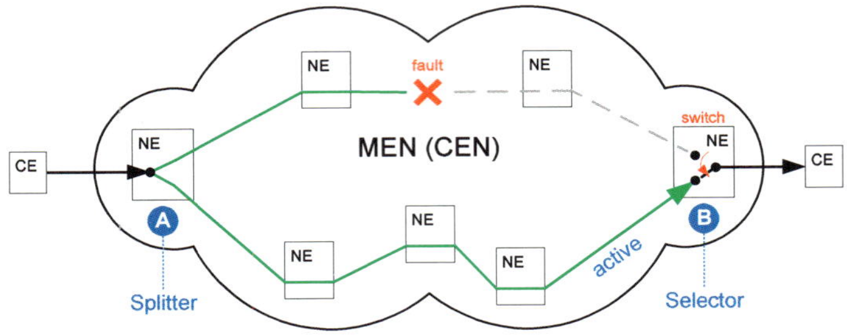

To protect bidirectional traffic, this scheme is used in both directions.

Examples of 1+1 linear protection include:

- UPSR (unidirectional path-switched ring) protection used by SONET
- Och-DPRING (optical channel dedicated protection ring) used by OTN/WDM

11.4.2 1:1 Linear Protection

1:1 linear protection supports point-to-point connectivity only.

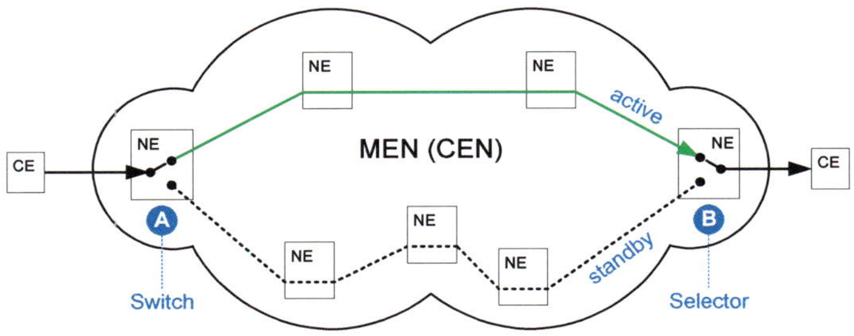

Note: *Because the standby path does not carry replicated traffic, it can potentially carry other traffic.*

Unidirectional traffic from A to B is protected as follows:

1. Two distinct network paths from A to B are provisioned: a working path and a protect path.
2. In normal operation, the working path is active. The protect path is on standby.
3. If network failure occurs, the protect path becomes active.

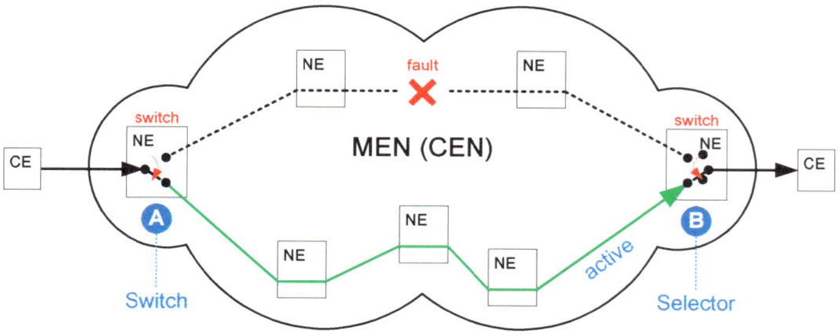

To protect bidirectional traffic, this scheme is used in both directions.

ITU-T G.8031 Ethernet Linear Protection Switching is an example of 1:1 linear protection.

11.4.3 xSTP-Based Protection

The network of bridging NEs that supports a multipoint service must be loop-free by design or must be controlled be loop-free (to avoid broadcast storms).

All variations of spanning tree protocol (xSTP)[36] artificially block links, such that the active topology is loop-free.

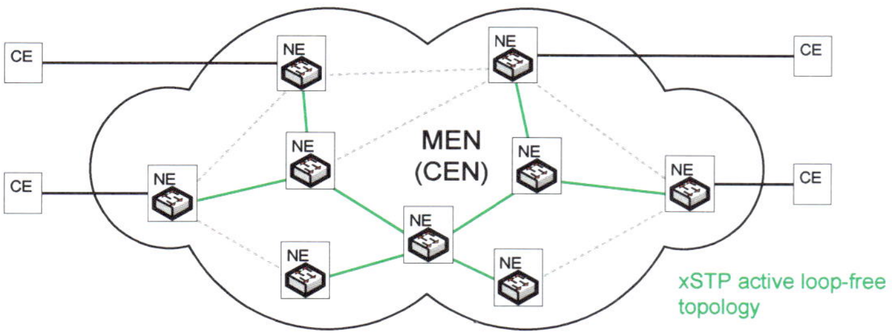

If a fault occurs in the active network, the xSTP algorithm converges to a new loop-free topology. However, re-convergence can take time.

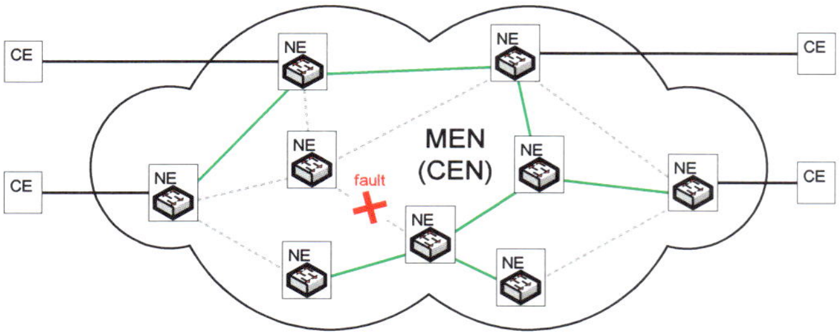

36 For example: STP (Spanning Tree Protocol per IEEE 802.1D), RSTP (Rapid Spanning Tree Protocol per IEEE 802.1w), and MSTP (Multiple Spanning Tree Protocol per IEEE 802.1s).

11.4.4
1:1 Ring Protection

1:1 ring protection (such as ITU-T G.8032) provides functionality similar to xSTP in that links are artificially blocked to produce an active topology that is loop-free.

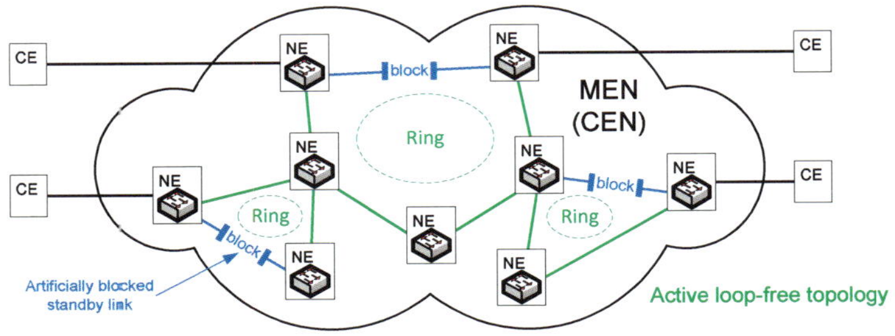

However, in contrast to xSTP, 1:1 ring protection applies only to ring-based network topologies.

If a fault occurs in an active link, the ring containing the faulty link opens its blocked link, producing a new loop-free topology.

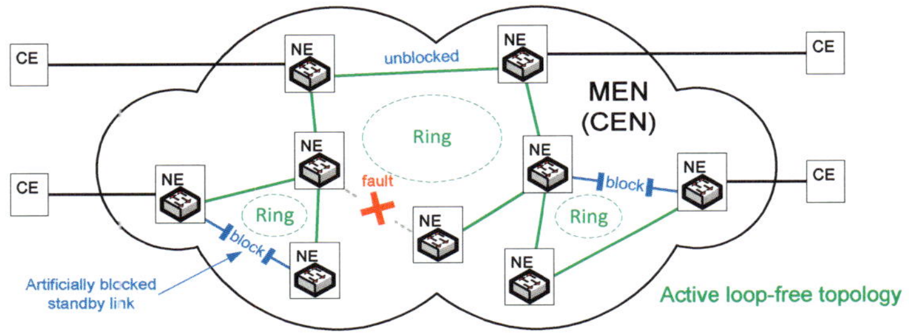

Protection switching for 1:1 ring protection is faster and more predictable than for xSTP protection.

11.4.5
Protection Mechanisms and Transport Technologies

Different transport technologies rely on different protection mechanisms as shown in the following table. Also note that an Ethernet service can be protected by several mechanisms (operating at different OSI layers, for example).

Table 32: Summary of Protection and Resiliency Mechanisms

Layer	Protection Mechanism	Description	Applicable Transport Technologies	50 ms Switchover Support
"2.5"	**FRR** (MPLS fast reroute)–per RFC 490 (IETF)	1:1 segment protection Creates local bypass to faulty node or link.	MPLS VPWS and MPLS VPLS	Yes
	MPLS-TP Ring Protection–under development (IETF) or ITUT-G.3132 (draft)	1:1 ring protection	MPLS-TP	Yes
	MPLS-TP Linear Protection–per RFC 6378 (IETF) or ITU-T G.8131	1:1 linear path protection	MPLS-TP	Yes
2	**xSTP**–any variant of spanning tree protocol: • STP–per IEEE 802.1D • RSTP–per IEEE 802.1w • MSTP–per IEEE 802.1s	Creates loop-free topology that automatically reconfigures in the event of node/port/link failure	Bridging, PB, or PBB over any topology	Depends*
	G.8032–ERPS (Ethernet ring protection switching, ITU-T)	1:1 ring protection Uses Y.1731 CCMs for fault detection.	Bridging, PB, or PBB over ring topology	Yes
	G.8031–ELPS (Ethernet linear protection switching, ITU-T)	1:1 linear path protection Uses Y.1731 CCMs for fault detection.	PBB-TE or Ethernet Tag Switching	Yes
1	**OCh-DPRING** (optical channel dedicated protection ring)	1+1 linear path protection	OTN or WDM	Yes
	BLSR (bidirectional line-switched ring)	1:1 ring protection	SONET	Yes
	UPSR (unidirectional path-switched ring)	1+1 linear path protection	SONET	Yes
0	**1:1 LAG**–link aggregation, per clause 43 of IEEE 802.3	1:1 link protection ENNIs and Type 2.2 UNIs require 1:1 LAG support.	All	Yes

* xSTP switchover time depends on the speed of spanning tree convergence.

11.5
Summary

This lesson has provided an overview of eleven transport technologies:

Table 33: Transport Technologies for Carrier Ethernet Services

OSI Model Layer	Transport Technology
Layer 1	• Synchronous Optical Network (SONET)/Synchronous Digital Hierarchy (SDH) • Optical Transport Network (OTN)—ITU-T G.709 • Wavelength Division Multiplexing (WDM)
Layer 2	• Bridging—IEEE 802.1D and 802.1Q–2005 • Provider Bridging (PB)—IEEE 802.1ad • Provider Backbone Bridging (PBB)—IEEE 802.1ah • Provider Backbone Bridging with Traffic Engineering extensions (PBB-TE)—IEEE 802.1Qay • Ethernet Tag Switching—Fujitsu implementation of connection-oriented Ethernet (COE) leveraging the IEEE 802.1ad frame format
Layer "2.5"	• Multiple Protocol Label Switching Virtual Private Wire Service (MPLS VPWS) • Multiple Protocol Label Switching Virtual Private LAN Service (MPLS VPLS) • Multiple Protocol Label Switching Transport Profile (MPLS-TP)

For the MEF-CECP exam, it is important to understand the advantages and limitations of each technology.

Note: *Different technologies rely on different protection mechanisms, as previously summarized in* Protection Mechanisms and Transport Technologies.

The following study aid arranges the transport technologies into three groups: Group A, Group B, and Group C.

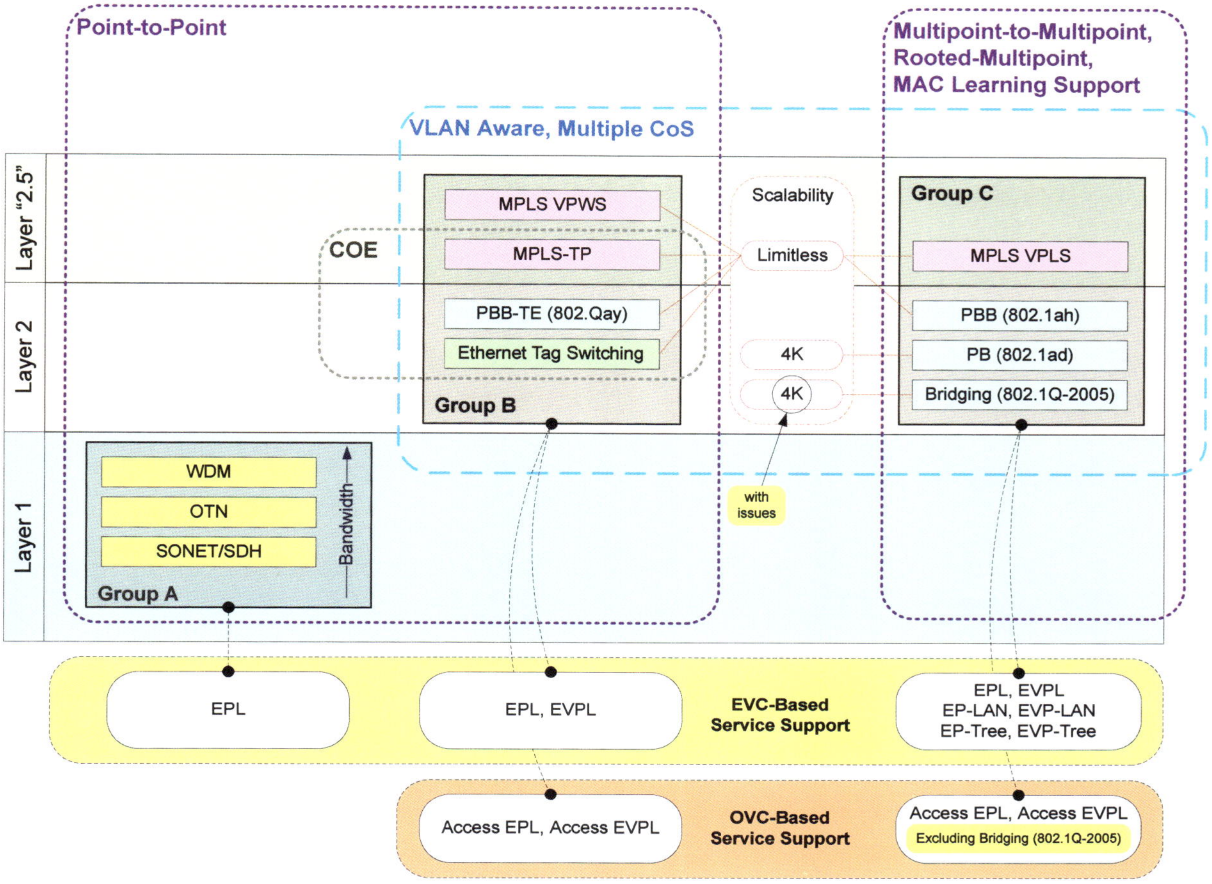

Figure 129: Transport Technology Study Aid

All Layer 1 transport technologies (SONET/SDH, OTN, and WDM) are placed in Group A. These technologies are point-to-point, are VLAN unaware, do not support multiple CoS, support only the EPL service, and differ from each other in bandwidth granularity.

Layer 2 and "2.5" technologies are divided into Groups B and C. Group B technologies (MPLS VPWS, MPLS TE, PBB-TE, and Ethernet Tag Switching) are point-to-point, are VLAN aware, support multiple CoS, and support EPL, EVPL, Access EPL, and Access EVPL services. All are COE technologies, except MPLS VPWS, which is not considered a COE technology. Scalability is unlimited for all Group B technologies.

Group C technologies (MPLS VPLS, PBB, PB, and Bridging) are VLAN aware and support multiple CoS, MAC learning, and all MEF services with one exception: 802.1Q-2005 bridging does not support E-Access services (due to lack of S-Tag support). Scalability for MPLS VPLS and PBB technologies is unlimited. The scalability of PB is limited by 4K unique S-VLAN IDs. The scalability of 802.1Q–2005 bridging technology is severely limited because the service provider has only 4K unique C-VLAN IDs, and they are shared among subscribers.

Note: *In the previous paragraph and figure, the number 4K is an approximation. The exact number of VLAN IDs is 4094. The 12-bit VID field allows 4096 values, but only 4094 values can be used for VLAN identification because, per IEEE standards, values 0 and FFF are reserved for other uses.*

11.6
Review

1. How many service instances can Multi-Protocol Label Switching (MPLS) support in one network?

    a. 16,777,216

    b. 16 million

    c. Unlimited

    d. 4094

2. Which two technologies forward traffic over an explicitly provisioned traffic-engineered path (similar to Layer 1) without bridging or a control plane? (Choose two.)

    a. Ethernet over MPLS-TP

    b. Ethernet over MPLS VPWS

    c. Ethernet over MPLS VPLS

    d. Ethernet over PBB-TE

    e. Ethernet over PBB

3. Which one of the following associations is <u>incorrect</u>?

    a. Provider Backbone Bridging – IEEE 802.1ah

    b. Bridging – IEEE 802.1Q–2005

    c. Ethernet Ring Protection Switching – ITU-T G.8032

    d. Ethernet Linear Protection Switching – ITU-T G.8031

    e. Provider Bridging – IEEE 802.1ag

    f. PBB-TE – IEEE 802.1Qay

4. Which two of the following statements describe an advantage of *Provider Bridging* compared to *Bridging*? (Choose two.)

    a. It allows customer VLAN Identifiers to be separated from service provider VLAN Identifiers.

    b. It can coexist in the same network with PBB-TE.

    c. It has unlimited scalability.

d. It supports ENNIs.

e. It can support multipoint-to-multipoint services and MAC learning.

5. Which three of the following technologies can support an EVPL service? (Choose three.)

    a. PB

    b. SONET/SDH

    c. PBB-TE

    d. WDM

    e. MPLS VPLS

    f. OTN

6. Which two of the following protection mechanisms use CCMs? (Choose two.)

    a. ITU-T G.8031, Ethernet linear protection switching

    b. RSTP (Rapid Spanning Tree Protocol)

    c. 1:1 link aggregation, per clause 43 of IEEE 802.3

    d. ITU-T G.8032, Ethernet ring protection switching

    e. FRR (MPLS Fast Reroute)

    f. SONET UPSR protection

7. What is the advantage of MPLS-TP compared to MPLS VPWS?

    a. It provides unlimited scalability.

    b. It supports multipoint-to-multipoint services, such as EVP-LAN and EVP-Tree.

    c. It supports traffic engineered paths, resulting in a highly predictable, connection-oriented service.

    d. It supports both EPL and EVPL services.

8. Which three of the following terms are related to a network's ability to detect and recover from faults without impacting subscribers? (Choose three.)

    a. Bandwidth

    b. Resiliency

    c. G.8031

   d. Scalability

   e. Protection

   f. VLAN awareness

9. Which three of the following technologies can support an Access EPL service? (Choose three.)

   a. PBB-TE

   b. OTN

   c. Bridging

   d. PBB

   e. PB

   f. SONET/SDH

10. Which two of the following technologies support MAC address learning? (Choose two.)

    a. MPLS VPWS

    b. PBB-TE

    c. PBB

    d. MPLS VPLS

    e. OTN

11. Which technology is also known at MAC-in-MAC?

    a. Provider Bridging

    b. MPLS VPLS

    c. Provider Backbone Bridging

    d. PBB-TE

    e. WDM

12. Which transport technology supports the highest bandwidth per optical fiber?

    a. OTN

    b. WDM

c. SONET

d. SDH

13. Which two of the following technologies can support multipoint-to-multipoint Ethernet services? (Choose two.)

    a. MPLS VPWS

    b. MPLS VPLS

    c. MPLS-TP

    d. PBB-TE

    e. PBB

    f. SONET/SDH

14. Which technology is also referred to as Q-in-Q?

    a. Provider Backbone Bridging

    b. OTN

    c. MPLS VPLS

    d. Provider Bridging

    e. PBB-TE

15. Which two of the following technologies support EP-LAN services? (Choose two.)

    a. MPLS-TP

    b. Provider Bridging

    c. PBB-TE

    d. SONET/SDH

    e. MPLS VPLS

16. Provider Backbone Bridging allows a single Carrier Ethernet network to accommodate a total of how many subscribers?

    a. Unlimited

    b. 4094

c. 16 million

d. 16,777,216

17. Which two of the following technologies support multicasting? (Choose two.)

a. SONET/SDH

b. Provider Bridging

c. MPLS-TP

d. MPLS VPLS

e. OTN over WDM

18. Which two of the following technologies can support EP-LAN services? (Choose two.)

a. PB

b. PBB-TE

c. MPLS-TP

d. MPLS VPWS

e. MPLS VPLS

19. Which one of the following technologies cannot support an Access EVPL service?

a. PBB-TE

b. Bridging

c. MPLS-TP

d. MPLS VPWS

Related Links

Answers: Transport Technologies

12
Applications

In this chapter:

This lesson provides a survey of applications that can be supported by MEF services.

Hint: *In preparing for the MEF-CECP 2.0 exam, your first objective should be to become deeply familiar with MEF service definitions in the abstract (the subject of previous lessons) because most MEF-CECP 2.0 exam questions are focused on that material. However, the material in this lesson is also very important because many MEF-CECP 2.0 exam questions are presented in the context of an application, which may be difficult to interpret if you are not familiar with it.*

12.1
Target Applications

In this section:

The following table lists target applications that are referenced in the MEF-CECP 2.0 exam. Become familiar with each application so that, given a scenario, you can choose a Carrier Ethernet solution that meets end-user specifications.

Table 34: Target Applications

Target Application	Description
Wholesale access services	First/last-mile Carrier Ethernet services enabling service providers to reach out-of-franchise (off-net) customer premises
Ethernet access to IP services	Ethernet access to IP-managed IP VPNs
	Ethernet access to Internet
	Ethernet access to cloud services
	Ethernet backhaul of IP video from DSLAM, GPON, or CMTS aggregator
Retail commercial/ business services	Connectivity between enterprise campuses/locations/sites
	Ethernet access to IP services
	Can involve a mixture of many requirements, including: • Multipoint-to-multipoint connectivity between enterprise sites (L3 VPN, L2 VPN) • Hub and spoke connectivity of branch offices with headquarters • Connectivity to partner businesses (supplier, credit agency, etc.) • Ethernet access to IP services
Mobile backhaul services	Interconnection of mobile operator's cell sites and on-net aggregation sites through Carrier Ethernet infrastructure
Support for legacy services	Support for TDM private line services, such as T1/E1 or SONET/SDH, over a Carrier Ethernet infrastructure
	Replacement for Frame Relay services

12.1.1 Wholesale Access Service

A wholesale access service is a first/last-mile service that enables service providers to reach out-of-franchise (off-net) customer premises. Typically, a retail service provider operates a carrier network and makes business agreements with one or more access providers to provide Ethernet connections between the carrier network and customer premises, as required, to support the Carrier Ethernet services that they offer. The MEF defines this type of provider as an Ethernet Access Provider (MEF 33).

The following figure illustrates MEF service implementation involving one retail service provider and two wholesale access providers.

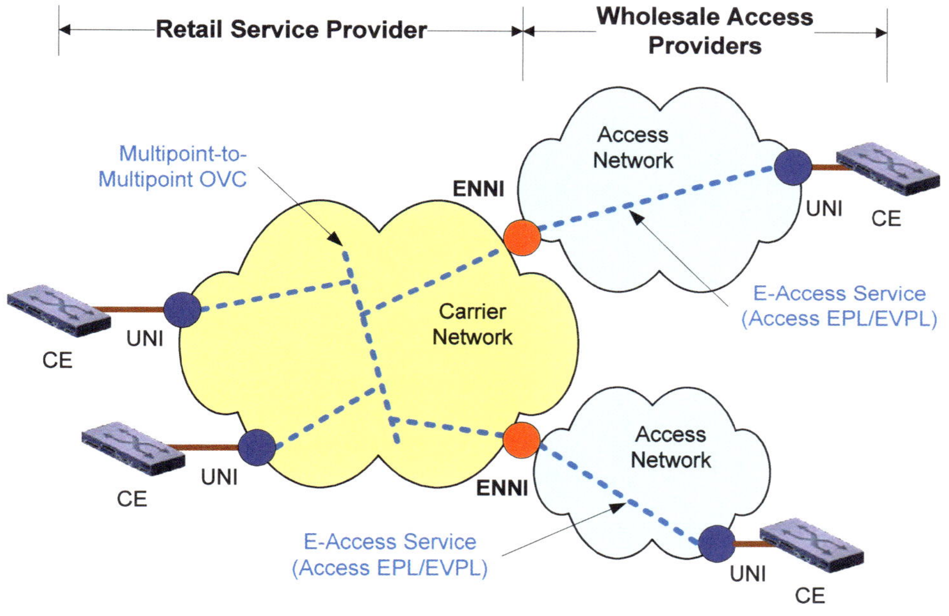

Figure 130: Wholesale Access Providers Supporting a Multipoint-to-Multipoint EVC

In this figure, the wholesale access providers furnish OVC-based E-Access services, as required, to support EVCs offered by the retail service provider. The figure shows only one E-Access service in each access network. Each ENNI would normally support many E-Access services, one for each service that the retail service provider orders from the wholesale access provider.

12.1.2 Ethernet Access to IP Services

Carrier Ethernet services can be used as a method for accessing IP services, such as Hosted Voice over IP (VoIP), the Internet, or IP virtual private networks (VPNs). This access method is analogous to using a T1 private line or Frame Relay. One of the key benefits of using Carrier Ethernet is much higher bandwidth. Ethernet's role in IP access applications is to provide connectivity to the IP service or ISP (Internet Service Provider) point of presence (POP) with appropriate bandwidth and performance attributes.

E-Line, E-Tree, and E-Access service types are appropriate for this application. For E-Access services, handoff to the ISP is through an ENNI. For EVC-based services (E-Line and E-Tree services), handoff is through a UNI.

Note: *Each of the following solutions ensures that subscribers cannot see each other's traffic. Communication between subscribers using applications such as e-mail is supported, but each e-mail must go to the ISP before it is forwarded to the appropriate subscriber.*

Port-Based Solutions

For port-based customer access, three services can be considered: EPL, EP-Tree, and Access EPL.

EPL Support for IP Access

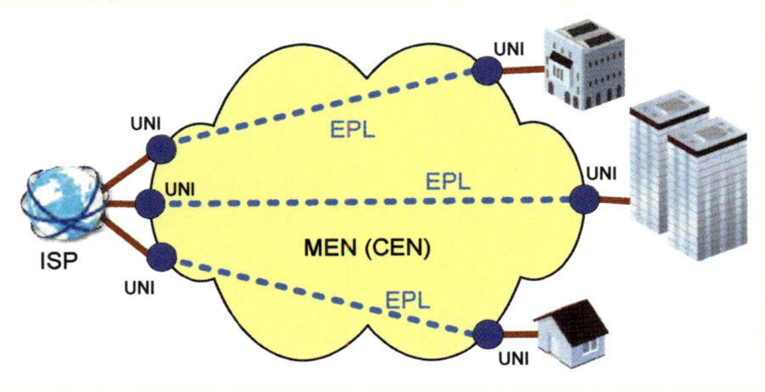

EPL service supports this application, but requires a separate ISP interface (UNI) for each IP service connection.

EP-Tree Support for IP Access

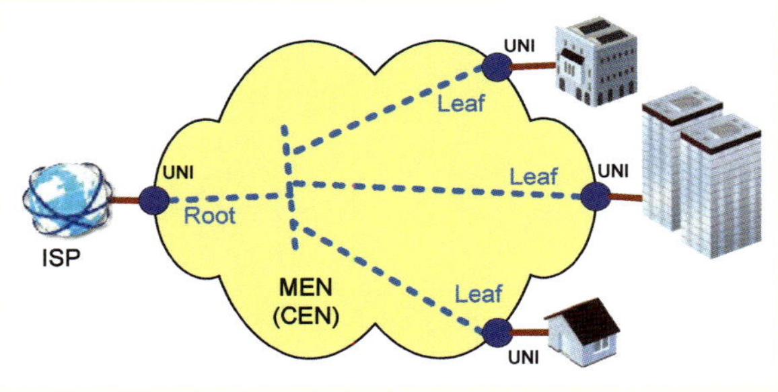

The EP-Tree service supports multiple IP service connections through a single ISP interface (UNI) and provides traffic separation for each connection (leaf-to-root path).

Access EPL Support for IP Access

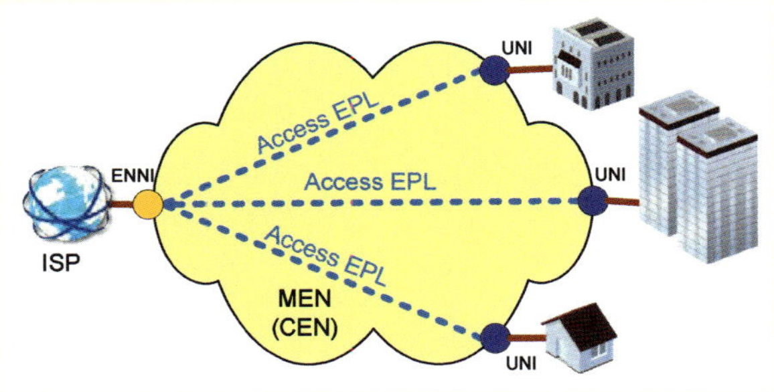

Access-EPL supports this application using one ISP interface (ENNI) to support multiple independent IP service connections. At the ENNI, each Access EPL is identified by an S-VLAN ID.

VLAN-Based Solutions

For VLAN-based customer access, three services can be considered: EVPL, EVP-Tree, and Access EVPL.

EVPL Support for IP Access

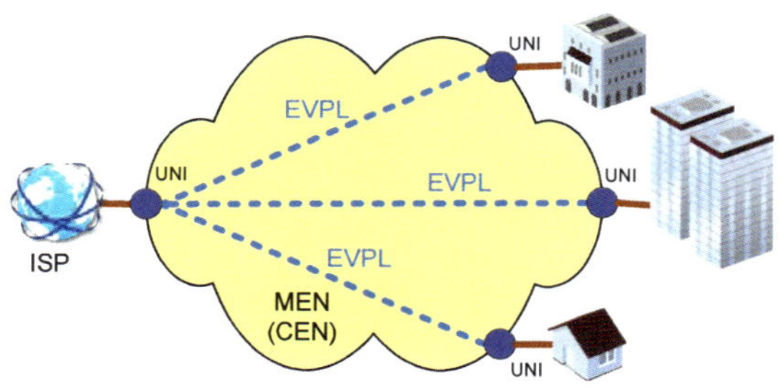

EVPL service supports this application and allows one ISP interface (UNI) to support multiple independent IP service connections. However, coordination is required because each EVPL is identified by CE-VLAN ID (not S-VLAN ID) at the ISP UNI.

Note: *CE-VLAN ID values for EVPLs cannot overlap at the ISP UNI. The values used by each EVPL must be distinct from those used by every other EVPL multiplexing to the ISP UNI.*

EVP-Tree Support for IP Access

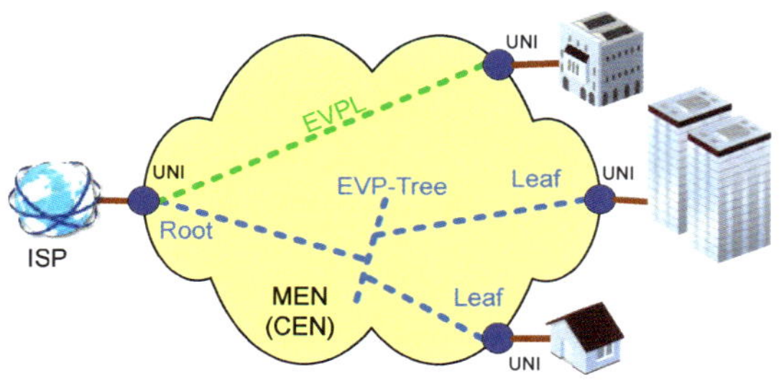

EVP-Tree support for IP services is similar to EP-Tree support for IP services, except that the service is VLAN-based, rather than port-based. Unlike the EP-Tree service, the EVP-Tree service can be multiplexed with other VLAN-based services.

Access EVPL Support for IP Access

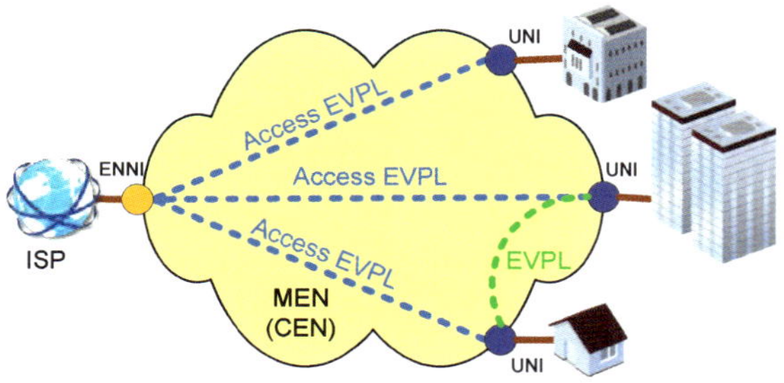

Access EVPL support for IP services is similar to Access EPL support for IP services, except that the service is VLAN-based, rather than port-based. Again, one ISP interface (ENNI) supports multiple independent IP service connections, each identified at the ENNI by an S-VLAN ID. However, as a VLAN-based service, the Access EVPL can be multiplexed with other VLAN-based services at the subscriber UNIs.

Note: *Each Access EVPL can support any set of CE-VLAN ID values at the ENNI. CE-VLAN ID values need not be distinct, as is required for EVPLs multiplexing to an ISP UNI.*

12.1.3 Retail Commercial/Business Services

Carrier Ethernet provides scalable, easily managed solutions for a wide variety of business applications. Requirements vary widely, but often include one of more of the following goals:

- Support for wide area networking (virtual networking) between sites
- Cost reduction by consolidating services (new and/or legacy) over a shared Ethernet-based infrastructure

- Data center storage replication for disaster recovery
- Performance guarantees for certain types of traffic
- Security requirements

Some business applications naturally map to a single MEF service and are straightforward to define. Others can involve multiple requirements that can be satisfied in more than one way by different combinations of MEF services.

Hint: *Deep familiarity with MEF service definitions is your best resource for creating good solutions and judging which one is best.*

The following examples illustrate some Carrier Ethernet solutions for business applications.

Example Business Application 1:

An enterprise has a headquarters and two branch offices and requires:

- Transparent, wide-area networking functionality between all three locations
- Connection to the Internet for all three locations through a firewall located at the headquarters

The following figure shows two potential solutions. Which solution best meets requirements?

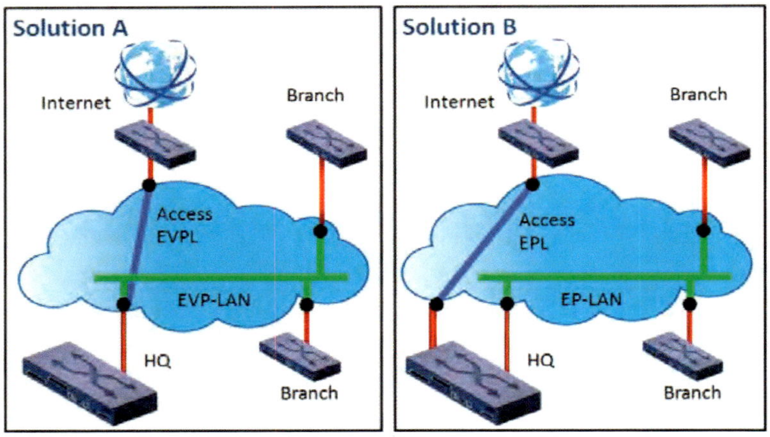

Solution B, shown with the firewall in the following figure, is the better solution for the following reasons:

- Internet connection through a dedicated port (UNI) with firewall provides excellent security.
- The EP-LAN service requires no CE-VLAN ID coordination between service provider and subscriber.
- The EP-LAN service supports transparency, including tunneling of BPDUs for STP.

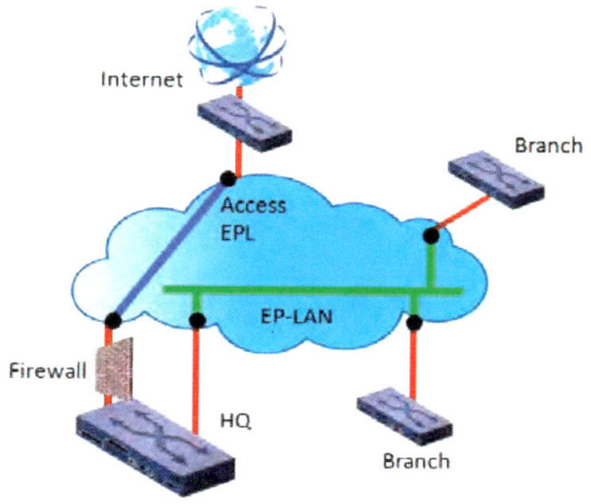

Example Business Application 2

Turbo 2000, an Internet service provider (ISP), uses the following architecture:

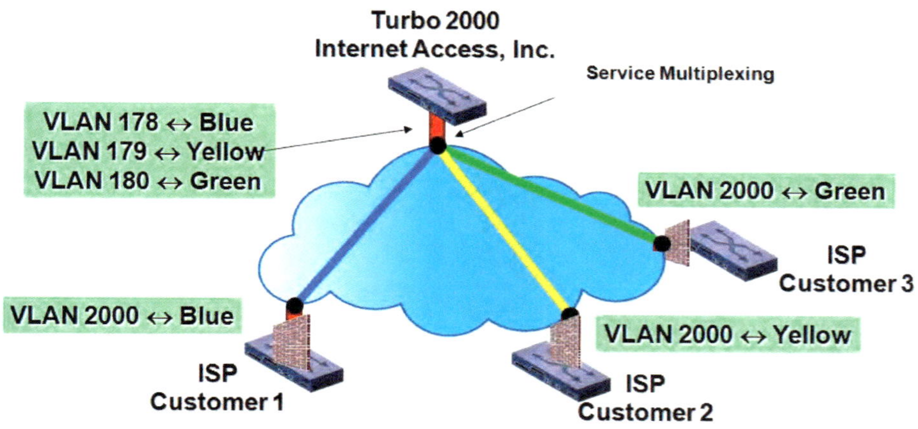

Questions:

1. *Why does Turbo 2000 support customers using EVPLs, rather than EPLs?*
2. *Why is VLAN ID set to 2000 at each customer site?*
3. *What value is required for EVC service attribute CE-VLAN ID Preservation?*
4. *Does this architecture allow customers to use the same port (UNI) for other services?*

Answers:
1. *So that Turbo 2000 (using service multiplexing) can support multiple customers with a single router port (UNI), rather than one router port (UNI) per customer.*
2. *To standardize configuration of customer sites.*
3. *CE-VLAN ID Preservation must be No to allow translation of the CE-VLAN ID.*
4. *Yes, this architecture allows service multiplexing at customer UNIs.*

Alternatively, Turbo 2000 might support customers using a single EVP-Tree service. In this case, the Turbo 2000 UNI would be designated as the only root, and each customer UNI would be designated as a leaf (to ensure that customers cannot see each other's traffic). This EVP-Tree service option would offer improved scalability (one service, rather than many EVPLs). However, it would be more complex to implement and would share bandwidth between subscribers, potentially allowing customers to impact each other's quality of service.

Example Business Application 3

Credit Check Inc., uses the following architecture to offer a high-availability credit checking service to customers:

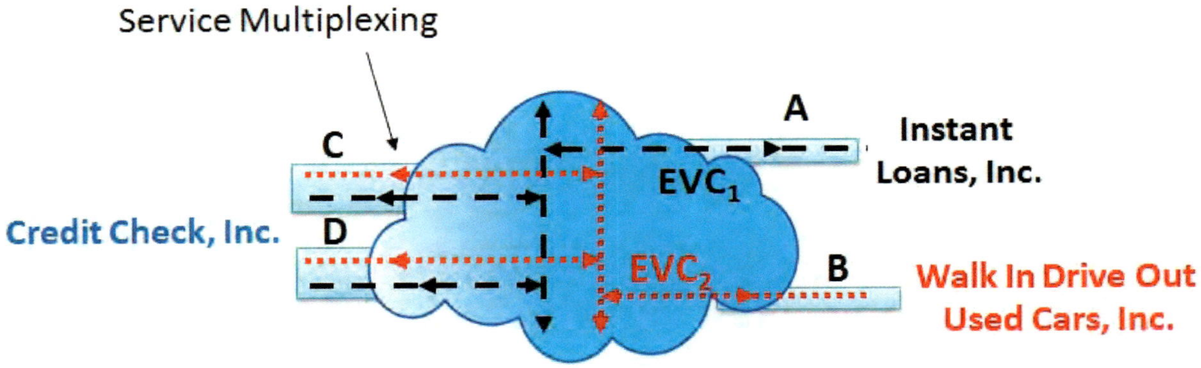

Questions:
1. *What type of MEF service are EVC1 and EVC2?*
2. *How does this architecture support high-availability for Credit Check Inc.?*
3. *How does this architecture make efficient use of router ports at Credit Check Inc.?*
4. *Does this architecture provide security for customers?*

Answers:
1. *Both are EVP-LAN services.*
2. *Each EVC provides redundant points of access to Credit Check, Inc.*
3. *EVP-LANs allow service multiplexing at both routers (C and D).*
4. *Yes, customers cannot see each other's traffic.*

12.1.4
Mobile Backhaul

Mobile communication (cellular/wireless communication) involves connections through several networks.

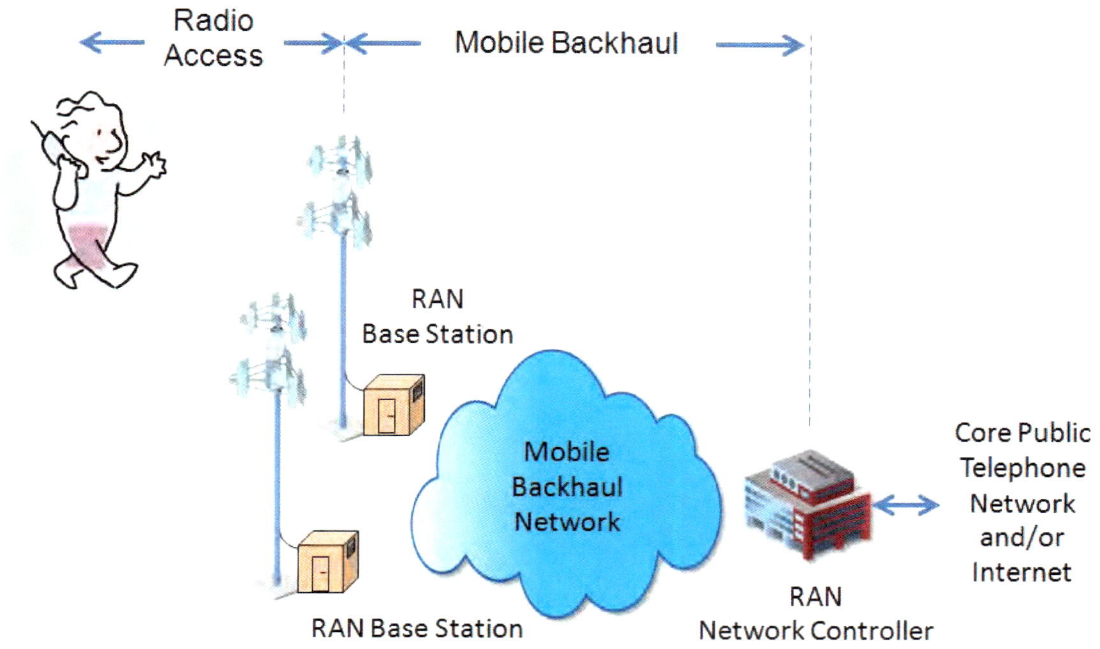

The first link, by radio wave, is between the smart phone (or similar device) and the cell tower. The second link is through a mobile backhaul network connecting one or more cell towers (RAN base stations) to a RAN network controller. The RAN network controller, in turn, provides a gateway to the core public phone network and/or Internet.

Mobile backhaul refers to the network that interconnects RAN base stations with RAN network controllers.

A **mobile backhaul service** is something that an access provider offers to a mobile/cellular service provider (also known as a wireless operator). It supports numerous end users. It is not offered to individual end users (cell phone users).

Mobile backhaul services are described in more detail later in this lesson. For now just recognize the role of mobile backhaul in cellular/wireless communication.

Related Links

Mobile Backhaul Services

12.2
Comparing and Positioning Carrier Ethernet Services with Legacy Services

In this section:

12.2.1
Support for TDM Private Lines

Time division multiplexing (TDM) services have dominated voice and data telecommunications for decades.

Table 35: Example Carrier Signals for TDM Services

System	Signal Type (approximate bandwidth)	
	US and Canada	Rest of the World
PDH (Plesiochronous Digital Hierarchy)	T1 (1.5 Mb/s)	E1 (2.0 Mb/s)
	T3 (45 Mb/s)	E3 (34 Mb/s)
SONET/SDH	OC-3 (150 Mb/s)	STM-1 (150 Mb/s)
	OC-12 (600 Mb/s)	STM-4 (600 Mb/s)

Traditionally, telecommunication service providers have supported TDM services through a TDM network, as shown by example in the following figure.

Note: *DS designations (DS1 and DS3) are used in connection with T1 and T3. Strictly speaking, a DS1/DS3 is the data carried on a T1/T3 circuit, but in practice the terms are used interchangeably.*

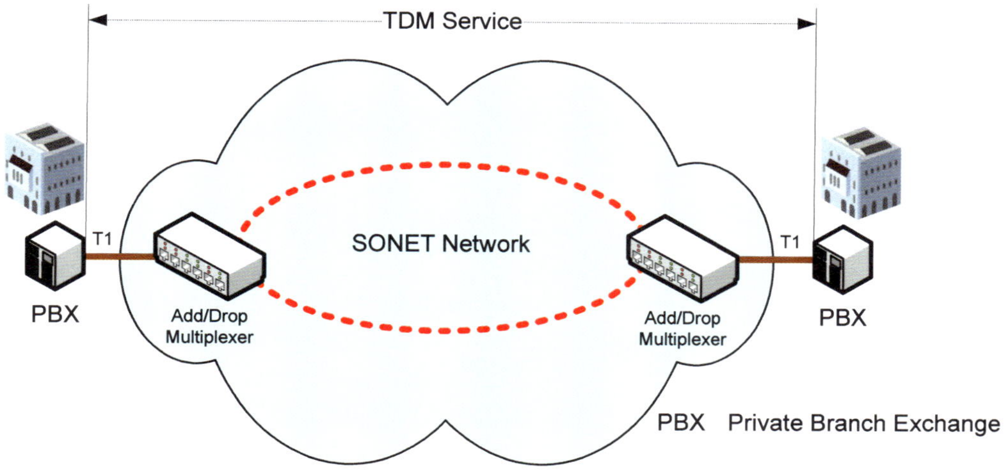

Figure 131: Example PBX Service Connection through a SONET Network

In this example, the service provider uses a SONET network to support a T1 private line between two PBXs (private branch exchanges), which are owned and operated by the subscriber.

The following figure shows the Carrier Ethernet solution for this service.

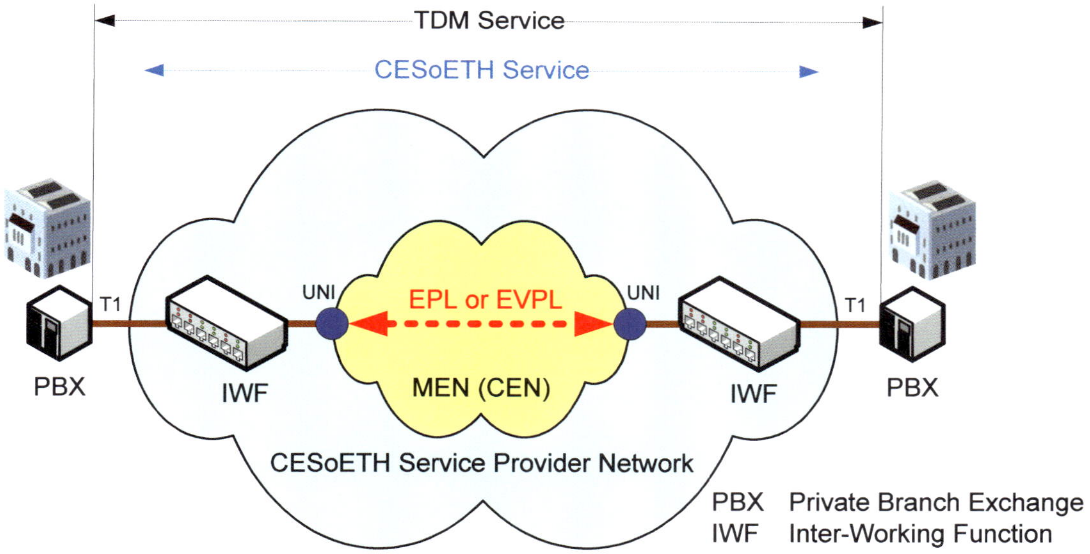

Figure 132: Example PBX Service Connection through Carrier Ethernet Network

This solution includes two parts:

- A point-to-point service (EPL or EVPL) providing Ethernet transport through the MEN (CEN)
- Two IWFs (Inter-Working Functions) located on both ends (to mediate between the TDM network and the Carrier Ethernet network)

The MEF calls this combination of technologies **Circuit Emulation Service over Ethernet (CESoETH)**.

Requirements for CESoETH are presented elsewhere in this study guide. For now, just recognize that the Carrier Ethernet service (EPL or EVPL) is subsumed within the CESoETH service. In the TDM domain, CESoE provides a service that is equivalent to a switched-circuit TDM service, but the service is implemented by "tunneling" TDM traffic through the Carrier Ethernet Network. The EPL/EVPL service functions like a virtual wire between IWFs within the CESoETH service.

Overall, CESoE enables TDM services to be transported across a Carrier Ethernet network, recreating the TDM circuit in the TDM network, as shown in the following figure.

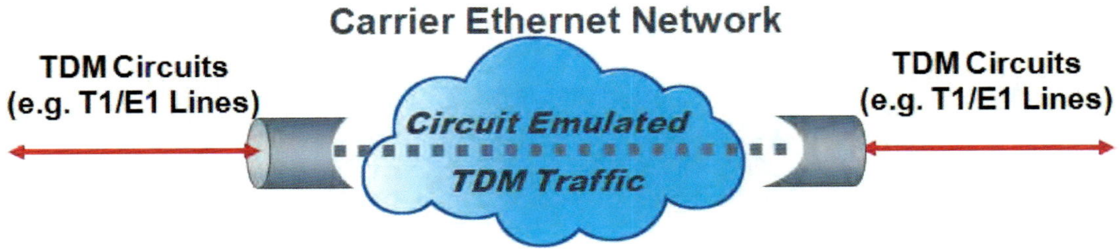

Related Links

Circuit Emulation Services over Ethernet

12.2.2 Replacement of Frame Relay Service

Frame Relay services are widely deployed by telecommunications service providers to support wide area networking (WAN).

Frame Relay

The following figure shows an example deployment of Frame Relay service equipment.

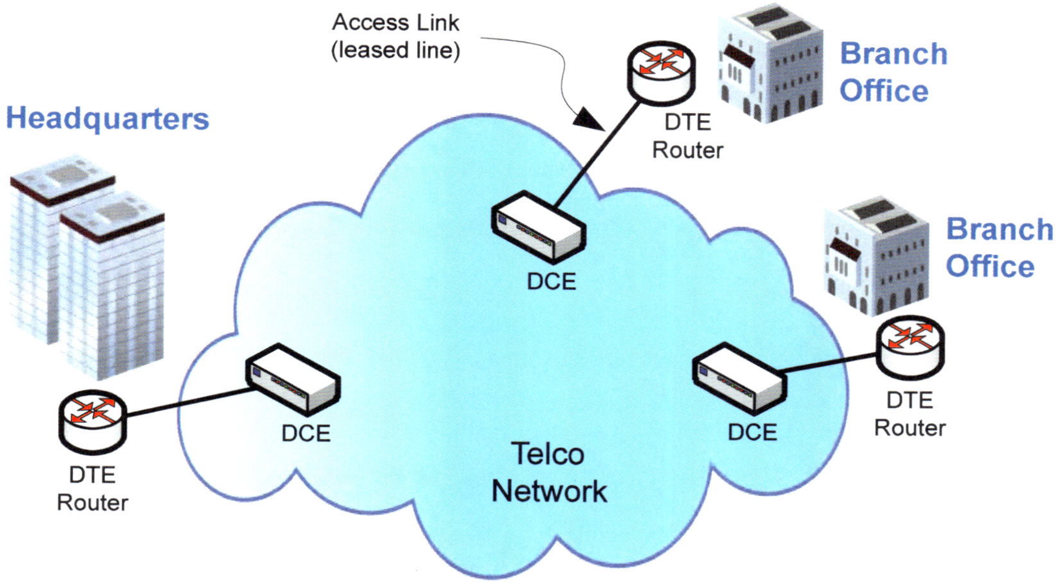

The following equipment is involved:

- **DTE (data terminal equipment)**–Frame Relay router located at the subscriber site
- **DCE (data communications equipment)**–Frame Relay switch located in the provider network

The access links, connecting DTEs to DCEs, are standard leased lines (for example, T1/ E1 service lines).

After DTE routers are deployed and connected, WAN connectivity between sites is established using **PVCs (permanent virtual circuits)**. PVCs are point-to-point virtual connections between sites that the subscriber purchases as needed.

The subscriber in this example might purchase two PVCs, creating a hub-and-spoke WAN topology:

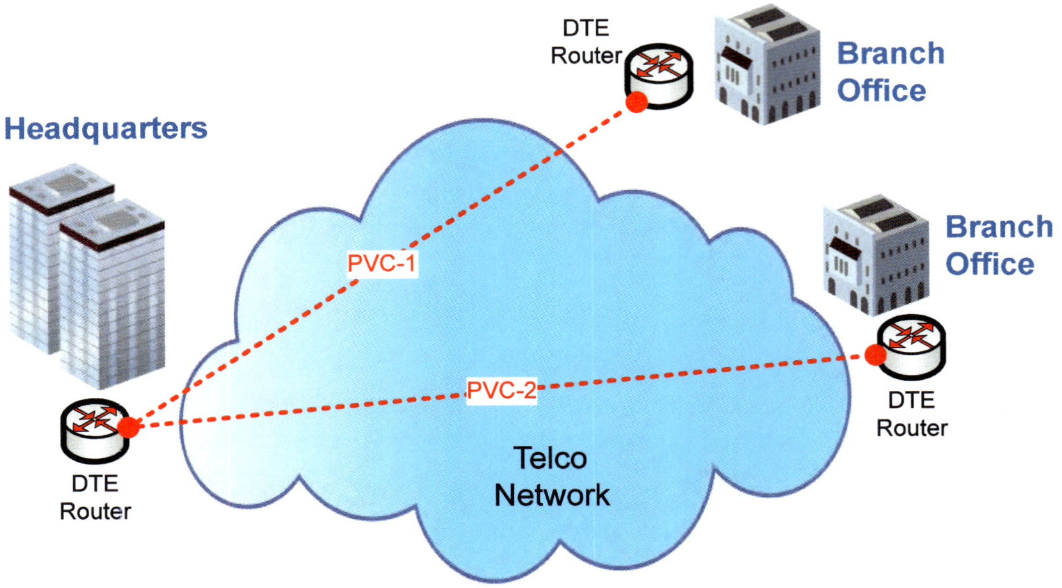

Or, the subscriber might purchase three PVCs, creating a fully meshed WAN topology:

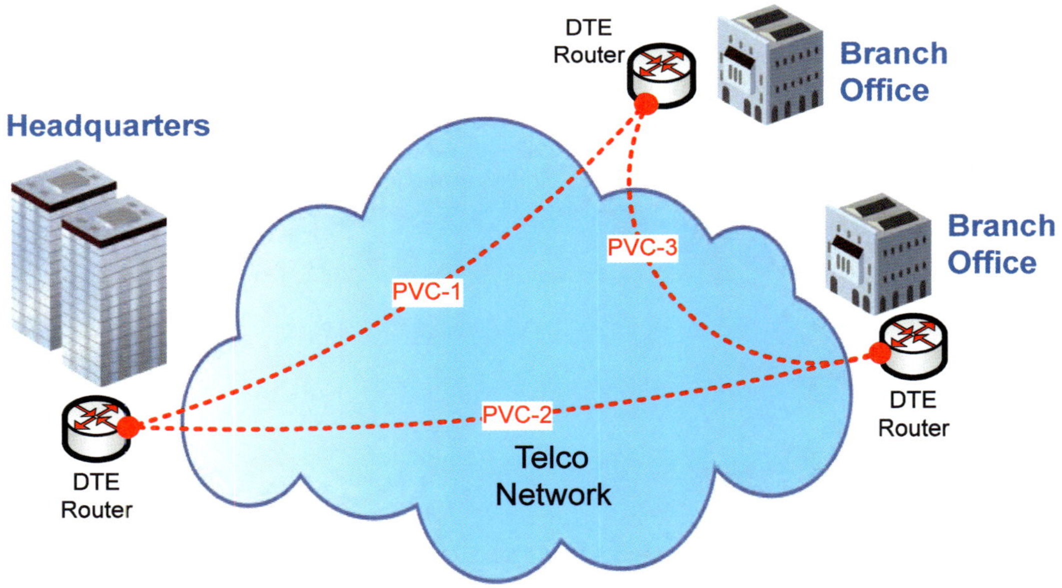

After PVCs are established, the subscriber provisions the DTE routers to associate service frames to PVCs. DTE routers insert a Frame Relay header into each service frame, and that header includes a **DLCI (data link**

connection identifier) that binds the service frame to a PVC. The DCE switches service frames based on the DLCI value.

A key benefit of Frame Relay over point-to-point leased lines (TDM private lines) is Frame Relay's ability to support multiple point-to-point links (PVCs) on a single interface (DTE router port).

From this description, do you see similarity between Frame Relay and MEF EVPL services?

Frame Relay Replacement by EVPLs

To make the analogy more explicit, compare the previous figures with the following MEF service figures.

The deployment of equipment looks similar, but different equipment is used:

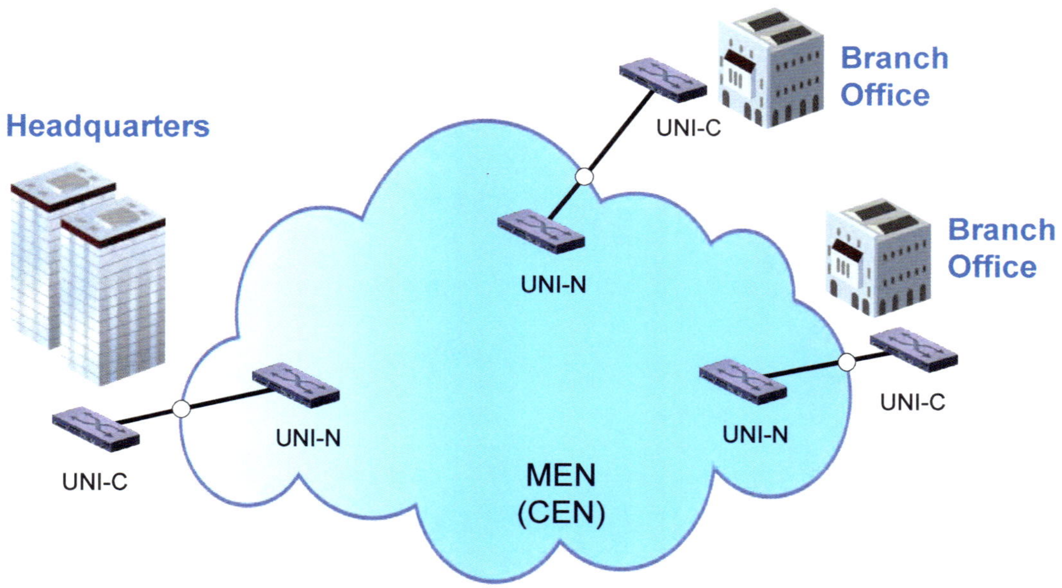

The UNI-C and UNI-N are Ethernet based and connect through an Ethernet link (rather than through a TDM leased line).

After the UNI-C and UNI-N equipment are deployed and connected, WAN connectivity is established using EVPL services that the subscriber purchases as needed (similar to Frame Relay PVCs).

The subscriber in this example might purchase two EVPLs, creating a hub-and-spoke WAN topology:

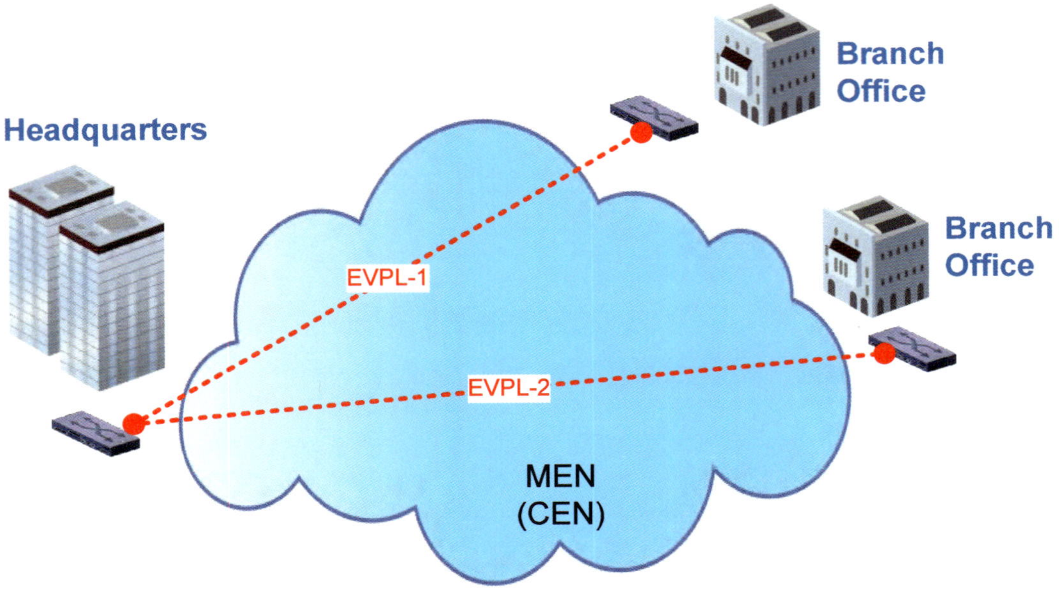

Or, the subscriber might purchase three EVPLs, creating a fully meshed WAN topology:

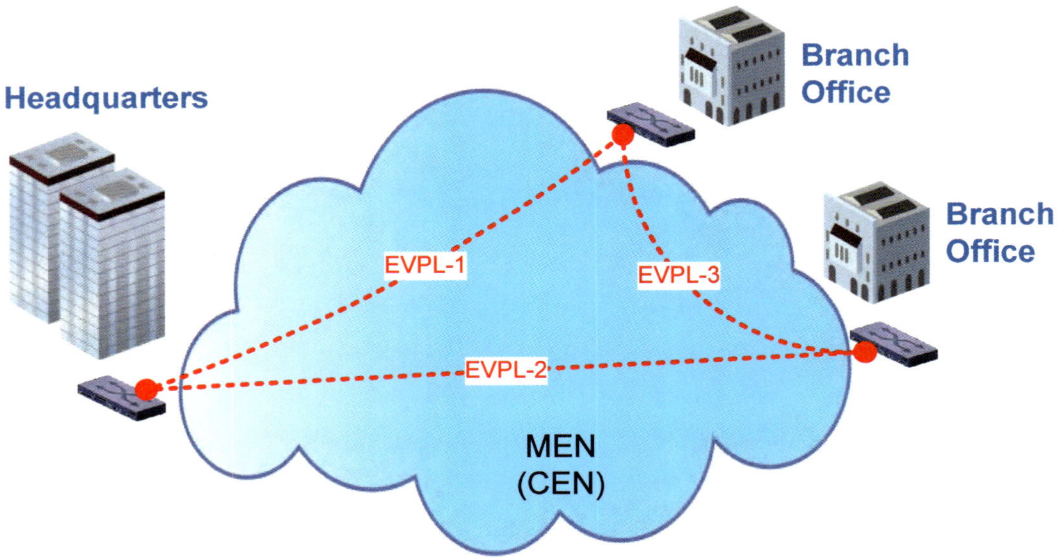

After EVPLs are established, the subscriber provisions the UNI-Cs (bridges or routers) to associate service frames on the customer premises equipment to EVPLs. The UNI-C inserts a CE-VLAN tag into the Ethernet header of each service frame. The CE-VLAN ID within the CE-VLAN tag binds the service frame to a specific EVPL.

Think again about the similarity between Frame Relay and MEF EVPL services:

- Both provide point-to-point connections.
- Both permit a single interface to support multiple point-to-point connections (unlike TDM private line services).

Frame Relay Replacement by an Ethernet Tree or Ethernet LAN Services

Although EVPL services closely resemble Frame Relay services, they are not necessarily the best option for service replacement. Multipoint-to-multipoint Ethernet Tree and LAN services should also be considered, especially for the support of multipoint services.

For example, if the service objective is to provide fully meshed WAN connectivity between three or more sites, an EP-LAN or EVP-LAN service may be better solution.

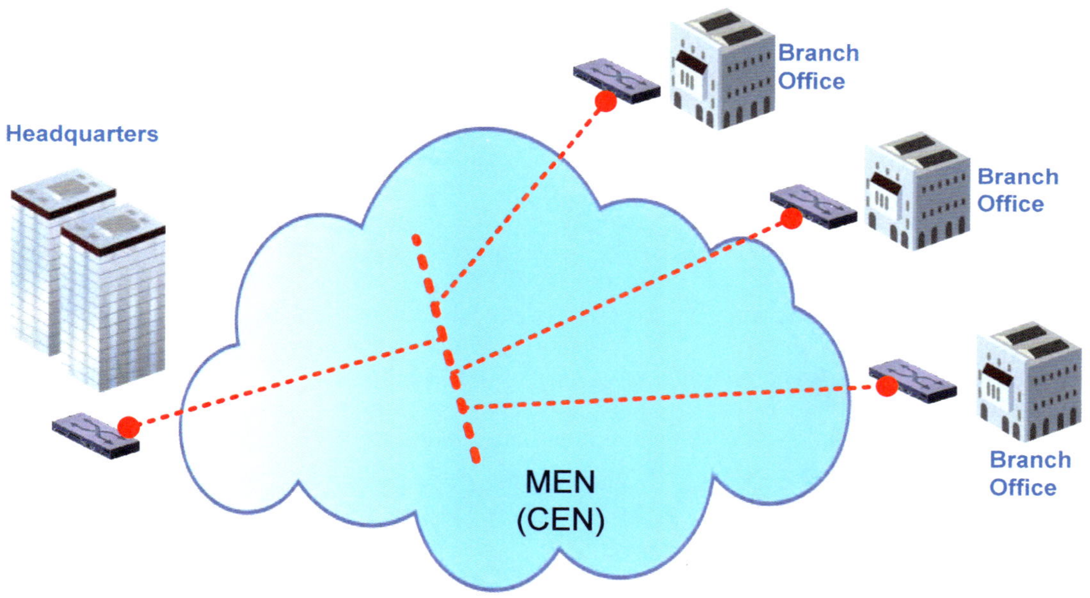

Figure 133: EP-LAN or EVP-LAN Service

Potential advantages include:

- EP-LAN and EVP-LAN services connect all participating sites without explicit point-to-point provisioning.

  Note: *Complete meshing of four sites would require six EVPLs (or six Frame Relay PVCs). Complete meshing of N sites would require N*(N-1)/2 EVPLs (or the same number of Frame Relay PVCs).*

- EP-LAN service requires no coordination of CE-VLAN IDs between the service provider and the subscriber.
- EP-LAN service allows tunneling of Layer 2 control protocol frames (including STP BPDUs).

If the application also requires some sites to be isolated from each other (as, for example, in a hub-and-spoke deployment of Frame Relay service), the EP-Tree or EVP-Tree service may be best.

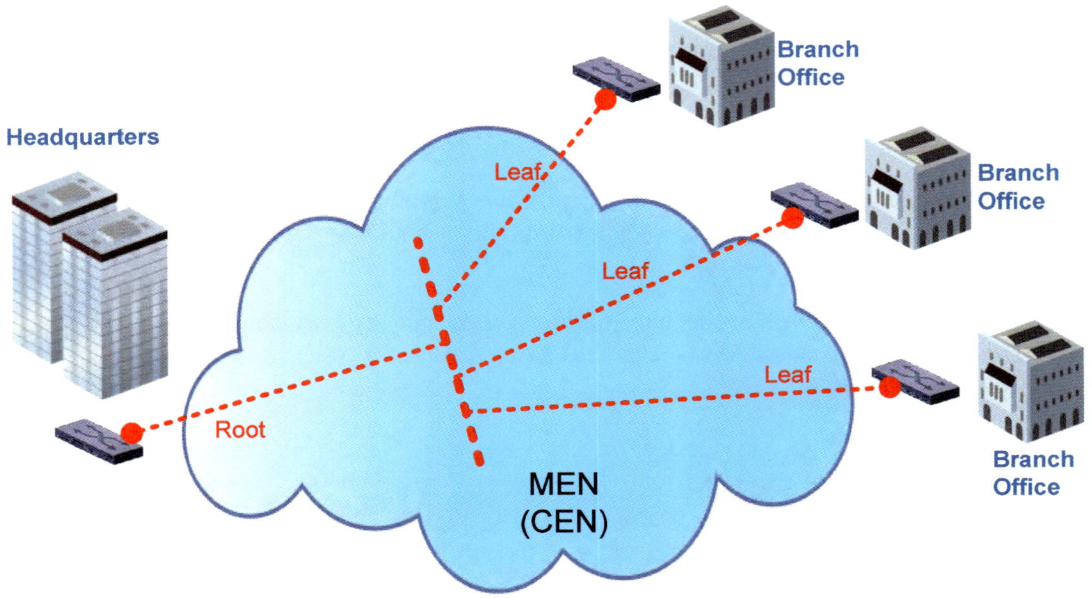

Figure 134: EP-Tree or EVP-Tree Service

Ethernet LAN and Tree services can also simplify subscriber-side provisioning compared to meshed EVPL services:

- Service frames are forwarded based on MAC learning, rather than CE-VLAN IDs.
- Customer-edge equipment (CE) does not have to be provisioned to assign CE-VLAN IDs, as required, to bind service frames to EVPLs.

12.2.3
Internet Access

TDM private lines, Frame Relay services, and other legacy methods for accessing the Internet can be replaced by Carrier Ethernet services, such as EVPL or E-Access services.

EVPL Support for Internet Access

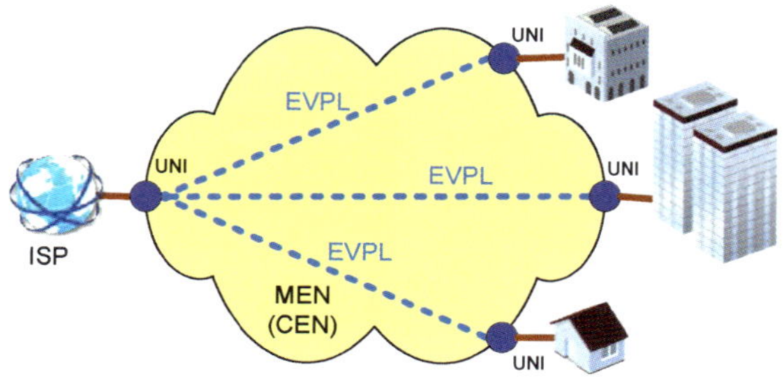

EVPL service supports this application and allows one ISP interface (UNI) to support multiple independent IP service connections. However, coordination is required because each EVPL is identified by CE-VLAN ID (not S-VLAN ID) at the ISP UNI.

Note: *CE-VLAN ID values for EVPLs cannot overlap at the ISP UNI. The values used by each EVPL must be distinct from those used by every other EVPL multiplexing to the ISP UNI.*

E-Access Support for Internet Access

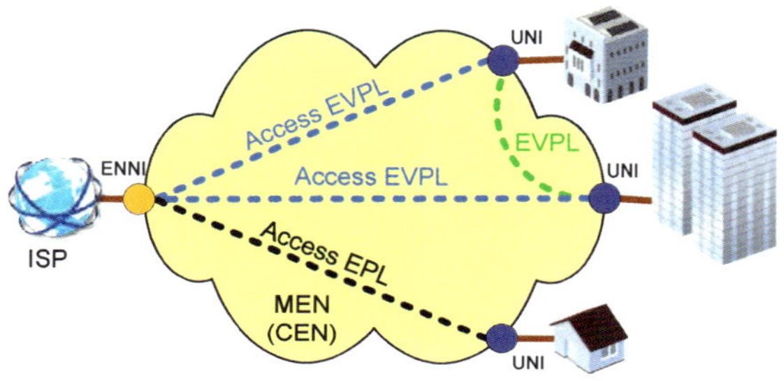

Access-EPL and Access-EVPL services support this application using one ISP interface (ENNI) to support multiple independent IP service connections. At the ENNI, each Access EPL/EVPL is identified by S-VLAN ID.

Note: *Each Access EVPL can support any set of CE-VLAN ID values at the ENNI. CE-VLAN ID values need not be distinct, as is required for EVPLs multiplexing to an ISP UNI.*

A key benefit of these solutions over TDM private lines is their ability to support multiple point-to-point connections at the shared interface between networks (UNI or ENNI). Frame Relay also has this ability, but at lower bandwidths compared to Carrier Ethernet service.

To support the E-Access solutions, the Ethernet service provider and the ISP must both be able to support S-VLAN tagging at the shared interface (ENNI). The EVPL solution requires support for C-VLAN tagging at the shared interface (UNI):

- A UNI supports multiple EVPLs (distinguished by CE-VLAN ID at the UNI).
- An ENNI supports multiple Access EPL/EVPLs (distinguished by S-VLAN ID at the ENNI).

Related Links

Ethernet Access to IP Services

12.2.4 Virtual Private Networks (VPNs)

A **virtual private network (VPN)** is a network formed by interconnecting two or more (geographically separate) networks into a single network through a public or third-party communication infrastructure.

If each network connects through an Ethernet bridge (Layer 2), the VPN is called a Layer 2 VPN.

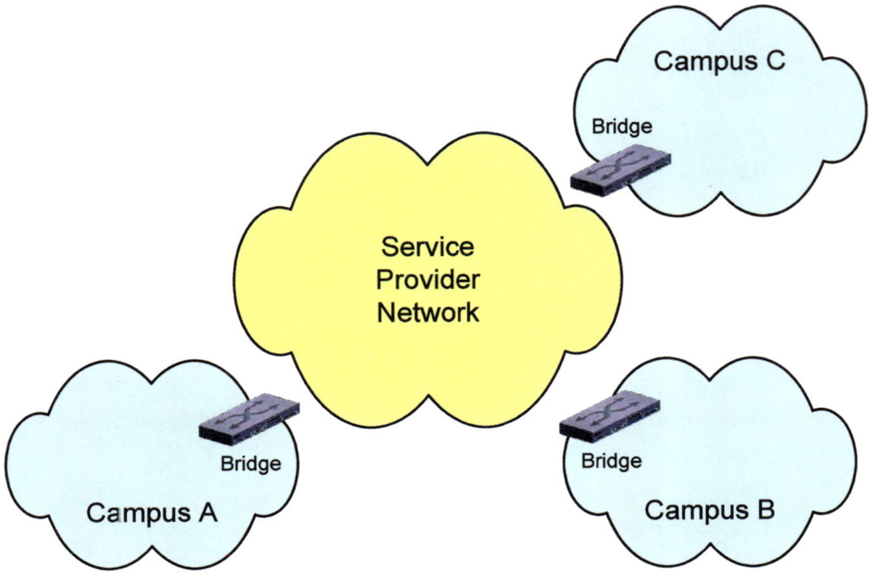

Figure 135: Layer 2 VPN

If each network connects through a router (Layer 3), the VPN is called a Layer 3 VPN.

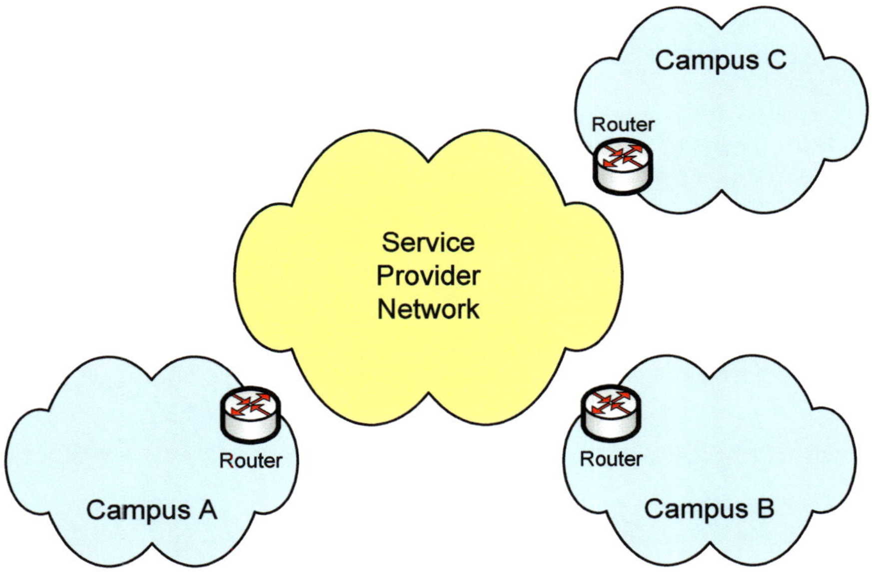

Figure 136: Layer 3 VPN

Service Provider Options for VPN Connectivity

For a Layer 2 VPN, the service provider can only offer connectivity at Layer 2:

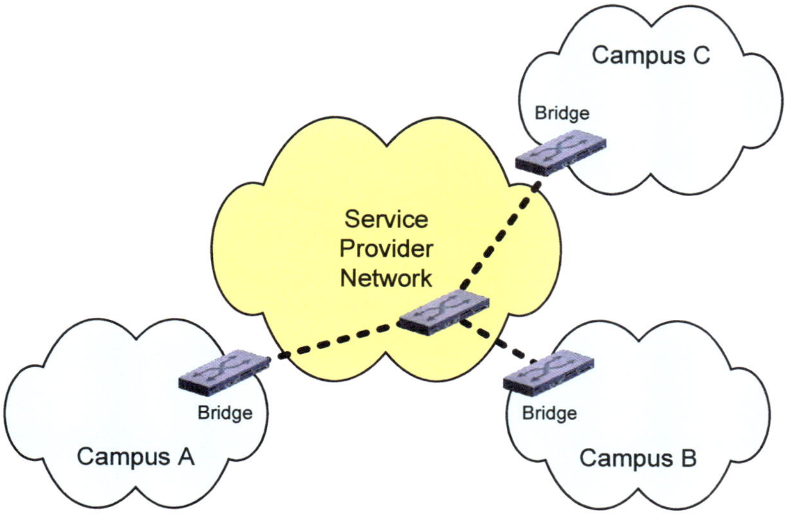

Figure 137: Layer 2 VPN with Layer 2 Connectivity

However, for a Layer 3 VPN, the service provider can offer connectivity at Layer 2:

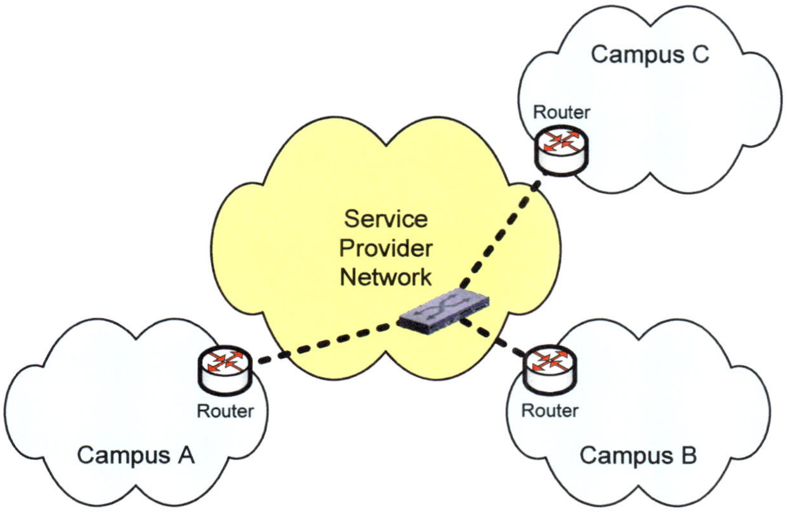

Figure 138: Layer 3 VPN with Layer 2 Connectivity

Or, the service provider can offer connectivity at Layer 3:

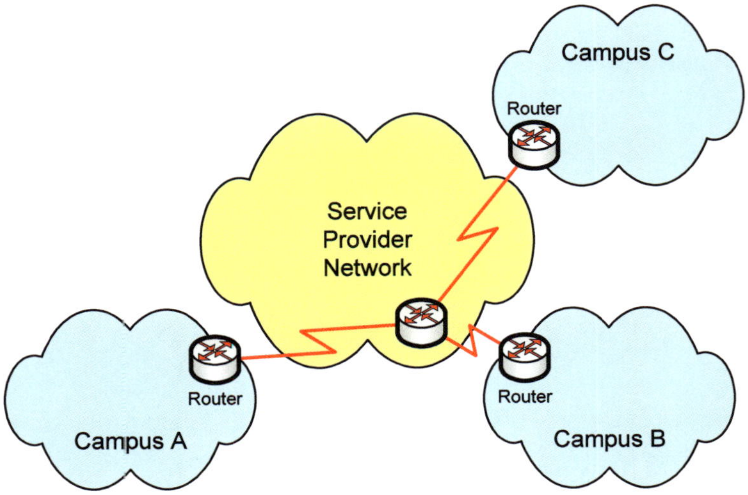

Figure 139: Layer 3 VPN with Layer 3 Connectivity

The last option (Layer 3 connectivity) is not considered further (going forward in this study guide) because Carrier Ethernet services do not support it (nor do Frame Relay or TDM Private Line services). It is included here for contextual completeness. If an enterprise uses this type of service, it must coordinate IP addressing within their network with the service provider.

Building a Layer 3 VPN with Layer 2 Service

An enterprise with three campuses wants to build a Layer 3 VPN with Layer 2 connectivity between the campuses.

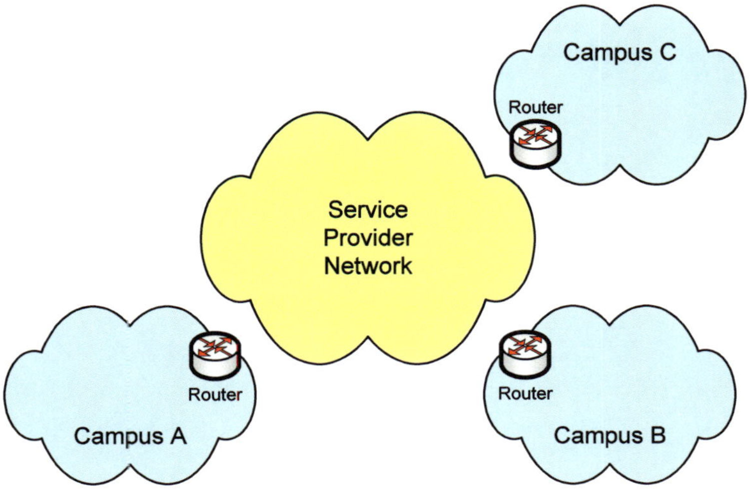

Figure 140: Layer 3 VPN

A service provider can offer a variety of Layer 2 services to support this application:

- Full mesh of Frame Relay PVCs
- Full mesh of EVPLs
- Full mesh of EPLs
- EP-LAN service
- EVP-LAN service

Each of these solutions involves only Layer 2 services: Frame Relay is a Layer 2 service as are all Carrier Ethernet services.

Note: *Because routers terminate STP bridging domains, Layer 3 VPN applications never require the service provider to tunnel BPDUs.*

Building a Layer 2 VPN with Layer 2 Service

The following figures show an enterprise that includes three campuses. Each campus has a managed network. The enterprise wants to build a Layer 2 VPN to interconnect the three campuses.

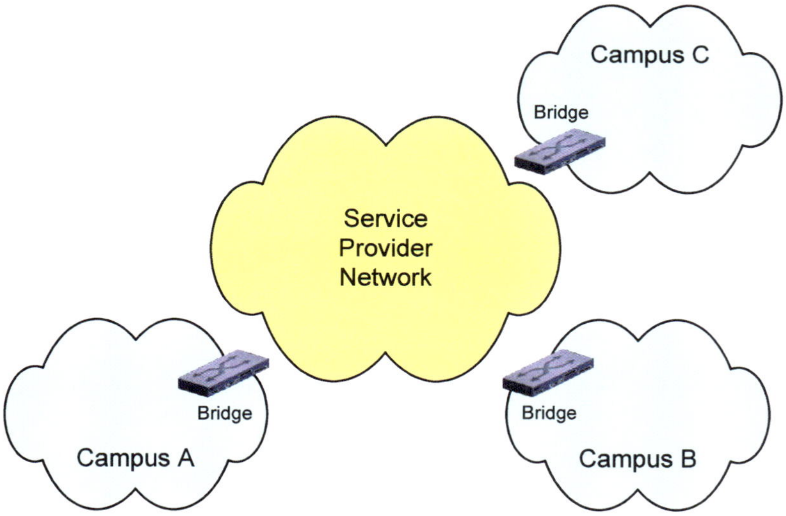

Figure 141: Layer 2 VPN

Frame Relay cannot be used in this application because the campuses interconnect through bridges (not routers).

Interconnection can be accomplished using the following Carrier Ethernet solutions:

- Full mesh of EVPLs
- Full mesh of EPLs
- EP-LAN
- EVP-LAN service

If transparent support of BPDUs is required, the port-based solutions (EPL and EP-LAN) are best.

The following figure illustrates a full mesh of EPLs.

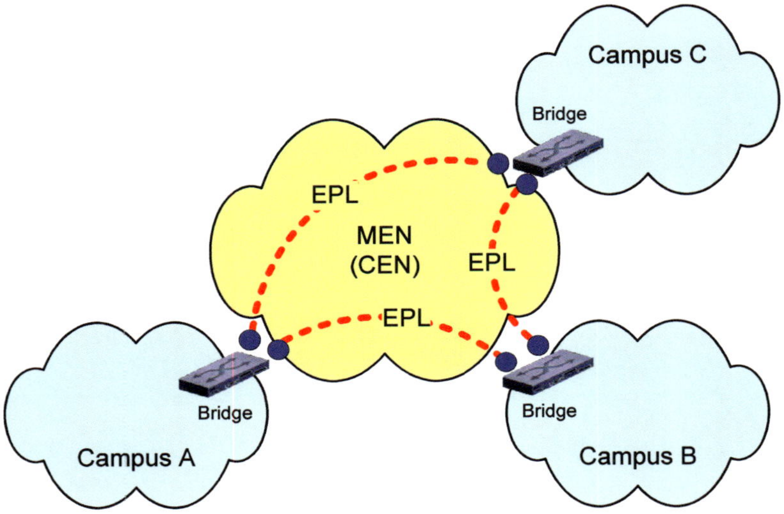

Figure 142: Layer 2 VPN Implemented with Full Mesh of EPL Services

12.3
Mobile Backhaul Services

In this section:

Mobile communication (cellular/wireless communication) involves connections through several networks.

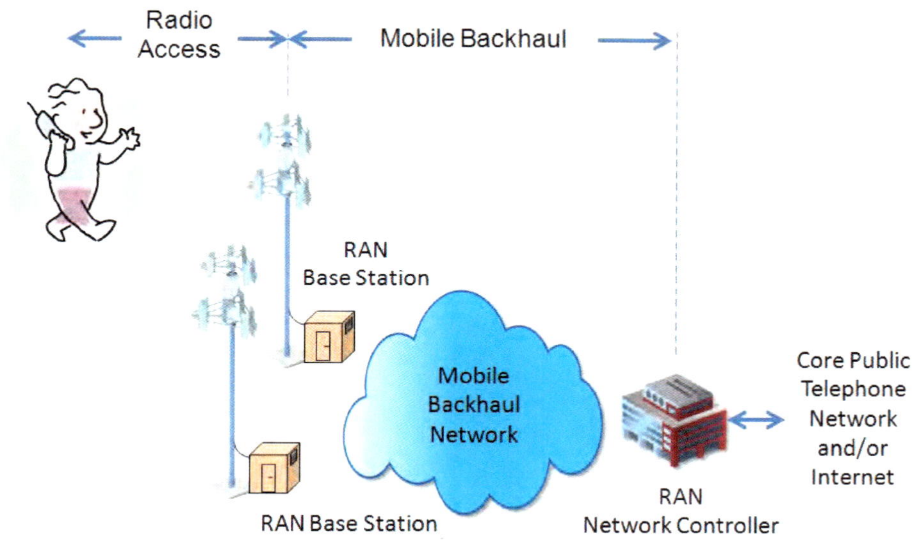

The first link, by radio wave, is between a smart phone (or similar device) and the cell tower. The second link is through a mobile backhaul network connecting one or more cell towers (RAN base stations) to a RAN network controller. The RAN network controller, in turn, provides a gateway to the core public phone network and/or Internet.

Mobile backhaul refers to the network that interconnects RAN base stations with RAN network controllers.

A **mobile backhaul service** is something that an access provider offers to a mobile/cellular service provider (also known as a wireless operator). It supports numerous end users. It is not offered to individual end users (cell phone users).

Mobile backhaul networks have traditionally used TDM and ATM technologies, but the latest generation of equipment and networks is based on Ethernet.

Table 36: Mobile Backhaul Technologies

Generation	Technologies
2G	TDM
3G	ATM, Ethernet (GSMA/HSPA, CDMA/EVDO)
4G	IP, Ethernet (LTE, WiMAX)

This trend toward Ethernet technology makes Carrier Ethernet a natural fit for mobile backhaul applications going forward.

MEF 22.1, Mobile Backhaul Phase 2 Implementation Agreement specifies requirements for mobile backhaul.

12.3.1 Mobile Backhaul Reference Model

MEF 22.1 uses the following basic reference model and terminology:

Reproduced with permission of the Metro Ethernet Forum

Term		Description
RAN	Radio Access Network	The wireless access network, including the mobile backhaul network connecting RAN CEs
RAN CE	RAN Customer Edge	A generic term that identifies a mobile network node/site (RAN NC or RAN BS)
RAN BS	RAN Base Station	A cell tower/site
RAN NC	RAN Network Controller	A node that aggregates traffic from multiple base stations and provides a gateway to the public phone network and/or Internet

Similar to the MEF reference model for EVC-based service, this reference model involves two stakeholders:

- **Service Provider**–The organization providing the mobile backhaul service
- **Mobile Operator (Subscriber/Customer)**–The organization purchasing the mobile backhaul service

However, the mobile backhaul service is not required to be purely Carrier Ethernet:

- The mobile backhaul service may involve a TDM-based (legacy) component operating in parallel with a Carrier Ethernet component.
- The Carrier Ethernet service component may connect indirectly to the RAN CE through a Generic Inter-working Function (GIWF).

For this reason, the previous figure does not include UNIs and does not indicate that the service provider network is a MEN (CEN).

12.3.2 Mobile Backhaul Use Cases

MEF 22.1 presents four uses cases to illustrate how Carrier Ethernet services can be applied in mobile backhaul applications.

Table 37: Mobile Backhaul Use Cases

Use Case	RAN CE Interface Type	In-Parallel Deployment with Legacy Network
Use Case 1a 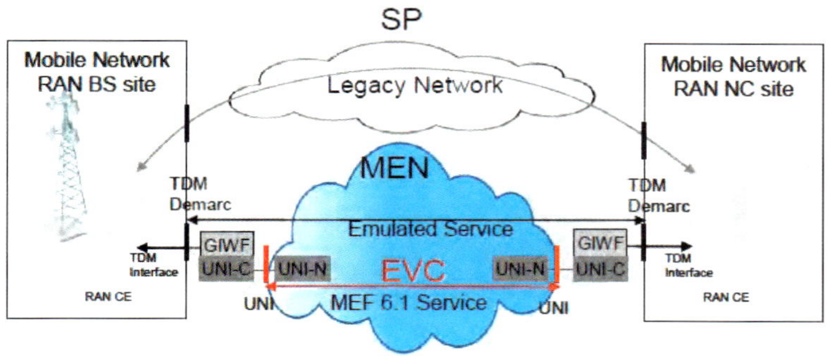 Reproduced with permission of the Metro Ethernet Forum	Non-Ethernet	Yes

Table 37: Mobile Backhaul Use Cases (Cont.)

Use Case	RAN CE Interface Type	In-Parallel Deployment with Legacy Network
Use Case 1b 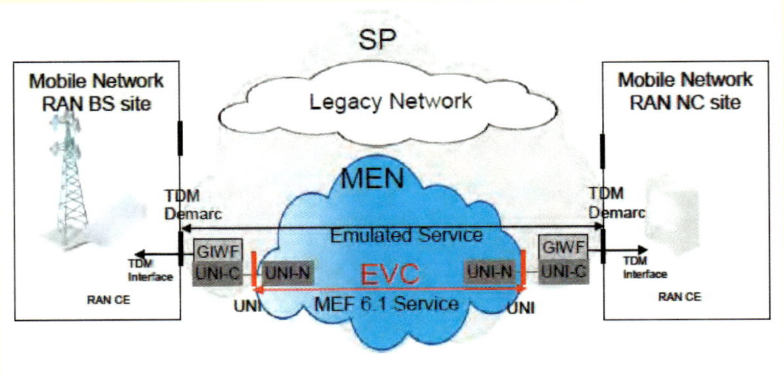 Reproduced with permission of the Metro Ethernet Forum	Non-Ethernet	No
Use Case 2a 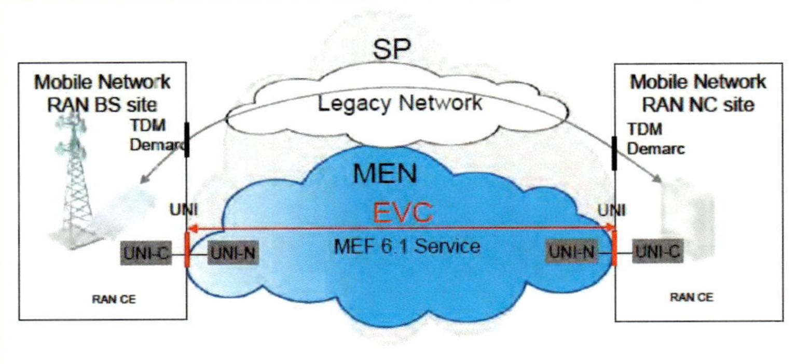 Reproduced with permission of the Metro Ethernet Forum	Ethernet	Yes
Use Case 2b 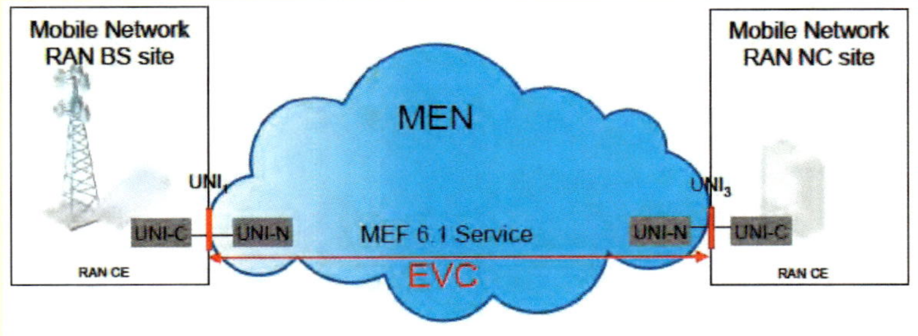 Reproduced with permission of the Metro Ethernet Forum	Ethernet	No

These four use cases outline a path of evolution from in-parallel deployment with legacy circuit-based equipment to solo deployment with Ethernet-based equipment.

In Use Cases 1a and 1b, the RAN CEs cannot be directly connected to UNIs because they have non-Ethernet–based interfaces, such as ATM or TDM. In these cases, the RAN CE connects to what the MEF calls a Generic Inter-Working Function (GIWF), which in turn connects to the UNI. Use Cases 2a and 2b illustrate deployments with RAN CE equipment that can be connected directly to the UNI through an Ethernet interface, eliminating the need for a GIWF.

Note: *The GIWFs encapsulate synchronous TDM circuits into Ethernet frames for transport across the asynchronous MEN (CEN) using Circuit Emulation Services over Ethernet (CESoETH). CESoETH is explained elsewhere in this study guide.*

Use Cases 1a and 2a show split access scenarios where the MEN (CEN) is deployed in parallel with a legacy network. These two use cases are used when an operator wants to continue to support low-bandwidth voice traffic with an existing legacy network but offload high-bandwidth data traffic to the MEN (CEN), thereby reducing demand on the legacy network. In Use Cases 1b and 2b, the MEN (CEN) supports all traffic between RAN CEs.

Related Links

Circuit Emulation Services over Ethernet

12.3.3 Mobile Backhaul for LTE

Note: *The initial MEF-CECP exam did not include this material. It was introduced in MEF 22.1.*

Mobile technologies such as GSM, WCDMA, and CDMA use a centralized *hub-and-spoke* network topology for mobile backhaul. As shown by examples in the following figure, multiple RAN base stations (BTS's or Node B's) connects to a single RAN network controller (BSC or RNC) in hub-and-spoke fashion.

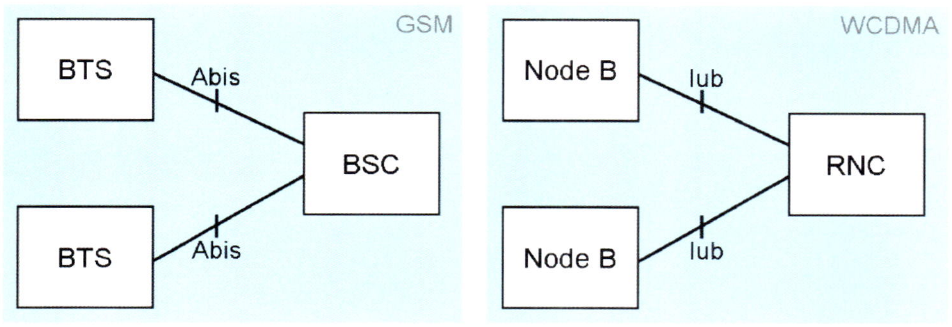

Figure 143: Example GSM and WCDMA Mobile Backhaul Networks (Fig 1, in MEF 22.1)

More recent mobile technologies such as LTE and WiMAX support decentralized control. LTE, for example, supports meshed connectivity between mobile backhaul network elements, as shown by example in the following figure.

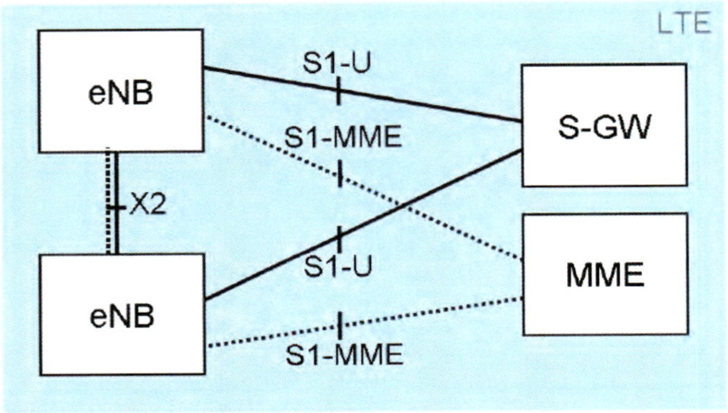

Reproduced with permission of the Metro Ethernet Forum

Figure 144: Example LTE Mobile Backhaul Network (Fig 2, in MEF 22.1)

This example includes four network elements: two RAN base stations (eNB's) and two RAN network controllers (one S-GW and one MME). The S-GW terminates data plane traffic (solid lines), while the MME terminates signal and control plane traffic (dashed lines). Each eNB has an X2 connection with a set of neighboring eNB's to support radio handovers.

The main point is that mobile backhaul for recent technologies, such as LTE, can involve meshed connectivity, in contrast to earlier technologies that involve only hub-and-spoke connectivity. As a consequence, recent technologies, such a LTE, tend to use a broader range of Carrier Ethernet services.

12.3.4 Mobile Backhaul CoS Requirements

The evolution in mobile backhaul toward Ethernet-based technology is driven by the explosive growth of high-bandwidth, best-effort data traffic, compared to low-bandwidth, real-time voice traffic. Under these conditions, support for differentiated quality of service (QoS) –supported by Carrier Ethernet– is key to efficient network utilization. With priority-based traffic management, the mobile backhaul service provider can realize better (more profitable) network utilization without compromising end-user experience. This is accomplished by forwarding voice traffic, synchronization traffic, and control traffic ahead of best-effort data traffic.

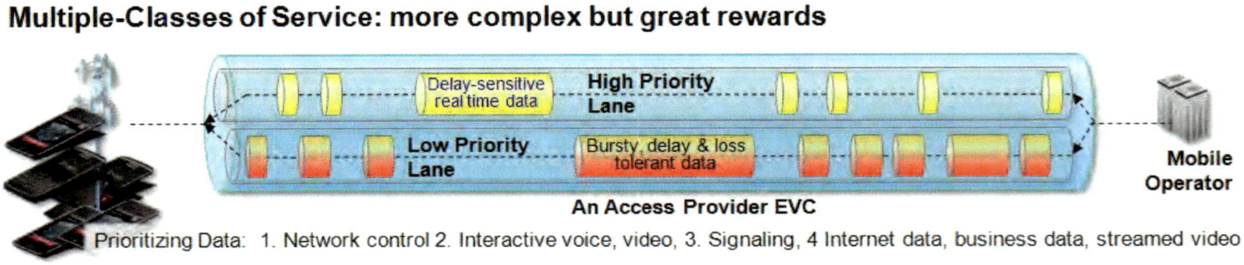

Reproduces with permission of the Metro Ethernet Forum

Figure 145: Example Mobile Backhaul Service with Multiple Classes of Service

Per MEF 22.1, a mobile backhaul EVC must support at least two classes of service (CoS). The following table, taken from MEF 22.1, provides examples on how mobile backhaul traffic classes might be mapped into four, three, or two CoS Names.

CoS Names	Generic Traffic Classes[2] mapping to CoS Names			
	4 CoS Names	3 CoS Names	2 CoS Names	2 CoS Names
Very High (H^+) Defined in this IA	Synchronization	-	-	-
High (H) Defined in [18]	Conversational. Signaling. Network Management and Control	Synchronization. Conversational. Signaling. Network Management and Control	Synchronization. Conversational. Signaling. Network Management Control. and Streaming media	Synchronization. Conversational. Signaling. Network Management. Control. and Streaming media
Medium (M) Defined in [18]	Streaming media	Streaming media	-	Interactive and Background
Low (L) Defined in [18]	Interactive and Background	Interactive and Background	Interactive and Background	

Figure 146: Examples of MBH Traffic Classes Mapping to Cos Names in MEN (Table 7 in MEF 22.1)

In this table, CoS Names H, M, and L correspond to CoS Labels defined in MEF 23.1. Each value (H, M, or L), used in combination with a performance tier (PT1, PT2, PT3, or PT4), implies specific values of CoS performance objectives (CPOs). Per MEF 22.1, a MEF-compliant mobile backhaul service should conform to performance tier PT1, but may conform to PT2 or PT3.

CoS Name H^+ is a special CoS Name defined to support packet-based synchronization traffic that requires more stringent performance than is offered by CoS Label H. The following table from MEF 22.1 includes CPOs for CoS Name H^+.

CoS Name	Ingress Bandwidth Profile**	One Way CPO for Mobile Backhaul Service {S, CoS ID, PT}							
		FD	MFD	IFDV	FDR	FLR	Availability	L	B
Very High (H^+)	*CIR>0* *EIR=0*	*≤10 ms*	*≤7 ms*	*N/S*	A_{FDR}	*≤.01 % (i.e., 10^{-4})*	$\geq A_{Avail}$	$\leq A_{HLI}$	$\leq A_{CHLI}$
High (H)	*CIR>0* *EIR≥0*	For CPO values across PT1 see Table 6 of MEF CoS IA [18].							
Medium (M)	*CIR>0* *EIR≥0*	For CPO values across PT1 see Table 6 of MEF CoS IA [18].							
Low (L)	*CIR≥0* *EIR≥0**	For CPO values across PT1 see Table 6 of MEF CoS IA [18].							

Notes:

- A_{FDR} *values and parameters for H^+ to be included in a future phase of this IA. Values for FD and MFD might change depending on values for FDR.*
- *For Synchronization traffic class (see Section 11.4.1) A_{IFDV} for H^+ = N/S since FDR is used. Also, either MFD or FD needs to be used in SLS.*
- *(*) both CIR = 0 and EIR = 0 is not allowed as this results in no conformant Service Frames. CIR=0 and EIR>0 results in non-specified objectives.*
- *(**) Ingress Bandwidth Profile for CoS Labels (H, M and L) are from Table 2 of MEF 23.1 [18] .*
- *CBS, EBS≥ 8xMTU per MEF 13 [11]*
- *See Table 5 of MEF CoS IA [18] for Parameters and values for H, M and L*

Figure 147: One Way CPOs for Point-to-Point Mobile Backhaul Service (Table 8 in MEF 22.1)

Related Links

CoS Labels and Performance Tiers
Mobile Backhaul Synchronization
Traffic Management and QoS

12.3.5 MEF Service Definitions for Mobile Backhaul

Per MEF 22.1, a case can be made for each EVC-based service (EPL, EVPL, EP-LAN, EVP-LAN, EP-Tree, or EVP-Tree) in mobile backhaul applications.

Table 38: MEF Service Application to Mobile Backhaul

MEF Service	Mobile Backhaul Application Notes	Example Illustration
Ethernet Private Line (EPL)	The EPL service might be preferred in cases where there is a desire for a 1:1 correspondence between RAN NC UNIs and RAN BS UNIs. However, this solution is not very scalable.	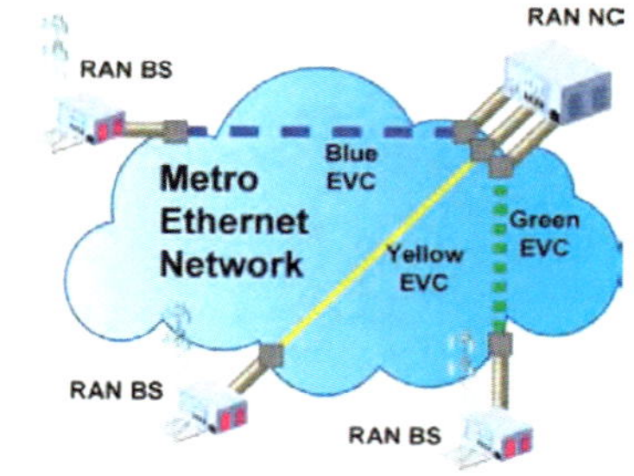 Reproduced with permission of the Metro Ethernet Forum
Ethernet Virtual Private Line (EVPL)	The Ethernet Virtual Private Line (EVPL) service may be used to emulate existing service offerings with a point-to-point relationship between each RAN NC site and each RAN BS site. EVPL supports multiplexing at the UNI. This allows services between the RAN BS and RAN NC to be multiplexed at the RAN NC UNI.	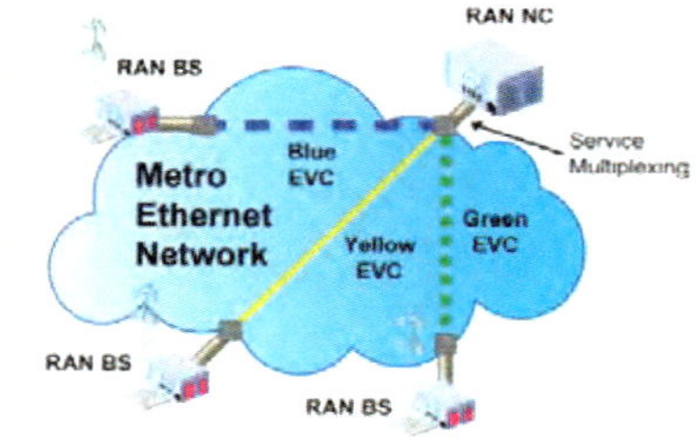
Ethernet Private LAN (EP-LAN)	Mobile operators with multiple RAN NC sites or deployments where inter RAN BS communication is permitted may want to interconnect them at high speeds so all sites appear to be on the same Local Area Network (LAN) and have equivalent performance. The EP-LAN service is defined to provide All to One bundling at each UNI, CE-VLAN ID preservation, CE-VLAN CoS preservation, and tunneling of key Layer 2 Control Protocols. A key advantage of this approach is that if the mobile operator has outsourced its backhaul network to another service provider or different company (for example, transport/transmission network organization), the mobile operator can configure VLANs at the RAN NCs and the RAN BSs without any need to coordinate with the other Service Provider.	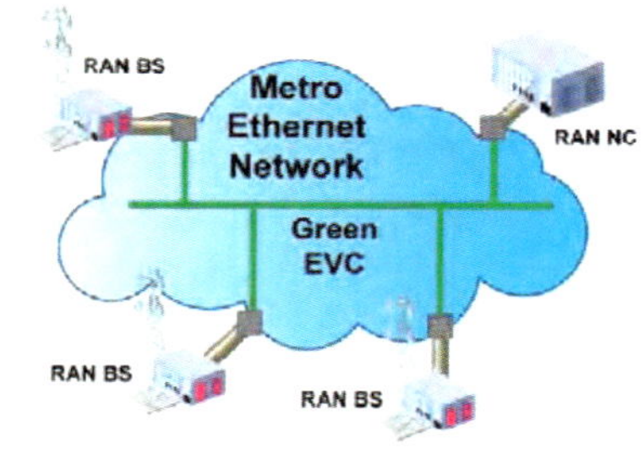

Table 38: MEF Service Application to Mobile Backhaul (Cont.)

MEF Service	Mobile Backhaul Application Notes	Example Illustration
Ethernet Virtual Private LAN (EVP-LAN)	Some mobile operators commonly desire an E-LAN service type to connect their UNIs in a Carrier Ethernet network, while at the same time accessing other services from one or more of those UNIs. An example of such a UNI is a mobile operator site that has co-siting of RAN BS of different technologies, such as legacy GSM and WiMAX. Each technology may have a specific EVC assigned to transport mobile backhaul traffic and different UNI peers. The EVP-LAN service may provide similar transparency as the EP-LAN case. For example, bundling may or may not be used on the UNIs in the Multipoint-to-Multipoint EVC. As such, CE-VLAN ID preservation, CE-VLAN CoS preservation, and tunneling of certain Layer 2 Control Protocols may or may not be provided.	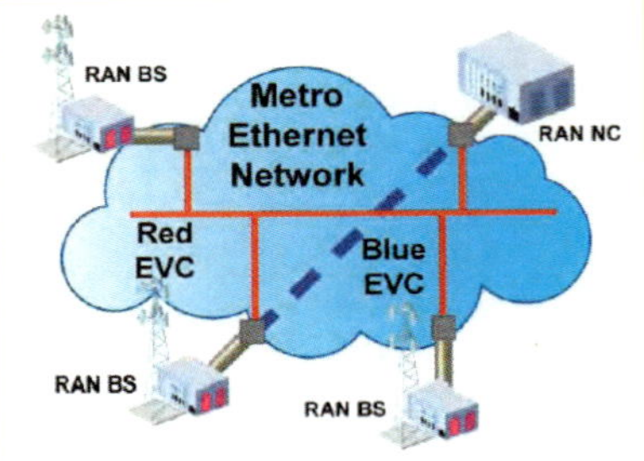 Reproduced with permission of the Metro Ethernet Forum
Ethernet Private Tree (EP-Tree)	Mobile operators with multiple sites may want to interconnect them to provide services other than those that resemble a LAN. These services may be distributed from a single or several centralized sites where the distribution sites are designated as roots and all the remaining sites are designated as leaves. Traditionally in mobile backhaul, the RAN BS sites only need to exchange Service Frames with the RAN NC sites and not with other RAN BSs. This behavior is possible in an Ethernet Private Tree (EP-Tree) service, where the RAN NC sites would be roots and the RAN BS sites would be leaves. The EP-Tree service is defined to provide All to One bundling, CE-VLAN ID preservation, CE-VLAN CoS preservation, and tunneling of key Layer 2 Control Protocols. A key advantage of this approach is that the mobile operator can configure VLANs across the sites without any need to coordinate with the Service Provider.	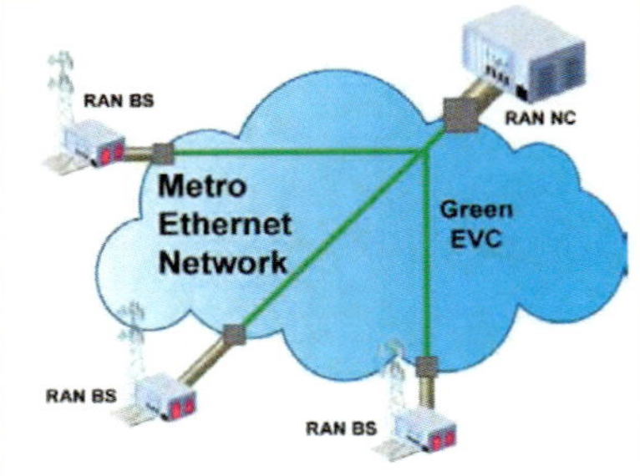 Reproduced with permission of the Metro Ethernet Forum
Ethernet Virtual Private Tree (EVP-Tree)	Some mobile operators desire to keep the root-leaf relationship between RAN NC and RAN BS sites, but also want to multiplex services at one or more of the interconnected UNIs. For such cases, the EVP-Tree service is used. Bundling may or may not be used on the UNIs in the Rooted-Multipoint EVC. As such, CE-VLAN ID preservation, CE-VLAN Cos preservation, and tunneling of certain Layer 2 Control Protocols may or may not be provided. In the example, an EVP-Tree service is used to transport mobile voice and data traffic, while the EVP-LAN service offers an inter-site connection for node and site management.	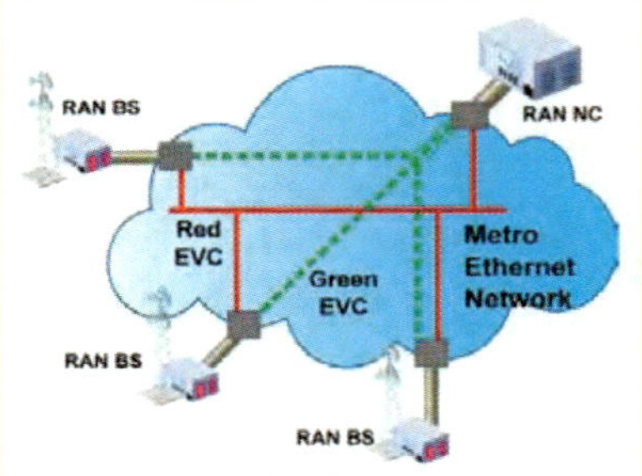 Reproduced with permission of the Metro Ethernet Forum

12.3.6 MEG Levels in Mobile Backhaul

Per MEF 22.1, if Service OAM (SOAM) is used to monitor a mobile backhaul service:

- The *Subscriber* MEG level is used by the Mobile Operator.
- The *EVC*, *Service Provider*, and *Operator* MEG levels are used by the Service Provider.
- The *UNI* MEG level is used to monitor connectivity between the UNI-C and the UNI-N.

12.3.7 Mobile Backhaul Synchronization

Clock synchronization is required to support voice traffic in mobile backhaul networks. Unlike TDM networks (which are synchronous by design), Ethernet-based networks are asynchronous (nodes have no common clock). So, to support mobile backhaul voice traffic, Carrier Ethernet services must be deployed in combination with a mechanism to distribute timing for clock synchronization.

Frequency, Phase, and Time Synchronization

Clock synchronization deals with the problem of ensuring that internal clocks across the network keep the same time in some sense (at least to ensure that clocks tick at the same frequency). Requirements vary with application. Three levels of clock synchronization are broadly recognized: frequency synchronization, phase synchronization, and time synchronization (sometime called time-of-day synchronization).

Frequency synchronization only ensures that clocks tick at the same rate. They can be out of phase.

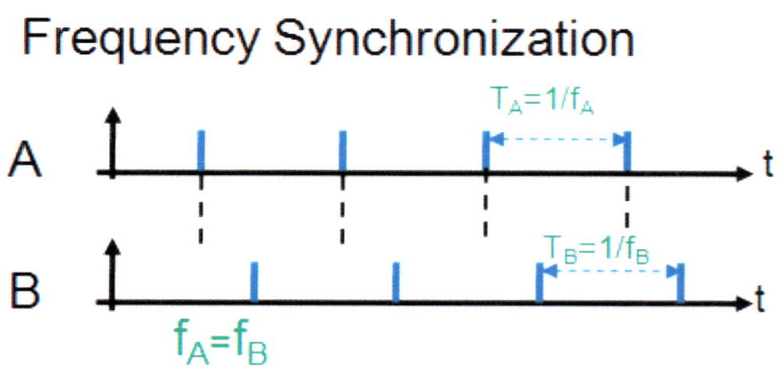

Reproduced with permission of the Metro Ethernet Forum

Phase synchronization ensures that clocks tick at the same rate and are in phase. In other words, all clocks across the network tick at the same time. However, it does not imply that clocks across the network assign the same value of time (or any value of time) to each clock tick.

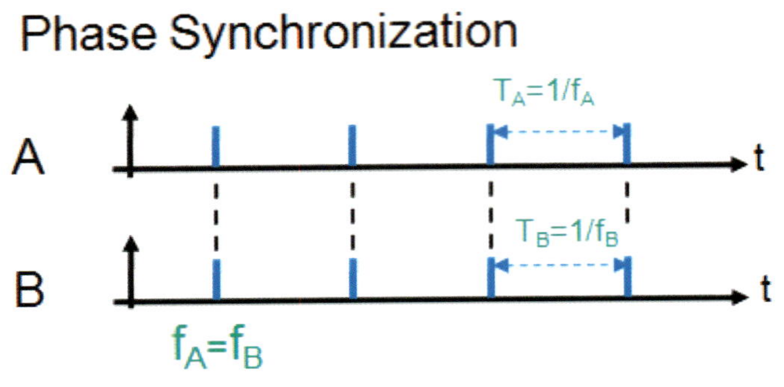

Reproduced with permission of the Metro Ethernet Forum

Note: *Phase synchronization implies frequency synchronization.*

Time (or time-of-day) synchronization ensures that clocks tick at the same rate, are in phase, and assign the same value of time to each clock tick.

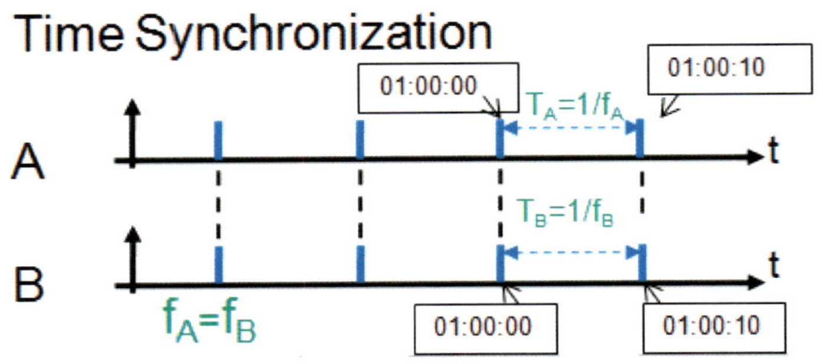

Reproduced with permission of the Metro Ethernet Forum

Note: *Time synchronization implies frequency synchronization and phase synchronization.*

Mobile Backhaul Synchronization Requirements

As shown in the following figure, synchronization requirements for mobile backhaul depend on which technology is used.

Mobile Network Architecture	Frequency Sync	Time-of-day / Phase Sync
CDMA2000		✓
GSM	✓	
UMTS-FDD	✓	
LTE-FDD	✓	
UMTS-TDD	✓	✓
LTE-FDD with MBMS-Single Freq. Network	✓	✓
LTE-TDD	✓	✓
Mobile WiMAX	✓	✓
TD-SCDMA	✓	✓

Reproduced with permission of the Metro Ethernet Forum

Mobile Backhaul Synchronization Methods

MEF 22.1 lists four methods for distributing timing from a primary reference clock to slave clocks at RAN BS sites.

Table 39: Timing Distribution Methods

Scenario	Description	Notes
1	Use GPS at RAN BS sites	MEF 22.1 mentions this as a possible method.
2	Distribute timing using a legacy TDM network.	Only viable for mobile backhaul Use Cases 1a and 2a, which include a legacy TDM network operating in parallel with the MEN (CEN).
3	Distribute timing over the Ethernet physical layer using synchronous Ethernet (SyncE).	SyncE follows the same approach used for traditional TDM (PDH/SONET/SDH) synchronization. It uses physical layer line signals and is implemented with similar engineering rules and principles. SyncE delivers frequency, but not phase or time of day synchronization.
4	Distribute timing by a packet-based method, such as through Circuit Emulation Services over Ethernet (CESoE).	CESoE is required for mobile backhaul Use Case 1b. ***Note:*** *CESoE is explained elsewhere in this study guide.*

MEF 22.1 addresses frequency synchronization only. Time and phase synchronization are for further study.

CESoE is required to distribute timing for mobile backhaul Use Case 1b.

Reproduced with permission of the Metro Ethernet Forum

Figure 148: Mobile Backhaul Use Case 1b

In this use case, synchronous TDM circuits (which carry timing) are encapsulated into Ethernet frames and transported across the asynchronous MEN (CEN) using CESoETH.

Note: *CESoE (Circuit Emulation Services over Ethernet) is explained elsewhere in this study guide.*

Related Links

Ethernet-Based Clock Synchronization
Mobile Backhaul Use Cases
Circuit Emulation Services over Ethernet

12.4
Circuit Emulation Services over Ethernet

In this section:

Circuit Emulation Services over Ethernet (CESoETH) technology is designed to support TDM circuits, such T1 or E1 lines, over an Ethernet network (MEN/CEN).

CESoETH can also enable packet-based clock synchronization over a MEN (CEN).

As shown in the following figure, CESoETH uses an EPL or EVPL service in combination with two Inter-Working Functions (IWFs).

Note: *In some MEF specifications, the term GIWF (Generic Inter-Working Function) is used in place of IWF. They are essentially the same thing.*

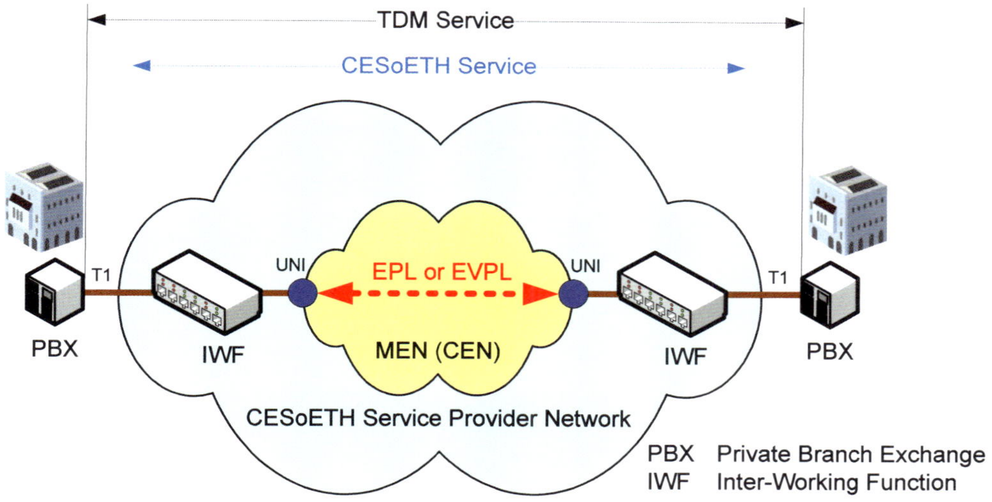

Figure 149: Example PBX Service Connection through Carrier Ethernet Network

The EPL or EVPL service carries the TDM traffic through the MEN (CEN).

The IWFs on each end mediate between the asynchronous Carrier Ethernet service (inside the MEN/CEN) and the synchronous TDM service (outside the MEN/CEN). The resulting service is called a CESoETH service.

CESoETH technology allows a MEN (CEN) service provider to support synchronous TDM traffic in the same MEN (CEN) with asynchronous Ethernet traffic.

Three MEF specifications are related:

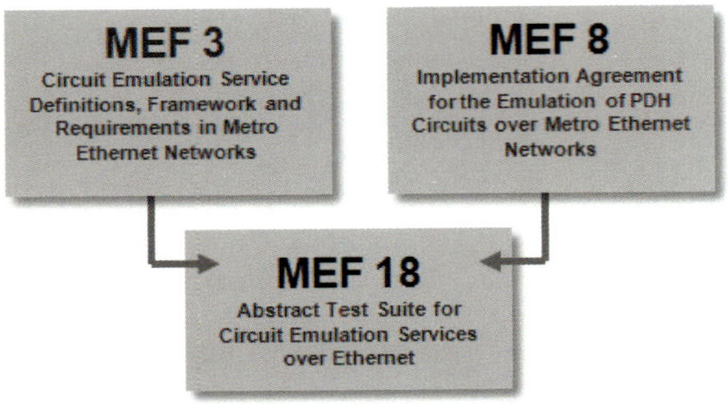

Reproduced with permission of the Metro Ethernet Forum

- MEF 3 – Circuit Emulation Service Definitions, Framework and Requirements in Metro Ethernet Networks
- MEF 8 – Implementation Agreement for the Emulation of PDH Circuits over Metro Ethernet Networks
- MEF 18 –Tests for compliance of CESoETH system – 334 tests for T1/E1 and DS3/E3, including tests of clock recovery

12.4.1 MEN/CEN Requirements for CESoETH

The Carrier Ethernet service that carries CESoETH must be a point-to-point service (either EPL or EVPL) and should deliver all service frames with high reliability and appropriate performance:

- Bandwidth profiles should include only CIR and CBS components (no EIR or EBS components).
- All service frame delivery attributes should be *Deliver Unconditionally*.
- Class of service performance objectives (CPOs) should be specified to meet TDM standards.

Bandwidth requirements for a TDM circuit, such as a T1 or E1 line, are well defined. The bit rate is constant, and the circuit is always on.

CESoETH service bandwidth is very granular and can be sized to fit any TDM service, large or small.

Related Links

CPOs (per CoS Label and Performance Tier)

12.4.2
Interworking Function

The Inter-Working Functions (IWFs) provide the "glue" between the TDM network and the MEN (CEN). The IWFs operate bidirectionally, as shown in the following figure.

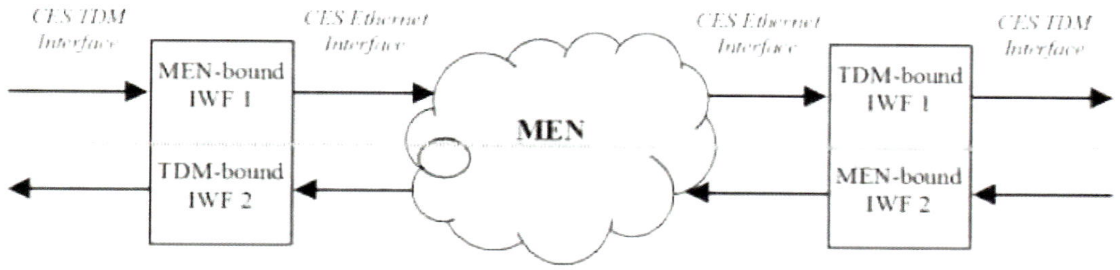

Reproduced with permission of the Metro Ethernet Forum

Figure 150: Bidirectional IWF Functionality (Figure 6-2 from MEF 8)

The MEN-bound IWF handles the packetization of the TDM data, encapsulation into Ethernet frames, and forwarding into the Ethernet network. The corresponding TDM-bound IWF extracts the TDM data from the Ethernet frames and recreates the TDM service.

Details of IWF functionality, provided in MEF 8, are very involved and mostly beyond the scope of this study guide. MEF 8 decomposes IWF functionality into four components, as shown in the following figure.

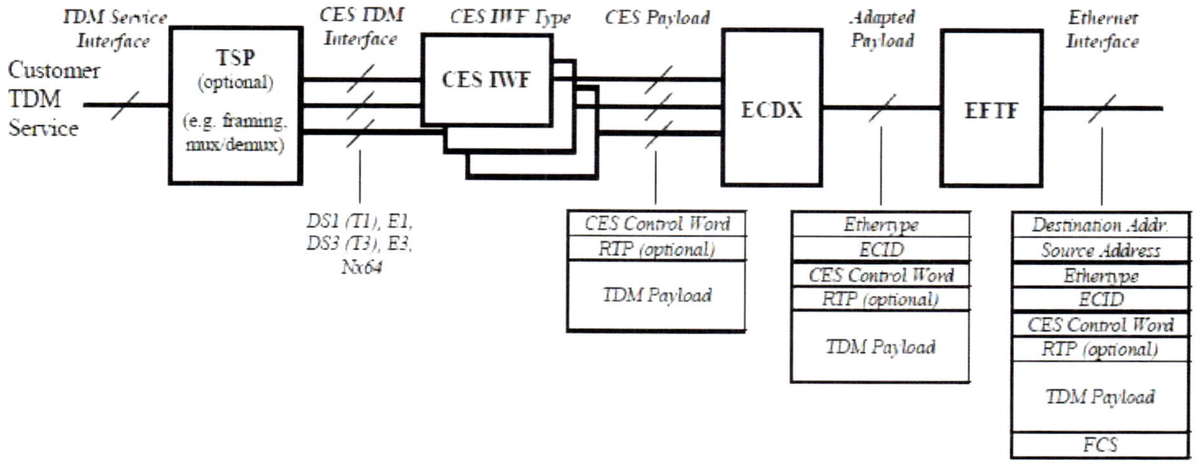

Reproduced with permission of the Metro Ethernet Forum

Figure 151: Functional Components of the IWF (Figure 6-3 from MEF 8)

- **TSP (TDM Service Processor)**–Is an optional component that is used to groom the customer signal(s), as required, for circuit emulation. The TSP might terminate framing overhead, demultiplex a customer signal into subcomponents (as shown in the figure), or multiplex several customer signals into an aggregate TDM signal. MEF 8 does not describe TSP operation in detail because it is considered to be part of an equipment vendor's own value-added function.
- **CES IWF (Circuit Emulation Inter-Working Function)**–The core component is responsible all of the functions required for circuit emulation, including data packetization and depacketization, sequencing, synchronization, TDM signaling, alarms, and performance monitoring.
- **ECDX (Emulated Circuit Demultiplexing Function)**–Associates an Emulated Circuit Identifier (ECID) and an Ethertype field to each packet.
- **EFTF (Ethernet Flow Termination Function)**–Add MAC addresses, VLAN tag (if required), padding, and frame check sequence (FCS) to complete the Ethernet frame.

12.4.3 Synchronization

Synchronization is the responsibility of the CES IWF (Circuit Emulation Inter-Working Function) within the IWF (or GIWF).

Put simply, synchronization requires that the egress clock (used to play out the data at the TDM-bound IWF) should be in sync with the ingress clock (used to input the data at the MEN/CEN-bound IWF). Otherwise, frame slips will occur over time.

The following figure illustrated four basic options for synchronizing the TDM-bound IWF.

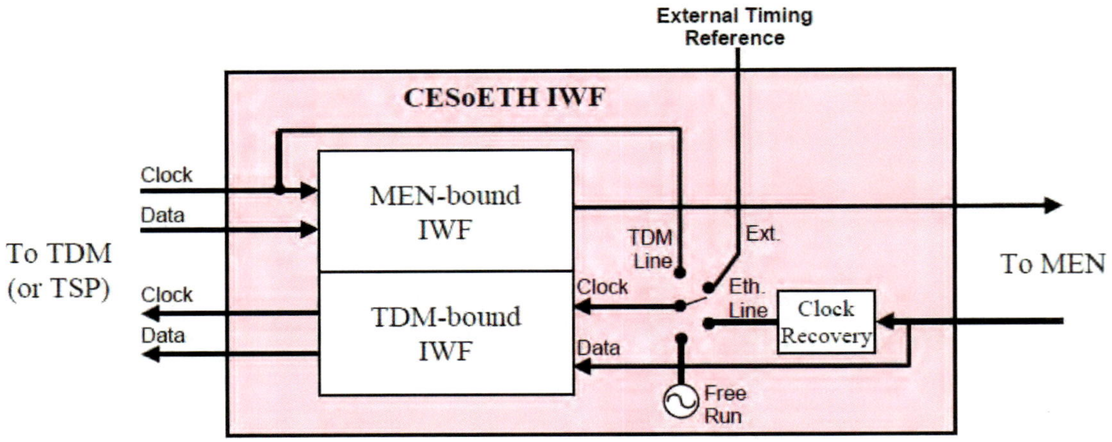

Figure 152: Synchronization Options for the TDM-Bound IWF (Figure 6-7 from MEF 8)

The four options are:

- **TDM line timing**—Use the clock from the incoming TDM line
- **External timing**—Use an external reference clock source
- **Free run timing**—Use a free-running oscillator
- **Ethernet line timing**—Recover the clock from the Ethernet interface

The following section describes methods for accomplishing the last option (Ethernet line timing).

12.4.4 Ethernet-Based Clock Synchronization

MEF 22.1 references two Ethernet-based approaches to distributing timing over a network:

1. Distribute timing over the Ethernet physical layer using **SyncE** (Synchronous Ethernet)
2. Distribute timing by a packet-based method using one or more of the following protocols:
   - **PTP** (Precision Time Protocol)—IEEE 1588v2 time stamping
   - **NTP** (Network Time Protocol)—IETF RFC 1305 time stamping
   - **ACR** (Adaptive Clock Recovery)—IETF RFC 4197
   - **RTP** (Real Time Protocol)—IETF RFC 3550

Both PTP and NTP include the following elements:

- A dedicated stream of data (such as can be provided by CESoETH)
- A time stamp exchange protocol (either PTP or NTP)
- One of the following clock recovery techniques:
  - **Adaptive timing clock recovery**—whereby the clock is recovered from data in the CESoETH frames and the arrival time of the frames
  - **Differential timing clock recovery**—whereby the clock is recovered from a combination of data contained in the CESoETH frames and knowledge of a reference clock

Related Links

Mobile Backhaul Synchronization Methods

12.5 Review

1. Why is CESoETH sometimes used in Carrier Ethernet mobile backhaul networks?

   a. For performance monitoring

   b. To support voice traffic and/or to distribute timing

   c. To multiplex traffic from multiple RAN base stations to the RAN network controller

   d. To support CoS

2. A mobile operator approaches a MEN (CEN) Service Provider to provide mobile backhaul service using a hub-and-spoke topology with a single Ethernet UNI at the hub site where the RAN Network Controller is located. Which of the following is most appropriate?

   a. Use one EPL service between each RAN Base Station and the RAN Network Controller.

   b. Use one EVP-LAN service interconnecting all RAN Base Stations.

   c. Use one EVPL service between each RAN Base Station and the RAN Network Controller.

   d. Use one EVP-Tree service with the RAN base station designated as root and the RAN network controllers designated as leaves.

3. A customer wants to replace a T1 private line with a Carrier Ethernet service. Which of the following bandwidth profiles is most appropriate?

   a. CIR=1.5 Mb/s, CBS=0 bytes, EIR=0, EBS=0

   b. CIR=2 Mb/s, CBS=10,000 bytes, EIR=2 Mb/s, EBS=10,000

   c. CIR=2 Mb/s, CBS=10,000 bytes, EIR=0, EBS=0

   d. CIR=0, CBS=0 bytes, EIR=2 Mb/s, EBS=10,000

4. Which two of the following functions are required to support CESoETH? (Choose two.)

   a. ITU-T G.8031 Ethernet linear protection switching

   b. Synchronization

   c. Internet access

   d. Link OAM (LOAM)

   e. Packetization and de-packetization

5. Which two of the following services can be used, in conjunction with CESoETH, in the replacement of a TDM private line service? (Choose two.)

    a. EP-LAN

    b. EPL

    c. EP-Tree

    d. EVPL

    e. EVP-LAN

6. LANs at three sites are to be interconnected. Which two of the following service scenarios can be configured to tunnel bridge protocol data units (BPDUs) so that spanning tree protocol (STP) works across the aggregation of connected LANs? (Choose two.)

    a. Each site has one UNI. All three sites are interconnected by an EP-LAN service.

    b. Each site has two UNIs. Each pair of sites is interconnected by an EVPL service.

    c. Each site has one UNI. Each pair of sites is interconnected by an EVPL service.

    d. Each site has one UNI. All three sites are interconnected by an EVP-LAN service.

    e. Each site has two UNIs. Each pair of sites is interconnected by an EPL service.

7. Which two of the following clock synchronization methods transmit timing packets across the Carrier Ethernet network? (Choose two.)

    a. Free run timing

    b. IETF RFC 1305 NTP (Network Time Protocol)

    c. SyncE (Synchronous Ethernet)

    d. TDM line timing

    e. IEEE 1588v2 PTP (Precision Time Protocol)

    f. External timing

8. A service provider wants to offer very transparent LAN connectivity, while limiting broadcast traffic. Which service solution is best?

    a. EVP-LAN service with EVC attribute *Broadcast Service Frame Delivery* set to *Deliver Conditionally*

    b. EP-LAN service with EVC attribute *Broadcast Service Frame Delivery* set to *Deliver Conditionally*

c. EVP-LAN service with EVC attribute *Broadcast Service Frame Delivery* set to *Deliver Unconditionally*

d. EP-LAN service with EVC attribute *Broadcast Service Frame Delivery* set to *Deliver Unconditionally*

9. A hospital has three clinics that need to connect to it. The three clinics are independent companies that must not be allowed to share information with each other. Which of the following service solutions is most suitable to connect the hospital to the three clinics if security is the primary concern?

   a. EP-LAN service

   b. EP-Tree service

   c. An EVPL service from each clinic to a service multiplexed UNI at the hospital

   d. An EVP-LAN service

10. Which EVC-based Ethernet services can be used for mobile backhaul? (Choose the <u>one</u> best answer.)

   a. Any

   b. EVPL or EVP-Tree

   c. EP-LAN or EP-Tree

   d. EVP-Tree , EVP-LAN, or EVPL

11. A sporting goods retailer has two data centers and numerous stores throughout the country. Each store needs to regularly exchange inventory, sales, and payroll information with both data centers. Stores should not exchange data with each other. Which of the following service solutions is best? (Choose the <u>one</u> best answer.)

   a. A full mesh of EVPL services interconnecting all sites

   b. A single EP-LAN service interconnecting all sites

   c. A single EP-Tree service with root UNIs at the data centers and leaf UNIs at the stores

   d. A single EVP-LAN service interconnecting all sites

   e. One EPL service connecting the two data centers and one EVPL service connecting each store to one of the data centers

12. An IP service provider wants to connect up to 100 customers through a single 1 Gb/s UNI with each customer allowed up to 10 Mb/s of bandwidth. None of the customers can be allowed to see traffic from any other customer. Which two the following service solutions are viable? (Choose two.)

   a. One EVPL service between each customer and the IP service provider

   b. One EVP-LAN service connecting all UNIs

c. One EVP-Tree service connecting all UNIs with the IP service provider UNI designated as a Root and all other UNIs designated as leaves

d. One EP-LAN service between each customer and the IP service provider

13. MEF 22.1 defines four use cases for transporting mobile backhaul traffic over Carrier Ethernet services. When is CESoE (Circuit Emulation Services over Ethernet) required?

 a. When RAN CE can connect to the Ethernet UNI through an Ethernet interface, eliminating the need for a Generic Inter-Working Function (GIWF)

 b. When the RAN CE cannot connect to the Ethernet UNI and there is no legacy network operating in parallel to deliver timing

 c. When timing is distributed over the Ethernet physical layer using synchronous Ethernet (SyncE)

 d. When the operator needs to support voice traffic via an existing legacy network and offload data to a Carrier Ethernet network

14. A customer wants to replace a T1 private line service between two PBXs (private branch exchanges) with a Carrier Ethernet service. Which technology is required to carry the customer's traffic over a Carrier Ethernet Network (CEN)?

 a. SONET/SDH

 b. CESoE

 c. ATM

 d. Frame Relay

15. An enterprise has a headquarters (HQ) and four branch offices. Each branch office requires access to the Internet through a firewall at HQ, but no access to each other. In addition, HQ requires connection to the Internet service provider (ISP). Which two of the following service arrangements supports these requirements using only one physical connection at each site? (Choose two.)

 a. One EP-Tree service with HQ and ISP designated as roots and branch offices designated as leaves

 b. Four EVPL services connecting branch offices to HQ, plus one EVPL service connecting HQ with ISP

 c. One EVP-LAN service interconnecting the branch offices and HQ, plus one EVPL service between HQ and ISP

 d. One EVPL service between HQ and ISP, plus one EVP-Tree service with HQ designated as the root and branch offices designated as leaves

16. The E-LAN service type supports which two of the following? (Choose two.)

 a. CESoETH

 b. MAC address learning

c. Leaf UNIs

d. G.8031 protection

e. Multipoint-to-multipoint EVC

17. A subscriber needs a single EVP-LAN service to connect numerous sites to three applications. All three applications use a different VLAN. All other VLANs should be blocked. Which two attribute settings are required? (Choose two.)

   a. Service Multiplexing = No

   b. Bundling = Yes

   c. Service Multiplexing = Yes

   d. CE VLAN ID Preservation = Yes

   e. Bundling = No

18. A subscriber requires an EP-LAN service between 6 sites: 3 in Los Angeles and 3 in Miami. The LA sites are on-net (reachable by the Service Provider's network). The Miami sites are off-net and only reachable through an Ethernet Access Provider that has no connectivity to the Service Provider. To deliver this service, the Service Provider purchases three Access EPLs from the Miami Ethernet Access Provider and one point-to-point OVC from an Interstate Operator. The Interstate Operator shares one ENNI with the LA Service Provider and one ENNI with the Miami Ethernet Access Provider. Which of the following statements is <u>not</u> consistent with this service scenario?

   a. The OVC in the Interstate Operator network uses end point map bundling.

   b. The OVC in the Service Provider network is multipoint-to-multipoint.

   c. The OVC in the Interstate Operator network provides hairpin switching.

   d. Three OVC end points in the Ethernet Access Provider network interconnect (through the ENNI) to a single OVC endpoint in the Interstate Operator network.

   e. Three OVC end points in the Service Provider network interconnect (through the ENNI) to a single OVC endpoint in the Interstate Operator network.

19. An Ethernet Mobile Backhaul Service provides:

   a. Radio access between cell phone users and cell tower base stations

   b. Temporary Ethernet services through a mobile WiMAX network

   c. Connectivity between two or more mobile phone users

   d. Connectivity between RAN base stations and RAN network controllers

   e. A gateway between the public mobile phone network and the Internet

20. What does MEF 18 contain?

   a. Test cases for compliance to functionality of Ethernet service traffic management

   b. Tests cases for compliance of CESoETH functionality

   c. Test cases for compliance of UNI Type 2 functionality

   d. Test cases for compliance to functionality of Ethernet services at the UNI

21. A service provider uses Carrier Ethernet as the access network technology to deliver an IP service. What functionality does Ethernet access provide for the service provider?

   a. A Layer 3 (IP) VPN

   b. Layer 2 connectivity to a Layer 3 service or an Internet Service Provider (ISP)

   c. Connection to the Internet through a Layer 2 (Ethernet) firewall

   d. Layer 2 emulation of a Layer 3 (IP) access service

   e. A virtual private gateway to the Internet

22. A company wants to connect two remote campuses to its corporate headquarters, extending all corporate VLANs across all three locations. Which two properties make the EP-LAN service a good choice for this application? (Choose two.)

   a. The service provider can forward frames without MAC address learning.

   b. The company does not have to coordinate CE-VLAN IDs with the service provider.

   c. The service can perform CE-VLAN ID translation if desired.

   d. The company can run STP across all locations if desired.

   e. The UNIs can support service multiplexing if desired.

23. Which two statements concerning Carrier Ethernet mobile backhaul are true? (Choose two.)

   a. A mobile backhaul service must support at least 2 classes of service.

   b. A mobile operator using SOAM, uses the Subscriber MEG level.

   c. If LTE technology is used, the mobile backhaul service topology must be "hub and spoke".

   d. CoS Label H^+ must be used for packet-based synchronization traffic.

24. A mobile operator requests mobile backhaul services for a 4G LTE network that includes base stations (called eNodeBs) with Ethernet interfaces. Which of the following options would you recommend for interconnecting the eNodeBs to the Carrier Ethernet mobile backhaul service?

a. Generic Inter-Working Function (GIWF)

b. Circuit Emulation Services over Ethernet (CESoE)

c. GIWF in combination with CESoE

d. None of the above

25. An enterprise needs Carrier Ethernet services that give two branch offices access to critical business applications located at corporate headquarters, but not to each other. The critical business applications use VLAN IDs 10 and 20. CE devices do not support translation of CE VLAN IDs. All other VLANs should be blocked. Both branch offices need access to both critical business applications. Which service arrangement will support this application?

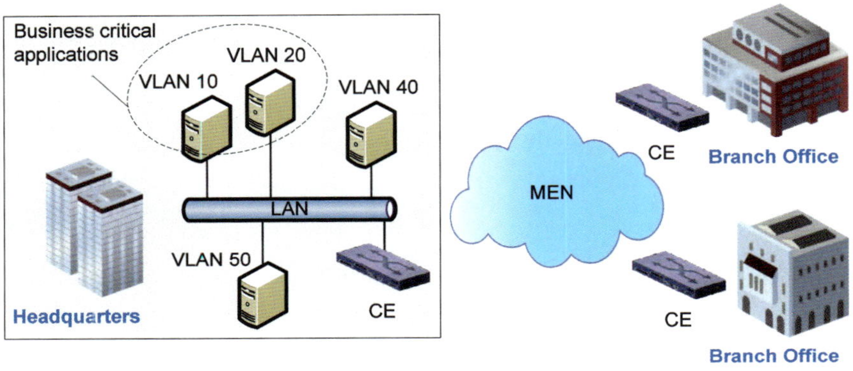

a. One Ethernet Private Line (EPL) service between each branch and the headquarters.

b. One Ethernet Virtual Private Line (EVPL) service between each branch and the headquarters. At the headquarters CE, both EVCs terminate on one UNI that is configured for service multiplexing.

c. One Ethernet Virtual Private Line (EVPL) service between each branch office and the headquarters. At the headquarters CE, the EVPLs terminate on different UNIs, which are both configured to support bundling.

d. One Ethernet Private Tree (EP-Tree) service with the headquarters UNI designated as the root and branch offices designated as leaves and with the root UNI configured to support bundling.

26. Which of the following is not required for a Carrier Ethernet service that carries CESoETH?

a. Bandwidth profiles include only CIR and CBS components (no EIR or EBS components).

b. The EVC must tunnel BPDUs.

c. The EVC is point-to-point.

d. All service frame delivery attributes should be set to "Deliver Unconditionally."

27. A subscriber has 10 sites with one UNI at each site. The 10 sites are interconnected by an EVP-LAN service that supports multiple VLANs between the sites. Each UNI also supports one EVPL service that connects the site to an Internet service provider (ISP). Which one of the following configuration options is not required?

a. Configure all UNIs with *Bundling* set to *Yes*.

b. Configure *CE-VLAN ID preservation* for each EVPL service to *Yes*.

c. Configure *CE-VLAN ID preservation* for the EVP-LAN service to *Yes*.

d. Configure all UNIs with *Service Multiplexing* set to *Yes*.

Related Links

Answers: Applications

13
MEF Certification Program

In this chapter:

The MEF defines and administers a threefold certification program to confirm compliance with and/or knowledge of Carrier Ethernet specifications:

- **Equipment Certification**–Confirms that the equipment offered by an equipment vendor complies with MEF standards and is fully capable of supporting high-performance Carrier Ethernet services

- **Service Certification**–Confirms that the service offered by a service provider conforms to MEF standards of quality and performance

- **Professional Certification**–Confirms that the individual awarded the certificate has proven knowledge and the skills to support the explosive growth of Carrier Ethernet across the world's markets

For both equipment and services, the MEF offers two categories of certification:

- **CE 1.0 Certification**[37]–Available since 2005. Certifies compliance with MEF 9 and/or MEF 14. Applicable to EPL, EVPL, and E-LAN[38] services only.

- **CE 2.0 Certification**– Introduced in 2013. Certifies compliance with *MEF Carrier Ethernet 2.0 Certification Blueprint*. Applicable to EPL, EVPL, EP-LAN, EVP-LAN, EP-Tree, EVP-Tree, Access EPL, and Access EVPL services.

For professional certification, the MEF now offers MEF-CECP 2.0 certification in place of the original MEF-CECP (MEF Carrier Ethernet Certified Professional) certification.

37 CE 1.0 certification is a rebranding of MEF 9 / MEF 14 certification.
38 EP-LAN together with EVP-LAN

13.1
CE 1.0 versus CE 2.0 Certification

In this section:

The *MEF Carrier Ethernet 2.0 Services Technical Foundation Document*, available from the MEF Web site (MetroEthernetForum.org), provides tables (such as the following) comparing CE 2.0 requirements to CE 1.0 requirements.

E-Line - EVC per UNI Service Attributes and SOAM Frames Handling

EVC per UNI Service Attributes	What's new in Carrier Ethernet 2.0	EPL 1.0	EPL 2.0	EVPL 1.0	EVPL 2.0
UNI EVC ID	Carrier Ethernet 1.0 and 2.0 require the UNI EVC ID to be a string formed by the concatenation of the UNI ID and the EVC ID.	✔	✔	✔	✔
CE-VLAN ID / EVC Map	Carrier Ethernet 2.0 requires that all Service Frames at the UNI map to a single point-to-point EVC for EPL and to specify mapping table of CE-VLAN IDs to EVC ID for EVPL.	✔	✔ Enh.	✔	✔ Enh.
Ingress BWP Per EVC	If ingress BWP per EVC is supported, Carrier Ethernet 2.0 requires configuration granularities per UNI Type 1 IA.	✔	✔ Enh.	✔	✔ Enh.
Ingress BWP Per CoS ID	If ingress BWP per CoS ID is supported, Carrier Ethernet 2.0 requires configuration granularities per UNI Type 1 IA.	✔	✔ Enh.	✔	✔ Enh.
Egress BWP Per EVC	Carrier Ethernet 2.0 requires to not specify Egress BWP per EVC.	✘	✘	✘	✘
Egress BWP Per CoS ID	Carrier Ethernet 2.0 requires to not specify Egress BWP per CoS ID.	✘	✘	✘	✘
SOAM Frames Handling	**What's new in Carrier Ethernet 2.0**	**1.0**	**2.0**	**1.0**	**2.0**
Connectivity Check messages	Carrier Ethernet 2.0 requires that all CCM frames at the default Test and Subscriber MEG levels be tunneled.	✘	✔ New	✘	✔ New
Linktrace messages with at least one MIP configured	Where at least one MIP is configured at the default Test or Subscriber MEG level, Carrier Ethernet 2.0 requires that Linktrace frames with the corresponding target MAC DA and MEG level be peered or discarded.	✘	✔ New	✘	✔ New
Linktrace messages without any MIPs configured	Where no MIPs are configured at the default Test or Subscriber MEG level, Carrier Ethernet 2.0 requires that Linktrace frames with the corresponding target MAC DA and MEG level be tunneled.	✘	✔ New	✘	✔ New
Unicast Loopback messages with at least one MIP configured	Where at least one MIP is configured at the default Test or Subscriber MEG level, Carrier Ethernet 2.0 requires that Loopback frames with the corresponding target unicast MAC DA and MEG level be peered or discarded.	✘	✔ New	✘	✔ New
Unicast Loopback messages with at least one MIP configured	Where at least one MIP is configured at the default Test or Subscriber MEG level, Carrier Ethernet 2.0 requires that Loopback frames with the corresponding MEG level and a target unicast DA not equal to the MAC address of any MIPs be tunneled.	✘	✔ New	✘	✔ New
Multicast Loopback messages	Carrier Ethernet 2.0 requires that all multicast Loopback frames at the default Test and Subscriber MEG levels be tunneled.	✘	✔ New	✘	✔ New

✔ New	New attribute introduced in CE 2.0	✔	Attribute introduced in CE 1.0 and still applicable as is in CE 2.0
✔ Enh.	Attribute introduced in CE 1.0 and enhanced in CE 2.0	✘	Attribute not specified

MEF 6

Reproduced with permission of the Metro Ethernet Forum

Figure 153: Example Table from CE 2.0 Technical Foundation Document

For each service, CE 2.0 certification includes all CE 1.0 requirements. So, CE 2.0 certification implies compliance with CE 1.0.

13.1.1
CE 1.0 Certification

CE 1.0, available since 2005, certifies that equipment (or service) complies with MEF 9 and/or MEF 14.

MEF Test Specification	Description
MEF 9 (Ethernet Services at the UNI)	244 Test Cases to verify that Carrier Ethernet EPL, EVPL, and E-LAN services comply with the functional specifications defined by the MEF for each service.
MEF 14 (Traffic Management Phase 1)	183 Test Cases to verify that Carrier Ethernet EPL, EVPL, and E-LAN services adhere to a strict set of MEF service performance objectives.

CE 1.0 certification is available for EPL, EVPL and E-LAN[39] services only.

CE 1.0 certification is based on technical specifications MEF 6 and MEF 10.

CE 1.0 service certification requires use of CE 1.0–certified equipment.

13.1.2
CE 2.0 Certification

CE 2.0 certification, introduced in 2013, certifies compliance with the *MEF Carrier Ethernet 2.0 Certification Blueprint*, available from the MEF Web site (MetroEthernetForum.org).

CE 2.0 certification is per service–The equipment manufacturer (or service provider) can apply for certification for any one or more of 8 Carrier Ethernet services: EPL, EVPL, EP-LAN, EVP-LAN, EP-Tree, EVP-Tree, Access EPL, and/or Access EVPL. There is no requirement to get certified for all eight services.

MEF-standardized classes of service are used in CE 2.0 Certification–CoS Labels with their associated performance objectives and performance tiers (as specified in MEF 23.1) are key elements of CE 2.0 service certification.

As shown in the following table, CE 2.0 Certification verifies that the equipment (or service) supports all applicable service attributes plus Service OAM frame handling.

39 EP-LAN together with EVP-LAN

Table 40: Attributes and Requirements Verified by CE 2.0 Certification

EPL, EVPL, EP-LAN, EVP-LAN, EP-Tree or EVP-Tree	Access EPL or Access EVPL
1. UNI Service Attributes 2. EVC Per UNI Service Attributes 3. EVC Service Attributes 4. Service OAM Frame Handling	1. UNI Service Attributes 2. ENNI Service Attributes 3. OVC Per UNI Service Attributes 4. OVC End Point Per ENNI Service Attributes 5. OVC Service Attributes 6. Service OAM Frame Handling

For details, refer to the *MEF Carrier Ethernet 2.0 Certification Blueprint* available from the MEF Web site (MetroEthernetForum.org).

Note: *CE 2.0 certification is based on MEF 6.1, MEF 10.2, MEF 23.1, MEF 26.1, and MEF 33.*

CE 2.0 Certification Deliverables

CE 2.0 companies receive a set of marketing and technical materials designed to be included in product documentation, Web site content, press releases, product launches, and RFP responses.

Additional deliverables, per certification, include:

- **CE 2.0 Certificate**–Official statement of CE 2.0 certification principally used as proof of compliance in response to customer requests. Certificates are delivered in soft and hard copies.
- **CE 2.0 Test Report**–Provides a comprehensive technical description of the CE 2.0 services that are tested together with a full set of test results. Reports are the property of the CE 2.0 Company and are distributed at its discretion.
- **CE 2.0 Certified Compliant Logo**–Delivered in print and publishing formats, the logo provides instant recognition of CE 2.0 companies, services, and equipment.
- **MEF Certification Registry**–New certifications are added to the official on-line registry of MEF certified companies and services. Posted on the MEF Web site, this directory is widely consulted in the industry.

13.2
Equipment Certification

MEF Equipment Certification applies to products, ensuring that products conform to MEF global Carrier Ethernet specifications. Equipment vendors apply for equipment certification through an independent streamlined process, with testing conducted by **Iometrix**, the official MEF certification lab. If equipment achieves certification, the equipment vendor receives a certificate that can be used in product documentation, Web site content, press releases, product launches, and RFP responses. The MEF public Web site provides a complete listing of certified products, per vendor, so that customers can validate certification.

MEF equipment certification also benefits service providers, helping them to confidently select equipment and systems appropriate for MEF service implementation.

Two categories of equipment certification are available:

- **CE 1.0 Certification**[40]–Available since 2005. Certifies compliance with MEF 9 and/or MEF 14. Applicable to EPL, EVPL, and E-LAN[41] services only.
- **CE 2.0 Certification**– Introduced in 2013. Certifies compliance with *MEF Carrier Ethernet 2.0 Certification Blueprint*. Applicable to EPL, EVPL, EP-LAN, EVP-LAN, EP-Tree, EVP-Tree, Access EPL, and Access EVPL services.

In addition, equipment can be certified with respect to MEF 18[42] and/or MEF 21[43]. These equipment certifications are independent of CE 1.0 and CE 2.0 certification.

Prerequisite: *Any equipment manufacturer with a Carrier Ethernet product can apply for certification of its products providing it is a MEF member at the time of the actual certification testing.*[44]

CE 2.0 Equipment Certification:

- **Is per service**–The equipment manufacturer can apply for certification with respect to any one or any combination of 8 Carrier Ethernet services: EPL, EVPL, EP-LAN, EVP-LAN, EP-Tree, EVP-Tree, Access EPL, and/or Access EVPL. Equipment is not required to support all 8 services.
- **Does not require CE 1.0 equipment certification**– There is no requirement for equipment to already have CE 1.0 certification (MEF 9 and/or MEF 14 certification) in order to be tested for CE 2.0 equipment certification.
- **Ensures compliance** to the *MEF Carrier Ethernet 2.0 Certification Blueprint*, available from the MEF Web site (MetroEthernetForum.org).
- **Provides deliverables** to the equipment manufacturer, as previously explained for CE 2.0 Certification.

40 CE 1.0 certification is a rebranding of MEF 9 / MEF 14 certification.
41 EP-LAN together with EVP-LAN
42 MEF 18, *Circuit Emulation Services over Ethernet*, includes 334 Test Cases to validate successful transport of TDM traffic over Carrier Ethernet networks, including stringent synchronization tests based on ITU-T G.8261, G.823, and G.824.
43 MEF 21, *UNI Type 2 Part 1 Link OAM*, includes 371 Test Cases to verify Link Operations, Administration & Maintenance (clause 57 of IEEE 802.3) at the User Network Interface (UNI) split into the UNI-C and the UNI-N.
44 *Once equipment certification is granted, it remains forever valid, regardless of MEF membership status.*

Equipment certification is carried out by the MEF-approved test lab, Iometrix, according to the following three-phase process:

1. **Qualification**–The equipment manufacturer conducts a detailed review of the candidate equipment for certification by telephone and email with Iometrix. A thorough and effective Qualification phase reduces effort required by the equipment manufacturer for the subsequent Configuration phase.

2. **Configuration**–Once the Qualification phase is complete, the equipment manufacturer configures the equipment according to recommendations by Iometrix in order to prepare the equipment for live testing. Again, an effective Configuration phase minimizes the time and effort required by the equipment manufacturer for the Testing phase.

3. **Testing**–The equipment manufacturer sends the product accompanied by suitably trained staff to the Iometrix test lab in San Francisco. Testing continues until certification is achieved.

For more information about equipment certification, refer to the MEF Web site (MetroEthernetForum.org) and the Iometrix Web site (Iometrix.com).

Related Links

CE 2.0 Certification

13.3
Service Certification

MEF Service Certification is designed for service providers and validates compliance of Carrier Ethernet services to MEF specifications through an independent streamlined process. Testing is conducted on live services provisioned across the service provider production network by **Iometrix**, the official MEF certification lab. If a service achieves certification, the service provider receives a certificate that can be used in service documentation, Web site content, press releases, service launches, and RFP responses. The MEF public Web site provides a complete listing of certified services, per vendor, so that customers can validate certification.

Two categories of service certification are available:

- **CE 1.0 Certification**[45]—Available since 2005. Certifies compliance with MEF 9 and/or MEF 14. Applicable to EPL, EVPL, and E-LAN[46] services only.
- **CE 2.0 Certification**— Introduced in 2013. Certifies compliance with *MEF Carrier Ethernet 2.0 Certification Blueprint*. Applicable to EPL, EVPL, EP-LAN, EVP-LAN, EP-Tree, EVP-Tree, Access EPL, and Access EVPL services.

Prerequisite: *Any service provider with a Carrier Ethernet service can apply for certification of its service - providing it is a MEF member at the time of the actual certification testing.*[47]

CE 2.0 Service Certification:

- **Is per service**—The service provider can apply for certification with respect to any one or any combination of 8 Carrier Ethernet services: EPL, EVPL, EP-LAN, EVP-LAN, EP-Tree, EVP-Tree, Access EPL, and/or Access EVPL. The service provider is not required to get all 8 services certified.
- **Is per performance tier**—The service provider will indicate the performance tier(s) that the service under certification supports.
- **Does not require CE 1.0 service certification**—There is no requirement for a service to already have CE 1.0 certification (MEF 9 and/or MEF 14 certification) in order to be tested for CE 2.0 service certification.
- **Does not require CE 2.0 certified equipment**—As stated on the MEF web site: "*It is not mandatory to use CE 2.0 certified network equipment for the CE 2.0 certified service – however it is strongly recommended by the MEF in order to maximize the efficiency of the process of certifying the CE 2.0 service.*"
- **Ensures compliance** to the *MEF Carrier Ethernet 2.0 Certification Blueprint*, available from the MEF Web site (MetroEthernetForum.org).
- **Provides deliverables** to the service provider, as previously explained for CE 2.0 Certification.

45 CE 1.0 certification is a rebranding of MEF 9 / MEF 14 certification.
46 EP-LAN together with EVP-LAN
47 *Once service certification is granted, it remains forever valid, regardless of MEF membership status.*

Service certification is carried out by the MEF-approved test lab, Iometrix, according to the following three-phase process:

1. **Qualification**– The service provider conducts a detailed review of the candidate services for certification by telephone and email with Iometrix. A thorough and effective Qualification phase reduces effort required by the service provider for the subsequent Configuration phase.

2. **Configuration**– Once the Qualification phase is complete, the service provider configures its service according to recommendations by Iometrix in order to prepare the service for live testing. Again, an effective Configuration phase minimizes the time and effort required by the service provider for the Testing phase.

3. **Testing**–Iometrix provides two or more network probes to the service provider, which are then attached by the service provider in pre-defined locations to enable remote testing by Iometrix of the service to be certified.

Note: *Services are tested remotely, in situ in the service provider network wherever it is located.*

Note: *For EVC-based services, testing is UNI-to-UNI. For Access EPL and Access EVPL services, testing is UNI-to-ENNI.*

Note: *No requirements apply to core network equipment but Ethernet services offered on different transport technologies such as SONET/SDH, DWDM, MPLS are generally subject to separate certifications.*

For more information about service certification, refer to the MEF Web site (MetroEthernetForum.org) and the Iometrix Web site (Iometrix.com).

Related Links

CE 2.0 Certification
CPOs (per CoS Label and Performance Tier)

13.4
Professional Certification

Reproduced with permission of the Metro Ethernet Forum

The MEF professional certification program is designed for telecommunication professionals seeking to validate their expertise, skills, and knowledge of Carrier Ethernet technologies, standards, services, and applications.

MEF's first professional certification, the MEF Carrier Ethernet Certified Professional (MEF-CECP) certification, was updated and replaced in December 2013 by the MEF-CECP 2.0 certification. The MEF no longer offers MEF-CECP certification, but those who have earned it remain MEF-CECPs (the certification does not expire).

MEF-CECP 2.0 certification is awarded to individuals who pass the MEF-CECP 2.0 exam, which covers everything in the MEF-CECP exam, plus new concepts, capabilities, and MEF specifications added after January 2012 as explained in this study guide.

Individuals who pass the MEF-CECP 2.0 exam are awarded:

1. An official MEF-CECP 2.0 certificate
2. A listing, as being MEF-CECP 2.0 certified, on the online Directory of MEF Certified Professionals posted at the widely visited EthernetAcademy.net Web site
3. Qualification to use the MEF-CECP 2.0 logo (for example, in e-mail signatures, business cards, and CVs) for instant recognition of their status

Figure 154: MEF-CECP 2.0 Logo

The MEF-CECP 2.0 exam is open to anyone (you need not be an MEF member) and may be taken online from your home or office, or from an MEF Accredited Training Provider. Refer to the EthernetAcademy.net Web site for more details.

13.5 Review

1. Which three of the following services can include CE 2.0 certified support for CoS Label H at Performance Tier 2? (Choose three.)

   a. EPL

   b. EP-LAN

   c. EP-Tree

   d. EVPL

   e. Access EVPL

2. Which two of the following are requirements for an equipment vendor to participate in MEF certification testing? (Choose two.)

   a. The equipment vendor must supply the equipment to Iometrix (the official MEF certification lab).

   b. The equipment vendor must attend MEF quarterly member meetings.

   c. The equipment vendor must be ISO 9001 compliant.

   d. The equipment vendor must be a member of the MEF.

3. Which two MEF test specifications are included in the CE 1.0 certification program? (Choose two.)

   a. MEF 9, Abstract Test Suite for Ethernet Services at the UNI

   b. MEF 14, Abstract Test Suite for Traffic Management Phase 1

   c. MEF 18, Abstract Test Suite for Circuit Emulation Services over Ethernet

   d. MEF 21, Abstract Test Suite for UNI Type 2 Part 1 Link OAM

4. For MEF service certification, where does testing occur?

   a. At the Iometrix lab (the official MEF certification lab)

   b. On the service provider's production network

   c. On a representative network using MEF-certified networking equipment

   d. At an MEF testing facility

5. Which one of the following is true for MEF CE 2.0 service certification?

   a. The service provider must provide on-site engineering support at the Iometrix lab during of testing.

   b. The service provider must be a member of the MEF.

   c. The service must terminate on MEF-certified equipment.

   d. Certification is lost if MEF membership is not renewed annually.

6. Where is equipment tested for MEF equipment certification?

   a. At the Iometrix lab (the official MEF certification lab)

   b. On the service provider's production network

   c. On the MEF service certification network

   d. At the equipment vendor's facility

7. CE 2.0 Certification allows <u>equipment</u> to be certified with respect to:

   a. Link OAM (per clause 57 of IEEE 802.3) at the UNI

   b. Any one or more of 8 Carrier Ethernet services: EPL, EVPL, EP-LAN, EVP-LAN, EP-Tree, EVP-Tree, Access EPL, and/or Access EVPL

   c. CESoETH (per ITU-T G.8261, G.823 and G.824)

   d. The following service only: EPL, EVPL, and/or E-LAN (EP-LAN together with EVP-LAN)

8. To participate in the MEF equipment or service certification program, a company must:

   a. have a least one employee with MEF-CECP certification

   b. be a member of the MEF

   c. participate in MEF quarterly member meetings

   d. have an in-house testing lab

Related Links

Answers: MEF Certification

14

Answers to Review Questions

In this chapter:

14.1
Answers: Introduction and Overview

1. What does MEF stand for?
   - a. Metro Ethernet Framework
   - **b. Metro Ethernet Forum**
   - c. Model Ethernet Forum
2. According to the MEF, how many attributes distinguish Carrier Ethernet from familiar LAN-based Ethernet?
   - a. 4
   - b. **5**
   - c. 6
3. Which one of the following statements is not true?
   - a. This study guide assumes you have a basic knowledge of Ethernet.
   - b. The bulk of MEF-CECP Exam questions relate to MEF specifications 6.1 and 10.2.
   - c. MEF standards published after April 2013 are not included in the MEF-CECP 2.0 exam.
   - **d. This study guide is meant to be used in complete isolation.**

   Note: *This study guide is meant to be used in conjunction with materials available from the MEF Web site (MetroEthernetForum.org).*
4. If a bridge receives an Ethernet frame with an unknown unicast destination address, the frame is:
   - **a. Flooded**
   - b. Buffered until the address can be determined
   - c. Dropped

Note: *This study guide assumes that you already have this level of Ethernet knowledge. Refer to* Assumed Knowledge.

5. When a bridge learns a MAC address, that address is:

    a. The destination address within a frame received by the bridge

    b. Flooded to all peer bridges

    c. **The source address within a frame received by the bridge**

Note: *This study guide assumes that you already have this level of Ethernet knowledge. Refer to* Assumed Knowledge.

14.2 Answers: MEF Services

1. The interface between an Operator MEN (Operator CEN) and the subscriber is called a(an):

   a. ENNI

   b. OVC

   c. **UNI**

   d. EVC

   e. CE

   Note: *Refer to General Reference Model.*

2. Service multiplexing allows:

   a. Multiple services on one EVC

   b. **Multiple EVCs on one UNI**

   c. Multiple UNIs on one EVC

   d. Bundling

   e. Multiple CE-VLAN IDs to be mapped to an EVC

   Note: *Refer to Service Multiplexing.*

3. What is the MEF definition of a UNI?

   a. The logical representation of a service connection between two or more EVCs

   b. **The physical demarcation point between the responsibility of the Service Provider and the responsibility of the Subscriber**

   c. All functions required to connect the CE to the MEN (CEN)

   d. The interface between two Operator MEN/CENs

   Note: *This definition appears in MEF 6.1 and MEF 10.2. Refer to Basic Reference Model.*

4. Equipment on the subscriber side of the UNI that connects to the MEN (CEN) is called the:

   a. Carrier edge

   b. Subscriber edge

c. User network interface

d. Consumer edge

e. **Customer edge**

f. Network edge

Note: *Refer to* Basic Reference Model.

5. Which two Carrier Ethernet service combinations can be offered simultaneously at a given UNI?

   a. EP-LAN and EVPL

   b. EVP-Tree and EPL

   c. **EVPL and EVP-LAN**

   d. EPL and EP-Tree

   Note: *Refer to* Port-Based versus VLAN-Based Services.

6. A UNI is configured to support all-to-one bundling. Which three statements are true? (Choose three.)

   a. **Service multiplexing is not supported at the UNI.**

   b. Any number of EVCs can be bundled to this UNI.

   c. **Bundling is not supported at the UNI.**

   d. **At this UNI, all CE-VLAN IDs map to the same EVC.**

   Note: *Refer to* Bundling.

7. Which three service types are defined in MEF 6.1? (Choose three.)

   a. Ethernet Virtual Connection (EVC)

   b. **Ethernet Line (E-Line)**

   c. **Ethernet Tree (E-Tree)**

   d. Ethernet Access (E-Access)

   e. **Ethernet LAN (E-LAN)**

   f. TDM Line (T-Line)

   Note: *Refer to* Taxonomy of Carrier Ethernet Services.

8. An enterprise needs to connect three branch offices with its two payroll processing data centers. The two data centers require interconnectivity for data mirroring (protection). The service must ensure that the branch offices cannot see each other's sensitive payroll data. Which two of the following service arrangements supports this application? (Choose two.)

- **a.** An EP-Tree service with the payroll data center UNIs designated as leaves and the branch office UNIs designated as roots
- **b.** **An EP-Tree service with payroll data center UNIs designated as roots and branch office UNIs designated as leaves**
- **c.** One EP-LAN service connecting all five sites
- **d.** **One EVPL service from each branch office to each payroll data center (six EVPLs in all), plus an EPL between the two payroll centers**

Note: *(a) Allows branches to see each other's payroll data. (b) Works. (c) Does not prevent communication between branch offices. (d) Works. Requires two UNIs at each payroll center.*

9. A UNI is configured without bundling or all-to-one bundling and is also configured to support service multiplexing. How many EVCs can terminate at this UNI?

- **a.** 0
- **b.** 0 or 1
- **c.** Up to 64
- **d.** **Up to 4094**

Note: *There are 4094 distinct CE-VLAN IDs. Per IEEE 802.1Q–2005, the maximum number of VLANs that can be supported by the 12-bit VID field is 4094 rather than 4096 because VID values 0 and FFF are reserved. Refer to* *CE-VLAN ID**.*

10. Which kind of UNI-to-UNI connectivity is used for the Ethernet Private Line (EPL) service?

- **a.** Ethernet Private Circuit
- **b.** Point-to-point Ethernet Private Connection
- **c.** **Point-to-point Ethernet Virtual Connection**
- **d.** Point-to-point Ethernet Virtual Circuit

Note: *Refer to* *EPL Service* *and recall that **EVC** stands for **Ethernet Virtual Connection**. All three incorrect answers are red herrings: the MEF does not use the terms "Ethernet Private Circuit," "Ethernet Private Connection," or "Ethernet Virtual Circuit."*

11. A customer with three locations requests a Carrier Ethernet service that provides the most transparent connectivity for all PDU types between sites. Which service is most suitable?

    a. **A single Ethernet Private LAN (EP-LAN)**

    b. A single Ethernet Virtual Private LAN (EVP-LAN)

    c. A single Ethernet Private Tree (EP-Tree)

    d. A single Ethernet Virtual Private Tree (EVP-Tree)

    Note: *(a) Works and provides the most transparency. (b) Works, but with less transparency. (c) and (d) All E-Tree service types limit connectivity between sites with Leaf UNIs (one of the site UNIs must be designated as a Root UNI). Refer to* Matrix of Service Distinctions.

12. McBary's, a business enterprise with numerous independently-owned restaurant franchises, wants to connect each franchise to the corporate data center to obtain daily sales information. Which of the following service arrangements is most suitable?

    a. A single Ethernet Virtual Private LAN (EVP-LAN) service

    b. One Ethernet Virtual Private Line (EVPL) service between each franchise and the corporate data center

    c. **A single Ethernet Private Tree (EP-Tree) service with the corporate data center UNI designated as the root and franchise UNIs designated as leaves**

    d. One Ethernet Private Line (EPL) service between each franchise and the corporate data center

    Note: *(a) Works, but also interconnects franchises which is not desirable because each franchise is independently owned (they are different businesses). (b) Works, but requires VLAN coordination between the different franchise locations, corporate data center and the service provider service provider. (c) Works best. (d) Works, but requires a separate UNI at the corporate data center for each franchise which may result in higher equipment costs due to the many physical connections at headquarters.*

13. Which two statements about a VLAN-based (virtual private) service are true? (Choose two.)

    a. All UNIs must be configured to support service multiplexing.

    b. All UNIs must be configured to support bundling.

    c. All UNIs must be configured to support all-to-one bundling.

    d. **All UNIs may be configured to support service multiplexing.**

    e. **All UNIs may be configured to support bundling.**

    f. All UNIs may be configured to support all-to-one bundling.

    Note: *Refer to* Port-Based versus VLAN-Based Services.

14. A UNI is configured without bundling or all-to-one bundling and is also configured to support service multiplexing. What is the maximum number of CE-VLAN IDs that can be mapped to a given EVC at the UNI?

- **a.** 0
- **b.** 1
- **c.** 256
- **d.** 4094
- **e.** 4096

Note: *Without bundling or all-to-one bundling, only one CE-VLAN ID can be mapped to an EVC. Refer to* Bundling *and* Bundling versus Service Multiplexing.

15. Which Carrier Ethernet services support bridging (frame forwarding based on MAC address learning)? (Choose all answers that are correct.)

- **a.** EPL
- **b.** EVPL
- **c.** **EP-LAN**
- **d.** **EVP-LAN**
- **e.** **EP-Tree**
- **f.** **EVP-Tree**

Note: *Refer to* Matrix of Service Distinctions.

14.3 Answers: How MEF Services are Defined (Part 1)

1. The MEF refers to an Ethernet frame transmitted across a UNI towards the subscriber as a (an):

   a. L2CP frame

   b. Subscriber frame

   c. **Service frame**

   d. UNI frame

   Note: *An Ethernet frame transmitted across the UNI toward the MEN (CEN) is also called a service frame. Refer to* Service Frames.

2. For an EVPL service, which two options are valid for Link OAM processing? (Choose two.)

   a. **Discard**

   b. Tunnel

   c. **Peer**

   d. Peer and Tunnel

   e. Discard and Tunnel

   Note: *Refer to* Per Protocol, Service-Specific L2CP Processing Requirements.

3. An EVC is provisioned with a maximum transmission unit (MTU) size of 1526 bytes. A service frame containing 1530 bytes is transmitted to the UNI from the CE. Which statement is correct?

   a. The frame should be dropped.

   b. The frame should be marked yellow (drop-eligible) and forwarded.

   c. **The service level specification (SLS) does not apply to the frame.**

   d. The frame should be truncated to 1526 bytes and forwarded.

   Note: *Refer to* MTU Size Service Attributes.

4. A broadcast service frame is transmitted from the subscriber to a UNI that is provisioned to support service multiplexing. Where will the service frame be delivered, assuming that it is not discarded?

   a. To all UNIs of each EVC multiplexed at the UNI

   b. To all UNIs in the EVC

c. To all EVCs multiplexed at the UNI

d. To all UNIs in the EVC, except the ingress UNI

e. To all UNIs in the EVC, except the egress UNI

Note: *An Ethernet frame cannot be assigned to more than one EVC. Refer to* Assigning Ethernet Frames to EVCs.

5. Which two services are required to tunnel L2CP service frames containing Bridge Protocol Data Units (BPDUs) to support Spanning Tree Protocol (STP)? (Choose two.)

a. EVP-LAN

b. EP-Tree

c. EVP-Tree

d. EPL

Note: *Tunneling is the process by which an L2CP service frame is passed through the service provider network without being processed and is delivered unchanged to the proper UNI(s).*

Note: *All port-based services are required to tunnel L2CP frames containing BPDUs to support STP, so EP-LAN service would also be correct if it were an option. Refer to* L2CP Processing Requirements for Bridging.

6. Which two of the following service attributes are *EVC per UNI* attributes? (Choose two.)

a. Maximum number of EVCs

b. Ingress Bandwidth Profile per CoS Identifier

c. CE VLAN ID Preservation

d. CE-VLAN ID / EVC Map

e. Bundling

Note: *Refer to* Bird's-Eye View of Service Attributes.

7. For an EPL service, which option is valid for L2CP processing of bridge protocol data units (BPDUs) used for spanning tree protocol (STP)?

a. Discard

b. Tunnel

c. Peer

d. Peer and Tunnel

Note: *Refer to* L2CP Processing Requirements for Bridging.

8. A EVPL service has to support an application that uses IP packets of length up to 800 bytes. What is the smallest value of attribute *EVC MTU Size* that can be specified for this service?

   a. 800 bytes

   b. 818 bytes

   c. 822 bytes

   d. 1500 bytes

   e. 1518 bytes

   f. 1522 bytes

   Note: *A C-Tagged Ethernet frame contains 22 bytes more than its payload (*Size of Service Frame Compared to Payload *). In this scenario the largest payload is an IP packet of length 800 bytes. So the service must support frames up to length 800+22=822 bytes. Additionally, the MEF requires that* ***EVC MTU Size*** *≥1522 bytes (*MTU Size Service Attributes*). The smallest value meeting both requirements is 1522 bytes.*

9. Which is a requirement of bundling?

   a. All CE-VLAN IDs are mapped to a single EVC.

   b. The UNI must be configured to support service multiplexing.

   c. The service must be an EVPL, EVP-LAN, or EVP-Tree service.

   d. More than one EVC must terminate at the UNI.

   Note: *(a) False. (b) False. Service multiplexing is optional. If service multiplexing is off, the UNI can support only one EVC, but that EVC can be bundled. (c) True. Bundling is only supported on VLAN-based services. (d) False. Not required. If you chose this answer, you may be confused about the difference between bundling and service multiplexing. Refer to* Bundling versus Service Multiplexing.

10. A UNI is configured without bundling or all-to-one bundling and is also configured to support service multiplexing. What is the maximum number of CE-VLAN IDs that can be mapped to a given EVC at the UNI?

    a. 4095

    b. 4094

    c. 64

    d. 1

    Note: *Without bundling or all-to-one bundling, only one CE-VLAN ID can be mapped to an EVC. Refer to* Service Multiplexing, Bundling, and VLAN Tag Preservation Service Attributes.

11. A UNI supports service multiplexing. Which statement is <u>not</u> true?

- **a.** The UNI can support up to 4094 EVCs.
- **b.** Bundling is optional at the UNI.
- **c.** At the UNI, at least one CE-VLAN ID must be mapped to each EVC.
- **d.** **At the UNI, all-to-one bundling is permitted.**

Note: *(a) True. There are 4094 unique CE-VLAN IDs. 4094 EVCs can be supported if only one CE-VLAN ID is mapped to each. (b) True. (c) True. (d) False. All-to-one bundling and service multiplexing are mutually exclusive. Refer to* Service Multiplexing, Bundling, and VLAN Tag Preservation Service Attributes.

12. Which statement is <u>not</u> true for a UNI configured with all-to-one bundling?

- **a.** Service multiplexing is No.
- **b.** Bundling is No.
- **c.** CE-VLAN ID preservation is Yes.
- **d.** CoS Preservation is Yes.
- **e.** **The UNI can support up to 4094 EVCs.**

Note: *Refer to* Service Multiplexing, Bundling, and VLAN Tag Preservation Service Attributes.

13. An EPL service has to support jumbo frames with payload of up to 3000 bytes. What is the smallest value of attribute *EVC MTU Size* that can be specified for this service?

- **a.** 3000 bytes
- **b.** 3018 bytes
- **c.** **3022 bytes**
- **d.** 3026 bytes
- **e.** 3030 bytes

Note: *A C-Tagged Ethernet frame contains 22 bytes more than its payload (*Size of Service Frame Compared to Payload *). In this scenario the largest payload is 3000 bytes. So the service must support frames up to length 3000+22=3022 bytes. Additionally, the MEF requires that* ***EVC MTU Size*** *≥1522 bytes (*MTU Size Service Attributes*). The smallest value meeting both requirements is 3022 bytes.*

14. The *CE-VLAN ID for Untagged and Priority Tagged Service Frames* service attribute applies to which three of the following services? (Choose three.)

- **a.** EPL services
- **b.** **EVPL services**
- **c.** EP-LAN services
- **d.** **EVP-LAN services**
- **e.** EP-Tree services
- **f.** **EVP-Tree services**

Note: *Refer to* Service Frame Mapping Service Attributes.

15. Which statement about Layer 2 Control Protocol (L2CP) handling is not true?

- **a.** The UNI *L2CP Processing* attribute must specify *Peer*, *Discard*, *Pass to EVC*, or *Peer and Pass to EVC*.
- **b.** For EVPL services, Link OAM L2CP frames must be peered or discarded.
- **c.** The EVC *L2CP Processing* attribute must specify *Tunnel* or *Discard*.
- **d.** **For EVPL services, BPDUs for STP should be tunneled, or may be discarded.**

Note: *Refer to* L2CP Processing Service Attributes.

16. The smallest value allowed for the *EVC MTU Size* attribute is:

- **a.** 4096 bytes
- **b.** 4094 bytes
- **c.** 1526 bytes
- **d.** **1522 bytes**

Note: *Refer to* MTU Size Service Attributes.

14.4
Answers: How MEF Services are Defined (Part 2)

1. At UNIs, which three of the following are MEF-permissible CoS identifiers? (Choose three.)

    a. Ethernet DEI bit

    b. **Differentiated services code point (DSCP) in the IP header**

    c. **PCP bits in IEEE 802.1Q customer VLAN tag (CE-VLAN tag)**

    d. Ethernet CFI bit

    e. **EVC identifier**

    Note: *Refer to* CoS Identification at UNIs.

2. Performance monitoring estimates frame delivery performance for:

    a. All service frames admitted to the EVC

    b. Yellow service frames

    c. All service frames, except L2CP service frames

    d. **Green service frames, per CoS**

    e. Green and Yellow service frames

    Note: *Refer to* EVC Performance Service Attributes.

3. Which three performance objectives are defined using a sliding window? (Choose three.)

    a. **Availability**

    b. Frame Delay (FD)

    c. Mean Frame Delay (MFD)

    d. Inter-Frame Delay Variation (IFDV)

    e. **High Loss Intervals (HLI)**

    f. **Consecutive High Loss Intervals (CHLI)**

    Note: *Refer to* Availability Performance *and* Resiliency Performance.

4. What is the maximum number of ordered UNI pairs in an EP-Tree service that has four leaf UNIs and two root UNIs?

   a. 6

   b. 8

   c. 9

   d. 10

   e. **18**

   Note: *4*2*2 (root-leaf ordered UNI pairs) + 2 (root-root ordered UNI pairs) = 18. Refer to* Ordered UNI Pair.

5. Which one of the following statements is false?

   a. The SLS is considered part of the SLA.

   b. The SLS contains technical details in engineering terms.

   c. The SLA includes all business agreements related to the service.

   d. **The SLA is considered part of the SLS.**

   Note: *Refer to* The MEF Service Agreement Framework (SLA and SLS).

6. The one-way Frame Delay performance objective for an EP-LAN service is specified as 80 ms. Which statement best describes the meaning of this performance objective?

   a. During an agreed time period, green service frames on average should spend no more than 80 ms in transit between CEs.

   b. **For each ordered UNI pair in the pre-agreed subset of UNI pairs, the average time a green service frame spends in transit between CEs should not exceed 80 ms.**

   c. No service frame should spend more than 80 ms in transit between CEs.

   d. For each ordered UNI pair in the set of all ordered UNI pairs, the average time a green service frame spends in transit between CEs should not exceed 80 ms.

   e. No green service frame should spend more than 80 ms in transit between CEs.

   Note: *Refer to* Frame Delay Performance.

7. A Carrier Ethernet service for a bank needs to deliver critical data that cannot be dropped. Bandwidth profiles should be set with:

   a. CIR>0, CBS=0, EIR=0, EBS=0

   b. **CIR>0, CBS>0, EIR=0, EBS=0**

c. CIR>0, CBS>0, EIR>0, EBS>0

d. CIR=0, CBS>0, EIR=0, EBS=0

e. CIR=0, CBS=0, EIR>0, EBS>0

Note: *Refer to* Single-Rate versus Dual-Rate Bandwidth Profile Implementations.

8. FLR (Frame Loss Ratio) performance is determined by:

a. [1 + (green service frames delivered)/(green service frames sent)] ×100

b. [1 + (total service frames delivered)/(total service frames sent)] ×100

c. [1 – (total service frames delivered)/(total service frames sent)] ×100

d. [1 – (green service frames delivered)/(green service frames sent)] /100

e. **[1 – (green service frames delivered)/(green service frames sent)] ×100**

Note: *Refer to* Frame Loss Ratio Performance.

9. A Service Level Agreement (SLA) is which one of the following?

a. **A contract between the subscriber and service provider specifying a business agreement**

b. An agreement stating service-level commitments in engineering terms

c. A technical specification for the service

d. A specification of service levels offered by the service provider

e. An agreement about service performance attributes, including frame delay, frame delay variation, frame loss ratio, and availability

Note: *Refer to* The MEF Service Agreement Framework (SLA and SLS).

10. Given a continuous stream of service frames with rate greater than EIR + CIR bits per second, approximately how many bytes of service frames will the bandwidth profile algorithm declare to be Yellow over a time interval of T seconds.

a. EIR * T bytes

b. **EIR/8 * T bytes**

c. CIR/8 * T bytes

d. (CIR+EIR) * T bytes

e. (EIR-CIR)/8 * T bytes

Note: *Refer to* Bandwidth Profiles.

11. A Subscriber requires an EVC that supports three Classes of Service: *Silver*, *Gold*, and *Diamond*. Which one of the following schemes is acceptable for identifying CoS at UNIs using PCP bit values (0-7)?

    a. Silver (5); Gold (6); Diamond (7)

    b. **Silver (7, 6); Gold (5, 4); Diamond (3, 2, 1, 0)**

    c. Silver (6); Gold (4); Diamond (2)

    d. Silver (0,1, 2); Gold (5); Diamond (6)

    e. Silver (4); Gold (5); Diamond (4)

    Note: *Refer to* CoS Identification at UNIs.

12. Which one of the following statements is not true?

    a. A performance attribute can include more than one performance objective.

    b. Performance is judged by comparing the measured value of the performance metric to the performance objective.

    c. A performance attribute can include more than one performance metric.

    d. **A performance metric is defined by a set of performance objectives (possibly null) and one performance parameter.**

    Note: *Refer to* EVC Performance Attributes: Metrics, Objectives, and Parameters.

13. When a performance attribute is measured over an agreed subset of UNI pairs (more than one UNI pair), how is the final value of the performance attribute calculated?

    a. By taking the best case (highest performing) UNI pair measurement

    b. By averaging the UNI pair measurements

    c. **By taking the worst-case (lowest performing) UNI pair measurement**

    d. By determining the median value within the set of UNI pair measurements

    Note: *Refer to* EVC Performance Service Attributes.

14. Which set of parameter settings ensures that the bandwidth profile algorithm will never declare a service frame green?

    a. CIR>0, EIR=0

    b. EBS=0, CBS>0

c. **CBS=0, EBS>0**

d. EIR>0, EBS>0

e. EBS>0, CBS>0

Note: *Refer to* Explaining the Bandwidth Profile Algorithm.

15. Four EVCs are service multiplexed to a UNI that is configured with the following per-UNI ingress bandwidth profile.

CIR (Committed Information Rate) = 200 Mb/s
CBS (Committed Burst Size) = 20,000 bytes
EIR (Excess Information Rate) = 800 Mb/s
EBS (Excess Burst Size) = 25,000 bytes
CF (Coupling Flag) = N
CM (Color Mode) = Color-Blind

Which statement is accurate?

a. The rate of yellow frames for each EVC is limited to 200 Mb/s.

b. **The total rate of green frames for all EVCs is limited to 200 Mb/s.**

c. The size of green frames is limited to 25,000 bytes.

d. The rate of green frames for each EVC is limited to 50 Mb/s.

e. The size of yellow frames is limited to 20,000 bytes.

Note: *Refer to* Ingress Bandwidth Profile per UNI Service Attribute.

16. Availability for a particular one-way service between two UNIs is defined with a sliding window of n=13 small time intervals Δt. Which condition ensures that the first Δt in the sliding window is declared to be **available**? (No Δt is part of a maintenance interval.)

a. The frame loss ratio (*flr*) for at least 7 Δt's in the sliding window is greater than threshold *C*.

b. The frame loss ratio (*flr*) for at least 7 Δt's in the sliding window is less than threshold *C*.

c. **The frame loss ratio (*flr*) for all Δt's in the sliding window is less than threshold *C*.**

d. The frame loss ratio (*flr*) for all Δt's in the sliding window is greater than threshold *C*.

Note: *Refer to* Availability Performance.

17. What happens to an Ethernet service frame that is that is marked Green by the bandwidth profile algorithm?

    a. It is discarded.

    b. It is placed in a higher class of service.

    c. It becomes drop-eligible.

    d. **It is forwarded.**

    e. It is buffered.

    Note: *Refer to* Color-Blind Bandwidth Profile Operation.

18. If a service frame is marked Yellow at MEN (CEN) ingress, which one of the following statements is true?

    a. **The subscriber is exceeding CIR/CBS, but not EIR/EBS.**

    b. The subscriber is exceeding CIR/CBS and EIR/EBS.

    c. The subscriber is exceeding EIR/EBS, but not CIR/CBS.

    d. The subscriber is not exceeding EIR/EBS or CIR/CBS.

    Note: *Refer to* Bandwidth Profiles.

19. Which two performance metrics are defined as a percentage?

    a. ***Availability* and *Frame Loss Ratio (FLR)***

    b. *High Loss Interval (HLI)* and *Consecutive High Loss Interval (CHLI)*

    c. *Inter-Frame Delay Variation (IFDV)* and *Mean Frame Delay (MFD)*

    d. *Frame Delay (FD)* and *Mean Frame Delay (MFD)*

    Note: *Refer to* Availability Performance *and* Frame Loss Ratio Performance.

20. Which four services allow specification of per-CoS Egress Bandwidth Profiles at UNIs? (Choose four.)

    a. EPL

    b. **EP-LAN**

    c. **EP-Tree**

    d. EVPL

e. **EVP-LAN**

f. **EVP-Tree**

Note: *Refer to* Bandwidth Profile Service Attributes.

21. What is the maximum number of ordered UNI pairs in a EP-LAN service that includes four UNIs?

a. 4

b. 6

c. 8

d. 12

e. 16

Note: *A multipoint-to-multipoint EVC with n UNIs has n*(n-1) ordered UNI pairs. Refer to* Ordered UNI Pair.

14.5 Answers: UNI Requirements

1. A MEF Type 2.2 UNI is required to support which type of protection?

    a. Layer 2 Control Protocol (L2CP)

    b. Link OAM per clause 57 of IEEE 802.3

    c. G.8031 Ethernet Linear Protection Switching

    d. **Link aggregation per clause 43 of IEEE 802.3**

    e. G.8032 Ethernet Ring Protection Switching

    Note: *(a) Not a protection mechanism. (b) Not a protection mechanism. (c) Not required. (d) Correct. (e) Not required. Refer to* Link Protection Requirement for Type 2.2 UNIs.

2. Which of the following provides port protection?

    a. L2CP

    b. **Link aggregation**

    c. E-LMI

    d. Link OAM

    e. Type 1 UNI

    Note: *Refer to* Link Protection Requirement for Type 2.2 UNIs.

3. Which two of the following statements about the Ethernet Local Management Interface (E-LMI) are true? (Choose two.)

    a. E-LMI protects against port failure using link aggregation.

    b. E-LMI is required for Type 2.1 UNIs.

    c. **E-LMI notifies the UNI-C of changes to EVC status.**

    d. E-LMI allows the subscriber and the service provider to monitor and diagnose link-level connectivity issues between UNI-C and UNI-N.

    e. **E-LMI allows UNI-C to acquire configuration information from UNI-N so that it can automatically configure itself.**

    Note: *(a) False. (b) Required for Type 2.2, but not Type 2.1. (c) True. (d) False. Describes LOAM. (e) True. Refer to* E-LMI Requirement for Type 2.2 UNIs.

4. Which three of the following requirements are mandatory for Type 2.2 UNIs, but optional for Type 2.1 UNIs? (Choose three.)

    a. **Link aggregation support**

    b. Service OAM support

    c. **Link OAM support**

    d. Support of multiplexed UNI for services such as EVPL

    e. Backward compatibility with UNI Type 1

    f. **E-LMI support**

    Note: *(a) Correct. (b) Mandatory for Type 2.1. (c) Correct. (d) Mandatory for Type 2.1. (e) Mandatory for Type 2.1. (f) Correct.*

5. Which one of the following statements about Type 2 UNIs is <u>not</u> true?

    a. Type 2 UNIs support Service OAM.

    b. MEF 20 describes two types of Type 2 UNIs.

    c. **Type 2 UNI requirements apply only to the UNI-N (they are not applicable to the UNI-C).**

    d. Type 2 UNIs are backward compatible with Type 1 UNIs.

    e. Type 2 UNIs support L2CP handling as defined in MEF 6.1 and MEF 10.2.

    Note: *Refer to* Type 2 UNI Requirements.

6. UNI Type 2.2 required protection mechanisms protect against which three types of failure? (Choose three.)

    a. **Port failure on UNI-N**

    b. Loss of timing

    c. Loss of "green" service frames due to bandwidth oversubscription

    d. E-LMI denial of service (DOS) attack

    e. **Port failure on UNI-C**

    f. **Disconnection or failure of a cable connecting the CE and MEN (CEN)**

    Note: *Refer to* Link Protection Requirement for Type 2.2 UNIs.

14.6 Answers: Extending MEF Services over Multiple Operator MEN/CENs

1. Which entity delivers Ethernet frames between ENNIs and UNIs/ENNIs?

    a. **Operator Virtual Connection**

    b. Ethernet Virtual Circuit

    c. Intermediate Virtual Connection

    d. Operator Virtual Circuit

    e. Ethernet Virtual Connection

    Note: *Refer to OVC.*

2. Per MEF 26.1, which three values are recognized for the *Protection Mechanism* ENNI service attribute? (Choose three.)

    a. LACP

    b. **None**

    c. **Link Aggregation**

    d. E-LMI

    e. **Other**

    Note: *Refer to Protection at ENNI.*

3. The *OVC End Points* of an operator virtual connection (OVC) must associate to:

    a. Exactly two UNIs and any number of ENNIs

    b. Exactly one ENNI and any number of UNIs

    c. UNIs only

    d. **At least one ENNI**

    e. Only ENNIs

    Note: *Refer to OVC and OVC End Point.*

4. Two or more operators can:

   a. connect through an ENNI

   b. be involved in an OVC

   c. be responsible for an EVC

   d. be involved in an EVC

   e. be responsible for an OVC

   Note: *Refer to* Introduction *and* Terminology. *Answer (a) is wrong because an ENNI connects two operators, not two or more operators.*

5. Which three frame formats are permitted at an ENNI? (Choose three.)

   a. Untagged

   b. C-Tag

   c. S-Tag and no C-Tag

   d. S-Tag and C-Tag

   e. Stacked S-Tags

   Note: *Refer to* ENNI Tagging Requirements.

6. How many EPL services can an ENNI support?

   a. One

   b. Two

   c. Up to 4094

   d. About 16 million

   e. Unlimited

   Note: *An ENNI can support up to 4094 services of any type (including EPL services). The number of services is limited by the number of unique S-VLAN ID values. Refer to* Encoding Information in S-Tags.

7. Which statement about the External Network-to-Network Interface (ENNI) is true?

   a. The subscriber and service provider agree on its exact physical location so that responsibilities are clearly defined.

   b. It can only support one EPL service.

c. Service provider is responsible for the service from the UNI to the ENNI, and then the organization in charge of the Operator MEN/CEN is responsible.

d. **It is a physical demarcation point between two Operator MEN/CENs.**

Note: *Refer to* ENNI and ENNI-N.

8. An OVC requires which one of the following?

a. A UNI

b. Two or more UNIs

c. Two or more EVCs

d. **An ENNI**

e. Two or more ENNIs

Note: *Refer to* OVC.

9. An *OVC End Point Map* has *bundling*. Which statement is true about the associated OVC?

a. *S-VLAN ID Preservation* can be *Yes* or *No*.

b. *CE-VLAN ID Preservation* can be *Yes* or *No*.

c. *CE-VLAN CoS Preservation* can be *Yes* or *No*.

d. ***S-VLAN ID CoS Preservation* can be *Yes* or *No*.**

Note: *S-VLAN ID Preservation, CE-VLAN ID Preservation, and CE-VLAN CoS Preservation must all be Yes. Refer to* End Point Map Bundling.

10. In which category of operator service attributes is *Broadcast Service Frame Delivery*?

a. **OVC Service Attributes**

b. PVC End Point per ENNI Service Attributes

c. ENNI Service Attributes

d. UNI Service Attributes

e. OVC per UNI Service Attributes

Note: *Refer to* OVC Service Attributes.

11. An *OVC End Point Map* has *bundling*. Which statement is not true about the associated OVC?

- **a.** The OVC cannot support hairpin switching.
- **b.** All end points of the OVC have bundling.
- **c.** *S-VLAN ID Preservation* is set to *Yes*.
- **d.** **At least one end point of the OVC terminates at a UNI.**
- **e.** The OVC is point-to-point.

Note: *Refer to* End Point Map Bundling.

12. Two Operator CENs connect through an ENNI. What is the advantage of using an ENNI with Link OAM?

- **a.** It provides link protection using two physical links, one active and one standby
- **b.** It allows the ENNI to support hairpin switching.
- **c.** **It allows both operators to monitor and diagnose ENNI connectivity issues**
- **d.** It allows two different types of physical layers to be connected at the ENNI
- **e.** It allows class of service (CoS) propagation through IEEE 802.1ad Ethernet Frames

Note: *Refer to* LOAM Requirement for ENNIs.

13. Which OVC service attribute is incompatible with hairpin switching?

- **a.** CE-VLAN CoS Preservation = Yes
- **b.** **S-VLAN ID Preservation = Yes**
- **c.** S-VLAN CoS Preservation = Yes
- **d.** CE-VLAN ID Preservation = Yes

Note: *Refer to* Hairpin Switching.

14. Which statement is not true about the service provider in the MEF 26.1 service model?

- **a.** The service provider is responsible for the service to the subscriber.
- **b.** The service provider is responsible for coordinating operators.
- **c.** The service provider may or may not be an operator.
- **d.** **Two or more operators can collaboratively function as the service provider.**

Note: *Refer to* Service Provider and Operators.

15. The smallest maximum transmission unit (MTU) size allowed for an operator virtual connection (OVC) is:

   a. 1522 bytes

   b. 1526 bytes

   c. 4094 bytes

   d. 4096 bytes

   e. None of the above

   Note: *Refer to* ENNI and OVC MTU Size.

16. How many EVCs can be supported by one OVC?

   a. Exactly one

   b. Up to 4094 if the OVC is point-to-point between two ENNIs

   c. Up to 4094 if UNI is configured with *Service Multiplexing = Yes*

   d. Exactly one if the UNI is configured with *All to One Bundling = Yes*

   Note: *Normally an OVC represents a segment of one EVC. However, the OVC with end point map bundling is a special case. Refer to* End Point Map Bundling.

17. Which three of the following statements are true?
(Choose three.)

   a. At ENNIs, drop-eligibility can be encoded in the S-Tag DEI bit or in the S-Tag PCP value.

   b. At ENNIs, all frames require C-Tags.

   c. At an ENNI, both Operators are responsible for setting S-Tag values appropriately before transmitting frames across the ENNI.

   d. Ingress ENNI frames are mapped to OVC end points based on S-VLAN ID value.

   e. At ENNIs, CoS may be encoded in the C-Tag or in the S-Tag.

   Note: *Refer to* Encoding Information in S-Tags.

18. Which of the following is <u>not</u> an ENNI requirement?

   a. ENNIs are required to support E-LMI.

   b. At ENNIs, all bandwidth profiles must be color-aware.

   c. ENNIs must support MTU sizes at least as large as 1526 bytes.

d. If an ENNI includes two physical links, it must be capable of implementing link aggregation.

e. ENNIs are required to support Link OAM.

Note: *(b) Refer to* Color-Aware Bandwidth Profiles at ENNI. *(c) Refer to* ENNI and OVC MTU Size. *(d) Refer to* Protection at ENNI. *(e) Refer to* LOAM Requirement for ENNIs.

19. Which of the following is an ENNI service attribute?

a. Bundling

b. **End Point Map**

c. Service Multiplexing

d. All-to-one bundling

Note: *Refer to* ENNI Service Attributes.

20. The *End Point Map* attribute:

a. is inactive if the ENNI is provisioned with all-to-one bundling

b. **specifies how ingress ENNI frames are assigned to OVC end points based on S-VLAN ID**

c. lists the CE-VLAN IDs that map to the EVC

d. is an *OVC End Point per ENNI* service attribute

Note: *Refer to* End Point Map.

21. A rooted-multipoint OVC has 5 OVC end points: 2 are designated Root, 2 are designated Leaf, and 1 is designated Trunk. A broadcast frame ingresses at a Leaf OVC end point. How many frames does the OVC deliver if OVC attribute *Broadcast Service Frame Delivery* is configured to *Deliver Unconditionally*?

a. 1

b. 2

c. **3**

d. 4

e. 5

Note: *In this scenario, the broadcast frame must be delivered to both Root end points and to the Trunk end point, but not to the other Leaf end point. At the Trunk end point, the broadcast frame must be transmitted with S-VLAN ID set to the* ***Leaf S-VLAN ID*** *value, indicating that it originated from a Leaf UNI (*Trunk Identifiers*).*

22. Why is the minimum MTU size requirement at ENNIs (1526 bytes) four bytes larger than the minimum MTU size requirement at UNIs (1522 bytes)?

   a. Because ENNIs do not support all-to-one bundling.

   b. To allow hairpin switching.

   c. To account for the S-Tag.

   d. Because color-aware bandwidth profiles are required at ENNIs.

   Note: *Refer to* ENNI and OVC MTU Size.

23. An EVPL service is implemented across three Operator CENs. How many UNIs, ENNIs, EVCs, and OVCs are involved?

   a. 2 UNIs, 2 ENNIs, 3 EVCs, and 1 OVC

   b. 2 UNIs, 3 ENNIs, 1 EVC, and 3 OVCs

   c. 2 UNIs, 2 ENNIs, 1 EVC, and 3 OVCs

   d. 2 UNIs, 3 ENNIs, 1 EVC, and 2 OVCs

   Note: *One point-to-point EVC: UNI–OVC–ENNI–OVC–ENNI–OVC–UNI. Refer to* OVC.

24. An EPL service spans three Operator MENs and each OVC is configured with *S-VLAN ID Preservation = Yes*. Which of the following statements is true?

   a. The S-Tag VLAN identifier will match at both UNIs.

   b. The S-Tag VLAN identifier is preserved within each OVC, but can be translated at the ENNIs.

   c. The S-Tag VLAN identifier will match at both ENNIs.

   d. The S-Tag VLAN identifier will match at both UNIs and both ENNIs.

   Note: *(a) S-Tags are not defined at UNIs. (b) ENNIs do not support VLAN ID translation. (c) With S-VLAN ID Preservation = Yes, the OVC that connects the two ENNIs ensures that the S-Tag VLAN ID matches at both ENNIs. (d) S-Tags are not defined at UNIs. Refer to* OVC Service Attributes.

25. An EP-LAN service spans two Operator CENs that connect through one ENNI. At all UNIs, ingress bandwidth profiles are configured with CIR=0, CBS=0, EIR=20 Mb/s, and EBS=2,000 bytes. How should the service be configured to ensure that yellow frames are not "promoted" to green?

   a. At all OVC end points, configure ingress and egress bandwidth profiles with *Color Mode = Color Aware*.

   b. At those OVC end points that connect to the ENNI (not those connecting to UNIs), configure ingress and egress bandwidth profiles with *Color Mode = Color Aware*.

c. **For both OVCs, configure *Color Forwarding* = *Yes*.**

d. At those OVC end points that connect to UNIs (not those connecting to the ENNI), configure ingress bandwidth profiles with *Color Mode* = *Color Aware*. Also, do not configure any bandwidth profiles at the OVC end points that connect to the ENNI.

Note: *Refer to* Color Forwarding.

26. Which statement about CoS identification at ENNIs is not true?

a. CoS is determined by S-Tag PCP bit values.

b. **At a given OVC end point, an S-Tag PCP bit value can map to more than one CoS.**

c. For each OVC end point, a CoS must be associated with each S-Tag PCP value.

d. At a given OVC end point, two different S-Tag PCP bit values can map to the same CoS.

e. The S-Tag supports up to eight classes of service.

Note: *Refer to* Encoding Information in S-Tags.

27. A service is implemented over 3 Operator MEN/CENs. Which three statements are correct? (Choose three.)

a. All three operators are responsible to the subscriber for the service.

b. **One of the operators can be the service provider.**

c. The subscriber has a business relationship with each operator.

d. **Each Operator MEN/CEN includes at least one ENNI-N.**

e. **At least two of the operators have no business relationship with the subscriber.**

Note: *Refer to* Service Provider and Operators *and* ENNI and ENNI-N.

28. An ENNI is to support many EVCs, including some EVCs configured with *EVC MTU Size* as large as 1800 bytes. What is the smallest value of the ENNI MTU Size that will support these EVCs?

a. 1526 bytes

b. 1800 bytes

c. **1804 bytes**

d. 2000 bytes

Note: *To support the S-Tag, the ENNI must support MTUs that are at least 4 bytes larger than the largest EVC MTU. Refer to* ENNI and OVC MTU Size.

29. A point-to-point OVC-based service can be configured with OVC attribute *S-VLAN ID Preservation* set to *Yes* or *No*. Which two statements are correct? (Choose two.)

a. If *S-VLAN ID Preservation* = *Yes*, the OVC-based service involves exactly 1 S-VLAN ID value.

b. **If *S-VLAN ID Preservation* = *Yes*, the OVC-based service can involve up to 4094 S-VLAN ID values.**

c. If *S-VLAN ID Preservation* = *No*, the OVC-based service involves exactly 2 S-VLAN ID values.

d. **If *S-VLAN ID Preservation* = *No*, the OVC-based service involves 1 or 2 S-VLAN ID values.**

Note: *Answer (b) is correct, not (a), because S-VLAN bundling is permitted if the OVC is point-to-point between ENNIs with* ***S-VLAN ID Preservation=Yes*** *(refer to* End Point Map Bundling*). Answer (d) is correct, not (c), because* ***S-VLAN ID Preservation=No****, does not require S-VLAN ID translation, but only permits it.*

14.7
Answers: MEF-Standardized Classes of Service

1. How many classes of service does MEF 23.1 define for Performance Tier 2?

    a. 2 classes of service

    b. 3 classes of service

    c. 4 classes of service

    d. 6 classes of service

    e. 12 classes of service

    Note: *In each performance tier, MEF 23.1 defines 3 standardized classes of service denoted by CoS Labels L, M, and H (*CoS Labels and Performance Tiers*).*

2. An *Ingress Bandwidth Profile Per ENNI Class of Service Identifier* is required at an OVC end point to police ingress ENNI frames marked with *CoS Label M*. Which one of the following ingress bandwidth profiles satisfies all MEF requirements?

    a. CIR=0, CBS=0 bytes, EIR=200 Mb/s, EBS=2000 bytes, CF=Y, CM=Y

    b. CIR=200 Mb/s, CBS=2000 bytes, EIR=200 Mb/s, EBS=2000 bytes, CF=N, CM=N

    c. CIR=200 Mb/s, CBS=2000 bytes, EIR=200 Mb/s, EBS=2000 bytes, CF=N, CM=Y

    d. CIR=0, CBS=0 bytes, EIR=200 Mb/s, EBS=2000 bytes, CF=Y, CM=N

    Note: *MEF 23.1 requires CIR>0 for bandwidth profiles associated with CoS Label M or H (*Ingress Bandwidth Profile Constraints (per CoS Label)*). At ENNIs, bandwidth profiles (if used) must be color-aware (*Color-Aware Bandwidth Profiles at ENNI*).*

3. A point-to-point service uses *CoS Label H*. Which CPO <u>must</u> be specified in the SLS?

    a. Both Frame Delay (FD) and Frame Delay Range (FDR)

    b. Both Mean Frame Delay (MFD) and Inter Frame Delay Variation (IFDV)

    c. Frame Loss Ratio (FLR)

    d. High Loss Intervals (HLI)

    e. Consecutive High Loss Intervals (CHLI)

    Note: *Refer to* CPOs (per CoS Label and Performance Tier).

4. According to MEF 23.1, if CoS Label and Color are encoded in S-Tag PCP bits, PCP bit value 3 indicates...

   a. M, Yellow

   b. **M, Green**

   c. H, Green

   d. Nothing

   Note: *Refer to* CoS Label and Color Identification Using PCP Bits, DEI Bit, and/or EVC/OVC End Point.

5. Which two statements about Class of Service Performance Objectives (CPOs) are correct? (Choose two.)

   a. CPOs for *CoS Label L* are more demanding than CPOs for *CoS Label H*.

   b. CPOs for *Performance Tier 4* are more demanding than CPOs for *Performance Tier 3*.

   c. **CPOs for *CoS Label M* are more demanding than CPOs for *CoS Label L*.**

   d. **CPOs for *Performance Tier 1* are more demanding than CPOs for *Performance Tier 3*.**

   Note: *Refer to* CoS Labels and Performance Tiers.

6. How many performance tiers does MEF 23.1 define for CoS Label H?

   a. 3 performance tiers

   b. **4 performance tiers**

   c. 6 performance tiers

   d. 12 performance tiers

   Note: *For each CoS Label, MEF 23.1 defines 4 performance tiers (*CoS Labels and Performance Tiers*).*

7. For each of the following services, all service frames map to *CoS Label L*. Which service is subject to CPO values defined in MEF 23.1?

   a. *Performance Tier 1* EP-LAN service with ingress bandwidth profile CIR > 0.

   b. ***Performance Tier 4* EVPL service with ingress bandwidth profile CIR > 0.**

   c. *Performance Tier 1* EPL service with ingress bandwidth profile CIR = 0.

   d. *Performance Tier 4* EVP-LAN service with ingress bandwidth profile CIR = 0.

   Note: *MEF 23.1 does not define CPO values for multipoint-to-multipoint (or rooted-multipoint) EVCs and OVCs (*CPOs (per CoS Label and Performance Tier)*). MEF 23.1 allows CIR=0 for CoS Label L (*Ingress

Bandwidth Profile Constraints (per CoS Label)*), but if CIR=0, all service frames are yellow (not subject to CPOs).*

8. According to MEF 23.1, if CoS Label and Color is encoded in S-Tag PCP bits, PCP bit value 4 indicates...

a. M, Yellow

b. M, Green

c. H, Green

d. **Nothing**

Note: *Refer to* CoS Label and Color Identification Using PCP Bits, DEI Bit, and/or EVC/OVC End Point.

14.8
Answers: E-Access Services

1. Both the Access EVPL and the Access EPL... (Choose two.)

   a. Allow CE-VLAN ID translation

   b. Associate each service instance to one S-VLAN ID value at the ENNI

   c. Are point-to-point OVCs

   d. Connect one ENNI and one UNI

   Note: *(a) Wrong because both require CE-VLAN ID preservation. (b) Correct. (c) Wrong because they are services involving more than just the OVC. (d) Correct.*

2. Two access providers establish an ENNI connection between their networks. What types of EVC-based services can they offer subscribers using only Access EPL and Access EVPL services? Assume that one of the access providers takes the role of service provider.

   a. E-Line only

   b. E-Line and E-LAN only

   c. E-Line and E-Tree only

   d. E-Line, E-LAN, and E-Tree

3. An E-Access service must support *CM=color-blind* for which bandwidth profile?

   a. Ingress bandwidth profile per OVC end point at UNI

   b. Egress bandwidth profile per OVC end point at UNI

   c. Ingress bandwidth profile per OVC end point at ENNI

   d. Egress bandwidth profile per OVC end point at ENNI

   Note: *Refer to* Special Ingress Bandwidth Profile Requirements. *Also, E-Access services do not support egress bandwidth profiles (*Bandwidth Profiles for E-Access Service*).*

4. Access EVPL differs from EVPL in that...

   a. Access EVPL supports multiple CoS, while EVPL supports only one CoS.

   b. Access EVPL is VLAN-based, while EVPL is port-based.

c. **Access EVPL is OVC-based, while EVPL is EVC-based.**

d. Access EVPL supports CE-VLAN ID Preservation, while EVPL does not.

Note: *(a) Wrong because Access EVPL supports only one CoS, while EVPL supports multiple CoS. (b) Wrong because EVPL is EVC-based, not OVC-based. (c) Correct. (d) Wrong because both support CE-VLAN ID Preservation.*

5. An Ethernet access provider offers Access EPL service with a 1 Gb/s UNI. The value of CIR in service attribute ***Ingress Bandwidth Profile Per OVC End Point*** at the UNI must be configurable to...

a. 500 Mb/s and 1 Gb/s

b. 400 Mb/s and 800 Mb/s

c. **20 Mb/s and 300 Mb/s**

d. 250 Mb/s and 500 Mb/s

Note: *Refer to* Special Ingress Bandwidth Profile Requirements. *(a) Wrong because 1 Gb/s > 70% of UNI speed. (b) Wrong because 800 Mb/s > 70% of UNI speed. (c) Correct. (d) Wrong because 250 Mb/s is not included in the list of values that must be supported.*

6. Which service supports CE-VLAN ID translation?

a. Access EVPL

b. Access EPL

c. **EVPL**

d. EPL

Note: *EVPL supports CE-VLAN ID translation if UNIs are configured with Bundling = No.*

7. E-Access services and E-Line services both include at least one...

a. **UNI**

b. ENNI

c. OVC

d. EVC

8. A service provider has been using four Access EPLs to support an EP-LAN for a subscriber. The subscriber now wants to convert the EP-LAN service to an EP-Tree service. How are the Access EPLs impacted by this change?

   a. They can no longer be used.

   b. The must be converted to Access EVPLs.

   c. They must now filter traffic.

   d. They are not impacted.

   Note: *The multipoint-to-multipoint OVC in the service provider network (operator MEN) would change to a rooted-multipoint OVC. The Access EPLs remain unchanged.*

9. Both Access EVPL and Access EPL...

   a. Support hairpin switching

   b. Can connect a customer site to an Internet service provider

   c. Support multiple CoS

   d. Are EVC-based services

   Note: *(a)Wrong because the general OVC-based operator service can support hairpin switching, but Access EPL/EVPL cannot. (b) Correct. (c) Wrong because both support only one CoS. (d) Wrong because both are OVC-based services.*

10. If an E-Access service is applied to reach an off-net customer site in an EVC-based service application, the EVC-based service...

   a. Can support only one class of service

   b. Cannot support CE-VLAN ID translation

   c. Must be point-to-point

   d. Must involve at least one ENNI

11. As a service provider you have agreed to provide an EPL service supporting IP packets up to 2000 bytes in length. However, one customer site is off-net, so you need to purchase an Access EPL service to reach that customer site. The access provider only offers Access EPLs with ingress bandwidth profiles at the ENNI and only supports the following values of CBS. Which value do you choose if you want to minimize CBS?

   a. CBS = 2000 bytes

   b. CBS = 2010 bytes

   c. CBS = 2020 bytes

d. **CBS = 2030 bytes**

e. CBS = 2040 bytes

Note: *The Access EPL must support C-Tagged Ethernet frames at UNIs which have 22 bytes more than their unencapsulated payload (Size of Service Frame Compared to Payload). In this case the payload is an IP packet of length 2000 bytes. So the Access EPL must support frames of length 2000+22=2022 bytes at the UNI. At the ENNI, an S-Tag is added to the frame (4 more bytes). So the Access EPL must support frames of length 2026 bytes at the ENNI. Therefore, CBS for an ingress bandwidth profile at the ENNI must be at least 2026 bytes. CBS=2030 bytes is the smallest value supported by the access provider that is greater than or equal to 2026 bytes.*

12. Which of the following statements is false?

    a. A UNI can support up to 4094 Access EVPLs.

    b. **A UNI can support up to 4094 Access EPLs.**

    c. An ENNI can support up to 4094 Access EPLs.

    d. An ENNI can support up to 4094 Access EVPLs.

Note: *A UNI can support up to 4094 Access EVPLs (one CE-VLAN ID per Access EVPL), but only one Access EPL (port-based at the UNI). Refer to Access EPL and Access EVPL.*

13. Which two services are used to reach off-net customer sites? (Choose two.)

    a. **Access EVPL**

    b. EVPL

    c. **Access EPL**

    d. EPL

14. Access EPL is similar to EPL in that... (Choose two.)

    a. Both support multiple classes of service.

    b. **Both support untagged frames and C-Tagged frames at the UNI.**

    c. Both support CE-VLAN ID translation.

    d. **Both support CE-VLAN CoS ID preservation.**

    e. Both are OVC-based services.

Note: *(a) Wrong because Access EPL supports only one CoS. (b) Correct. (c) Wrong because neither support CE-VLAN ID translation. (d) Correct. (e) Wrong because EPL is EVC-based, not OVC-based.*

15. An E-Access service must support *CM=color-aware* for which bandwidth profile?

- **a.** Ingress bandwidth profile per OVC end point at UNI
- **b.** Egress bandwidth profile per OVC end point at UNI
- c. **Ingress bandwidth profile per OVC end point at ENNI**
- **d.** Egress bandwidth profile per OVC end point at ENNI

Note: *Refer to* Special Ingress Bandwidth Profile Requirements*. Also, E-Access services do not support egress bandwidth profiles (*Bandwidth Profiles for E-Access Service*).*

14.9
Answers: Ethernet OAM

1. Which one of the following statements describes how frame loss ratio (FLR) is measured for an EPL service?

   a. For each EVC CoS, count all green and yellow frames sent and received over an agreed time interval.

   b. For each EVC CoS, count all green frames sent and received over an agreed time interval.

   c. For an EVC, count all frames sent and received (regardless of CoS) over an agreed time interval.

   d. For an EVC, count all green and yellow frames sent and received over an agreed time interval.

   Note: *Refer to Single-Ended Service Loss and recall, from Key Caveats, that performance objectives apply to green service frames only (they say nothing about delivery of yellow service frames). Therefore, only green frames should be used for performance measurement.*

2. NE A and NE B are connected by an Ethernet link. Brian, an operator, has access only to NE B (NE B is in Brian's domain of operations, but NE A is not). Which two options below can be available (with correct provisioning) to Brian? (Choose two.)

   a. Use SOAM loopback to redirect all traffic at NE A back toward NE B.

   b. Use SOAM loopback to verify link connectivity.

   c. Use LOAM loopback to redirect all traffic at NE B back toward NE A.

   d. Use LOAM loopback to redirect all traffic at NE A back toward NE B.

   e. Use RDI to initiate LOAM link discovery.

   Note: *(a) SOAM loopback does not redirect traffic. Refer to SOAM Loopback. (b) Refer to SOAM Loopback. (c) LOAM loopback creates a port-based loopback on the far side, not the near side. Refer to Link OAM. (d) Refer to Link OAM. (e) This answer is nonsense.*

3. Which of the following is <u>not</u> an SOAM frame?

   a. Connectivity Check Message (CCM)

   b. Alarm Indication Signal (AIS)

   c. Remote defect indication (RDI)

   d. Loss Measurement Message (LMM)

   Note: *Refer to SOAM Frames. (c) RDI is not a SOAM frame. It is encoded in a SOAM frame, the CCM, but it is not itself a SOAM frame. Refer to Remote Defect Indication (RDI).*

4. Which two of the following statements are correct ? (Choose two.)

   a. LOAM support is required for Type 2.1 UNIs and at ENNIs.

   b. SOAM CCMs can support 50 ms protection switching.

   c. SOAM domains can overlap.

   d. An ME can include more than two MEPs.

   e. An ME can include any number of MIPs.

   Note: *(a) LOAM is required for Type 2.2 UNIs and at ENNIs. Refer to* Link OAM. *(b) Refer to* Continuity Check Message (CCM). *(c) False. Refer to* SOAM Domains. *(d) An ME includes only two MEPs. Refer to* Maintenance Entity (ME). *(e) Refer to* Maintenance Entity (ME).

5. If a MEP sends a Loss Measurement Message (LMM), where does it go?

   a. To all MEPs in every ME

   b. To peer MIPs in the same ME

   c. To all MIPs in every ME

   d. To all MEPs in the MEG

   e. To the peer MEP in the same ME

   Note: *Refer to* Single-Ended Service Loss.

6. Per MEF 30.1, to whom does the EVC MEG belong?

   a. Subscriber

   b. Service Provider

   c. Operator

   d. Subscriber and Operator

   e. Operator and Service Provider

   Note: *Refer to* SOAM MEG Levels.

7. MEP clock synchronization is required to measure which two of the following? (Choose two.)

   a. One-way delay with the Single-Ended Delay function

   b. Two-way delay with the Single-Ended Delay function

c. **One-way delay with the Dual-Ended Delay function**

d. Two-way delay with the Dual-Ended Delay function

Note: *Refer to* Single-Ended Delay *and* Dual-Ended Delay.

8. To isolate a fault within a point-to-point OVC, an operator can:

   a. Inspect all continuity check messages (CCMs) for the appropriate ME.

   b. Initiate SOAM Loopback over the appropriate ME.

   c. Send a frame loss measurement message (LMM) to the appropriate peer MEP.

   d. Process the most recent remote defect indicator (RDI) per IEEE 802.1Q.

   e. **Initiate SOAM Link Trace over the appropriate ME.**

   Note: *Refer to* SOAM Link Trace.

9. Which Maintenance Entity (ME) should the service provider use to monitor the performance of an Ethernet service that spans two Operator MENs?

   a. Subscriber ME

   b. **EVC ME**

   c. Operator ME

   d. ENNI ME

   e. UNI ME

   Note: *Refer to* Suggested MEs and Default MEG Levels.

10. Which two SOAM frames are used for performance monitoring? (Choose two.)

   a. CCM (Continuity Check Message)

   b. **LMM (Loss Measurement Message)**

   c. LBM (Loopback Message)

   d. LTM (Link Trace Message)

   e. **DMM (Delay Measurement Message)**

   Note: *Refer to* SOAM Frames.

11. One-way frame delay for a service frame in an EVC-based service... (Choose two.)

    a. Measures time of travel, CE-to-CE, for the first bit in the service frame.

    b. Includes the time of transmission between the UNI-C and the UNI-N at both UNIs.

    c. Includes the time that the service frame waits in queue at the CE to be scheduled for transmission.

    d. Depends on the length of the service frame.

    Note: *Refer to* Frame Delay Performance.

12. Which four of the following items are connectivity fault management (CFM) functions? (Choose four.)

    a. Remote defect indication (RDI)

    b. Link Trace

    c. Frame Delay Measurement

    d. Alarm Indication Signal (AIS)

    e. Loopback

    f. Frame Loss Measurement

    Note: *Refer to* SOAM Connectivity Fault Management. *(c) PM, not CFM. (f) PM, not CFM.*

13. How many ordered UNI pairs are in an EVC containing three UNIs?

    a. Two

    b. Three

    c. Six

    d. Nine

    Note: *Refer to* Ordered UNI Pair.

14. Which two statements about SOAM frames are true? (Choose two.)

    a. AIS frames are sent away from the direction of failure.

    b. AIS frames are sent toward the direction of failure.

    c. CCM frames containing RDI are sent away from the direction of failure.

    d. CCM frames containing RDI are sent toward the direction of failure.

    Note: *Refer to* Remote Defect Indication (RDI) *and* Alarm Indication Signal (ETH-AIS).

14.10
Answers: Access Technologies

1. What are three advantages offered by *Ethernet over Bonded Copper*? (Choose three.)

   a. Easy to deploy in rugged terrain

   b. Low cost

   c. No distance limitation

   d. Supports rates up to 1 Gb/s through bonding of additional copper pairs

   e. Resiliency through bonding copper pairs

   f. Widely available copper infrastructure

   Note: *Refer to* Summary.

2. An office, located 3 miles (4.8 km) from the service provider network, requires 2 Mb/s of symmetric connectivity to an EVP-LAN service. A single copper pair is available. Which of the following access technologies is most appropriate?

   a. Ethernet over DOCSIS

   b. G.SHDSL

   c. Ethernet over PDH

   d. 10PASS-TS

   Note: *Refer to* 2BASE-TL (G.SHDSL).

3. What is the advantage of having Ethernet service over WiMAX?

   a. Bandwidth split among multiple users

   b. No cabling

   c. Resilience

   d. No reach limitation

   Note: *Refer to* Ethernet over Wireless Network.

4. Which access technology can support Ethernet services in parallel with TV services as commonly deployed by MSO/Cable TV operators?

   a. Ethernet over GPON

   b. Ethernet over Bonded Copper

   c. **Ethernet over HFC**

   d. Ethernet over PDH

   e. Ethernet over Terrestrial Microwave

   Note: *Refer to* Ethernet over HFC.

5. A mobile backhaul operator needs to reach a cell tower located in an undeveloped rural area. Which two of the following technologies could be most quickly deployed? (Choose two.)

   a. Ethernet over PDH

   b. **Ethernet over Terrestrial Microwave**

   c. **Ethernet over Free Space Optics**

   d. Ethernet over SONET/SDH

   e. Ethernet over bonded copper pairs

   f. Ethernet over Fiber

   Note: *Refer to* Summary.

6. What are two advantages of *Ethernet over SONET/SDH*? (Choose two.)

   a. Supports multiple classes of service

   b. **Highly reliable**

   c. Supports multipoint services (EP-LAN, EP-Tree)

   d. **Provides dedicated bandwidth that is not impacted by other user traffic**

   e. Supports circuit emulation

   Note: *Refer to* Ethernet over SONET/SDH.

7. An incumbent provider (ILEC/PTT) wants to provide an Ethernet service at 6 Mb/s over its existing copper network infrastructure. The UNI will be 20 km (12.4 miles) away from the nearest switching office. Which technology do you recommend for this application?

   a. Ethernet over bonded copper pairs (2BASE-TL)

   b. Ethernet over HFC/DOCSIS

   c. **Ethernet over bonded T1/E1 PDH circuits**

   d. Ethernet over bonded T3/E3 PDH circuits

   Note: *Refer to* Ethernet over PDH *(also* Summary*).*

8. What is an advantage of *Ethernet over PDH* compared to *Ethernet over Copper*?

   a. Uses widely deployed infrastructure

   b. Easy, low-cost, and rapid to deploy

   c. Supports eight classes of service

   d. Supports resiliency through bonding

   e. **No distance limitation**

   Note: *Refer to* Ethernet over PDH *and* Ethernet over Copper.

9. Which two of the following access technologies support rates over 1 Gb/s? (Choose two.)

   a. **Ethernet over SONET/SDH**

   b. Ethernet over PDH

   c. Ethernet over HFC/DOCSIS

   d. **Ethernet over Active Fiber**

   Note: *Refer to* Summary.

10. Which wireless access technology uses licensed frequency spectrum in high-power applications?

    a. Ethernet over DOCSIS

    b. **Ethernet over terrestrial microwave**

    c. Ethernet over free space optical

    d. Ethernet over WiFi

    Note: *Refer to* Ethernet over Wireless Network.

11. Which of the following access technologies has the shortest reach?

    a. Ethernet over Terrestrial Microwave

    b. Ethernet over Bonded Copper

    c. Ethernet over PDH

    d. Ethernet over HFC/DOCSIS

    e. Ethernet over Active Fiber

    Note: *Refer to* Summary.

12. Which of the following is a disadvantage of *Ethernet over Active Fiber*?

    a. Distance limited to under 40 km

    b. Lack of widespread availability for fiber infrastructure

    c. Speed limited to 100 Mb/s

    d. Dated technology

    e. Poor bandwidth granularity

    Note: *Refer to* Ethernet over Active Fiber.

13. What are two advantages of using *Ethernet over GPON* versus *Ethernet over copper*? (Choose two.)

    a. The physical media is widely deployed.

    b. Higher bandwidth

    c. Supports efficient multicasting

    d. Resiliency through bonding

    Note: *Refer to* Ethernet over PON *and* Ethernet over Copper.

14.11
Answers: Transport Technologies

1. How many service instances can Multi-Protocol Label Switching (MPLS) support in one network?

    a. 16,777,216

    b. 16 million

    c. **Unlimited**

    d. 4094

    Note: *Refer to* Layer 2.5 Technologies (Multiprotocol Label Switching).

2. Which two technologies forward traffic over an explicitly provisioned traffic-engineered path (similar to Layer 1) without bridging or a control plane? (Choose two.)

    a. **Ethernet over MPLS-TP**

    b. Ethernet over MPLS VPWS

    c. Ethernet over MPLS VPLS

    d. **Ethernet over PBB-TE**

    e. Ethernet over PBB

    Note: *COE technologies forward traffic over explicitly provisioned traffic-engineered paths. Refer to* Summary.

3. Which one of the following associations is <u>incorrect</u>?

    a. Provider Backbone Bridging – IEEE 802.1ah

    b. Bridging – IEEE 802.1Q-2005

    c. Ethernet Ring Protection Switching – ITU-T G.8032

    d. Ethernet Linear Protection Switching – ITU-T G.8031

    e. **Provider Bridging – IEEE 802.1ag**

    f. PBB-TE – IEEE 802.1Qay

    Note: *Provider Bridging is IEEE 802.1ad. Refer to* PB (Provider Bridging).

4. Which two of the following statements describe an advantage of *Provider Bridging* compared to *Bridging*? (Choose two.)

   a. **It allows customer VLAN Identifiers to be separated from service provider VLAN Identifiers.**

   b. It can coexist in the same network with PBB-TE.

   c. It has unlimited scalability.

   d. **It supports ENNIs.**

   e. It can support multipoint-to-multipoint services and MAC learning.

   Note: *Refer to* PB (Provider Bridging).

5. Which three of the following technologies can support an EVPL service? (Choose three.)

   a. **PB**

   b. SONET/SDH

   c. **PBB-TE**

   d. WDM

   e. **MPLS VPLS**

   f. OTN

   Note: *Refer to* Summary.

6. Which two of the following protection mechanisms use CCMs? (Choose two.)

   a. **ITU-T G.8031, Ethernet linear protection switching**

   b. RSTP (Rapid Spanning Tree Protocol)

   c. 1:1 link aggregation, per clause 43 of IEEE 802.3

   d. **ITU-T G.8032, Ethernet ring protection switching**

   e. FRR (MPLS Fast Reroute)

   f. SONET UPSR protection

   Note: *Refer to* Protection Mechanisms and Transport Technologies.

7. What is the advantage of MPLS-TP compared to MPLS VPWS?

   a. It provides unlimited scalability.

   b. It supports multipoint-to-multipoint services, such as EVP-LAN and EVP-Tree.

   c. **It supports traffic engineered paths, resulting in a highly predictable, connection-oriented service.**

   d. It supports both EPL and EVPL services.

   Note: *Refer to* MPLS-TP (MPLS Transport Profile).

8. Which three of the following terms are related to a network's ability to detect and recover from faults without impacting subscribers? (Choose three.)

   a. Bandwidth

   b. **Resiliency**

   c. **G.8031**

   d. Scalability

   e. **Protection**

   f. VLAN awareness

   Note: *Refer to* Protection Mechanisms and Transport Technologies.

9. Which three of the following technologies can support an Access EPL service? (Choose three.)

   a. **PBB-TE**

   b. OTN

   c. Bridging

   d. **PBB**

   e. **PB**

   f. SONET/SDH

   Note: *Refer to* Summary.

10. Which two of the following technologies support MAC address learning? (Choose two.)

    a. MPLS VPWS

    b. PBB-TE

c. **PBB**

d. **MPLS VPLS**

e. OTN

Note: *Refer to* Summary.

11. Which technology is also known at MAC-in-MAC?

a. Provider Bridging

b. MPLS VPLS

c. **Provider Backbone Bridging**

d. PBB-TE

e. WDM

Note: *Refer to* PBB (Provider Backbone Bridging).

12. Which transport technology supports the highest bandwidth per optical fiber?

a. OTN

b. **WDM**

c. SONET

d. SDH

Note: *Refer to* WDM (Wavelength Division Multiplexing).

13. Which two of the following technologies can support multipoint-to-multipoint Ethernet services? (Choose two.)

a. MPLS VPWS

b. **MPLS VPLS**

c. MPLS-TP

d. PBB-TE

e. **PBB**

f. SONET/SDH

Note: *Refer to* Summary.

14. Which technology is also referred to as Q-in-Q?

- **a.** Provider Backbone Bridging
- **b.** OTN
- **c.** MPLS VPLS
- **d.** **Provider Bridging**
- **e.** PBB-TE

Note: *Refer to* PB (Provider Bridging).

15. Which two of the following technologies support EP-LAN services? (Choose two.)

- **a.** MPLS-TP
- **b.** **Provider Bridging**
- **c.** PBB-TE
- **d.** SONET/SDH
- **e.** **MPLS VPLS**

Note: *Refer to* Summary.

16. Provider Backbone Bridging allows a single Carrier Ethernet network to accommodate a total of how many subscribers?

- **a.** **Unlimited**
- **b.** 4094
- **c.** 16 million
- **d.** 16,777,216

Note: *Refer to* PBB (Provider Backbone Bridging).

17. Which two of the following technologies support multicasting? (Choose two.)

- **a.** SONET/SDH
- **b.** **Provider Bridging**
- **c.** MPLS-TP

d. **MPLS VPLS**

e. OTN over WDM

Note: *Multicasting requires multipoint-to-multipoint service. Refer to* Summary.

18. Which two of the following technologies can support EP-LAN services? (Choose two.)

a. **PB**

b. PBB-TE

c. MPLS-TP

d. MPLS VPWS

e. **MPLS VPLS**

Note: *Refer to* Summary.

19. Which one of the following technologies cannot support an Access EVPL service?

a. PBB-TE

b. **Bridging**

c. MPLS-TP

d. MPLS VPWS

Note: *Refer to* Summary.

14.12
Answers: Applications

1. Why is CESoETH sometimes used in Carrier Ethernet mobile backhaul networks?

    a. For performance monitoring

    b. To support voice traffic and/or to distribute timing

    c. To multiplex traffic from multiple RAN base stations to the RAN network controller

    d. To support CoS

    Note: *Refer to* Mobile Backhaul Synchronization *and* Circuit Emulation Services over Ethernet.

2. A mobile operator approaches a MEN (CEN) Service Provider to provide mobile backhaul service using a hub-and-spoke topology with a single Ethernet UNI at the hub site where the RAN Network Controller is located. Which of the following is most appropriate?

    a. Use one EPL service between each RAN Base Station and the RAN Network Controller.

    b. Use one EVP-LAN service interconnecting all RAN Base Stations.

    c. Use one EVPL service between each RAN Base Station and the RAN Network Controller.

    d. Use one EVP-Tree service with the RAN base station designated as root and the RAN network controllers designated as leaves.

    Note: *(a) Not appropriate because this solution would require multiple ports at the RAN network controller (hub). (b) Not appropriate because this solution only interconnects the RAN base stations and not the RAN network controller. (c) Most appropriate. (d) Not appropriate because the RAN base stations should be leaves and the RAN network controller should be a root.*

3. A customer wants to replace a T1 private line with a Carrier Ethernet service. Which of the following bandwidth profiles is most appropriate?

    a. CIR=1.5 Mb/s, CBS=0 bytes, EIR=0, EBS=0

    b. CIR=2 Mb/s, CBS=10,000 bytes, EIR=2 Mb/s, EBS=10,000

    c. CIR=2 Mb/s, CBS=10,000 bytes, EIR=0, EBS=0

    d. CIR=0, CBS=0 bytes, EIR=2 Mb/s, EBS=10,000

    Note: *T1 private line replacement requires CESoE (*Support for TDM Private Lines*). CESoE requires EIR=EBS=0 (*MEN/CEN Requirements for CESoETH*).*

4. Which two of the following functions are required to support CESoETH? (Choose two.)

   a. ITU-T G.8031 Ethernet linear protection switching

   b. Synchronization

   c. Internet access

   d. Link OAM (LOAM)

   e. Packetization and de-packetization

   Note: *Refer to* Interworking Function.

5. Which two of the following services can be used, in conjunction with CESoETH, in the replacement of a TDM private line service? (Choose two.)

   a. EP-LAN

   b. EPL

   c. EP-Tree

   d. EVPL

   e. EVP-LAN

   Note: *Refer to* Support for TDM Private Lines.

6. LANs at three sites are to be interconnected. Which two of the following service scenarios can be configured to tunnel bridge protocol data units (BPDUs) so that spanning tree protocol (STP) works across the aggregation of connected LANs? (Choose two.)

   a. Each site has one UNI. All three sites are interconnected by an EP-LAN service.

   b. Each site has two UNIs. Each pair of sites is interconnected by an EVPL service.

   c. Each site has one UNI. Each pair of sites is interconnected by an EVPL service.

   d. Each site has one UNI. All three sites are interconnected by an EVP-LAN service.

   e. Each site has two UNIs. Each pair of sites is interconnected by an EPL service.

   Note: *Refer to* L2CP Processing Requirements for Bridging *and* Building a Layer 2 VPN with Layer 2 Service.

7. Which two of the following clock synchronization methods transmit timing packets across the Carrier Ethernet network? (Choose two.)

   a. Free run timing

   b. IETF RFC 1305 NTP (Network Time Protocol)

c. SyncE (Synchronous Ethernet)

d. TDM line timing

e. **IEEE 1588v2 PTP (Precision Time Protocol)**

f. External timing

Note: *Refer to* Ethernet-Based Clock Synchronization.

8. A service provider wants to offer very transparent LAN connectivity, while limiting broadcast traffic. Which service solution is best?

   a. EVP-LAN service with EVC attribute *Broadcast Service Frame Delivery* set to *Deliver Conditionally*

   b. **EP-LAN service with EVC attribute *Broadcast Service Frame Delivery* set to *Deliver Conditionally***

   c. EVP-LAN service with EVC attribute *Broadcast Service Frame Delivery* set to *Deliver Unconditionally*

   d. EP-LAN service with EVC attribute *Broadcast Service Frame Delivery* set to *Deliver Unconditionally*

   Note: *Refer to* EP-LAN Service *and* Data Service Frame Delivery Service Attributes.

9. A hospital has three clinics that need to connect to it. The three clinics are independent companies that must not be allowed to share information with each other. Which of the following service solutions is most suitable to connect the hospital to the three clinics if security is the primary concern?

   a. EP-LAN service

   b. **EP-Tree service**

   c. An EVPL service from each clinic to a service multiplexed UNI at the hospital

   d. An EVP-LAN service

   Note: *(a) No security. (b) Best solution; hospital is root and clinics are leaves. (c) Some security risk: misprovisioning at hospital UNI might lead to unwanted information sharing. (d) Poor security.*

10. Which EVC-based Ethernet services can be used for mobile backhaul? (Choose the <u>one</u> best answer.)

    a. **Any**

    b. EVPL or EVP-Tree

    c. EP-LAN or EP-Tree

    d. EVP-Tree , EVP-LAN, or EVPL

    Note: *Refer to* MEF Service Definitions for Mobile Backhaul.

11. A sporting goods retailer has two data centers and numerous stores throughout the country. Each store needs to regularly exchange inventory, sales, and payroll information with both data centers. Stores should not exchange data with each other. Which of the following service solutions is best?
(Choose the one best answer.)

a. A full mesh of EVPL services interconnecting all sites

b. A single EP-LAN service interconnecting all sites

c. **A single EP-Tree service with root UNIs at the data centers and leaf UNIs at the stores**

d. A single EVP-LAN service interconnecting all sites

e. One EPL service connecting the two data centers and one EVPL service connecting each store to one of the data centers

Note: *(a) Stores should not be connected. (b) Stores should not be connected. (c) Best solution. (d) Stores should not be connected. (e) Could work, but stores are not directly connected to both data centers.*

12. An IP service provider wants to connect up to 100 customers through a single 1 Gb/s UNI with each customer allowed up to 10 Mb/s of bandwidth. None of the customers can be allowed to see traffic from any other customer. Which two the following service solutions are viable? (Choose two.)

a. **One EVPL service between each customer and the IP service provider**

b. One EVP-LAN service connecting all UNIs

c. **One EVP-Tree service connecting all UNIs with the IP service provider UNI designated as a Root and all other UNIs designated as leaves**

d. One EP-LAN service between each customer and the IP service provider

Note: *Refer to Ethernet Access to IP Services.*

13. MEF 22.1 defines four use cases for transporting mobile backhaul traffic over Carrier Ethernet services. When is CESoE (Circuit Emulation Services over Ethernet) required?

a. When RAN CE can connect to the Ethernet UNI through an Ethernet interface, eliminating the need for a Generic Inter-Working Function (GIWF)

b. When the RAN CE cannot connect to the Ethernet UNI and there is no legacy network operating in parallel to deliver timing

c. When timing is distributed over the Ethernet physical layer using synchronous Ethernet (SyncE)

d. When the operator needs to support voice traffic via an existing legacy network and offload data to a Carrier Ethernet network

Note: *Refer to Mobile Backhaul Use Cases.*

14. A customer wants to replace a T1 private line service between two PBXs (private branch exchanges) with a Carrier Ethernet service. Which technology is required to carry the customer's traffic over a Carrier Ethernet Network (CEN)?

- **a.** SONET/SDH
- **b.** **CESoE**
- **c.** ATM
- **d.** Frame Relay

Note: *Refer to* Circuit Emulation Services over Ethernet.

15. An enterprise has a headquarters (HQ) and four branch offices. Each branch office requires access to the Internet through a firewall at HQ, but no access to each other. In addition, HQ requires connection to the Internet service provider (ISP). Which two of the following service arrangements supports these requirements using only one physical connection at each site? (Choose two.)

- **a.** One EP-Tree service with HQ and ISP designated as roots and branch offices designated as leaves
- **b.** **Four EVPL services connecting branch offices to HQ, plus one EVPL service connecting HQ with ISP**
- **c.** One EVP-LAN service interconnecting the branch offices and HQ, plus one EVPL service between HQ and ISP
- **d.** **One EVPL service between HQ and ISP, plus one EVP-Tree service with HQ designated as the root and branch offices designated as leaves**

Note: *(a) Allows branches direct access to ISP without going through HQ firewall. (b) Works. At HQ, all four EVPLs terminate at one UNI. (c) Does not prevent communication between branches. (d) Works. At HQ, the EVPL and EVP-Tree terminate on the same UNI.*

16. The E-LAN service type supports which two of the following? (Choose two.)

- **a.** CESoETH
- **b.** **MAC address learning**
- **c.** Leaf UNIs
- **d.** G.8031 protection
- **e.** **Multipoint-to-multipoint EVC**

Note: *Refer to* E-LAN Service Type.

17. A subscriber needs a single EVP-LAN service to connect numerous sites to three applications. All three applications use a different VLAN. All other VLANs should be blocked. Which two attribute settings are required? (Choose two.)

- **a.** Service Multiplexing = No
- **b.** **Bundling = Yes**
- **c.** Service Multiplexing = Yes
- **d.** **CE VLAN ID Preservation = Yes**
- **e.** Bundling = No

Note: *Bundling is required to support three VLANs on a single service with all other VLANs blocked. If Bundling=Yes, CE-VLAN ID preservation must be Yes. Service Multiplexing can be either Yes or No. Refer to* Service Multiplexing, Bundling, and VLAN Tag Preservation Service Attributes.

18. A subscriber requires an EP-LAN service between 6 sites: 3 in Los Angeles and 3 in Miami. The LA sites are on-net (reachable by the Service Provider's network). The Miami sites are off-net and only reachable through an Ethernet Access Provider that has no connectivity to the Service Provider. To deliver this service, the Service Provider purchases three Access EPLs from the Miami Ethernet Access Provider and one point-to-point OVC from an Interstate Operator. The Interstate Operator shares one ENNI with the LA Service Provider and one ENNI with the Miami Ethernet Access Provider. Which of the following statements is <u>not</u> consistent with this service scenario?

- **a.** The OVC in the Interstate Operator network uses end point map bundling.
- **b.** The OVC in the Service Provider network is multipoint-to-multipoint.
- **c.** **The OVC in the Interstate Operator network provides hairpin switching.**
- **d.** Three OVC end points in the Ethernet Access Provider network interconnect (through the ENNI) to a single OVC endpoint in the Interstate Operator network.
- **e.** Three OVC end points in the Service Provider network interconnect (through the ENNI) to a single OVC endpoint in the Interstate Operator network.

Note: *Hairpin switching occurs in the Service Provider network (not in the Interstate Operator network). Refer to following figure.*

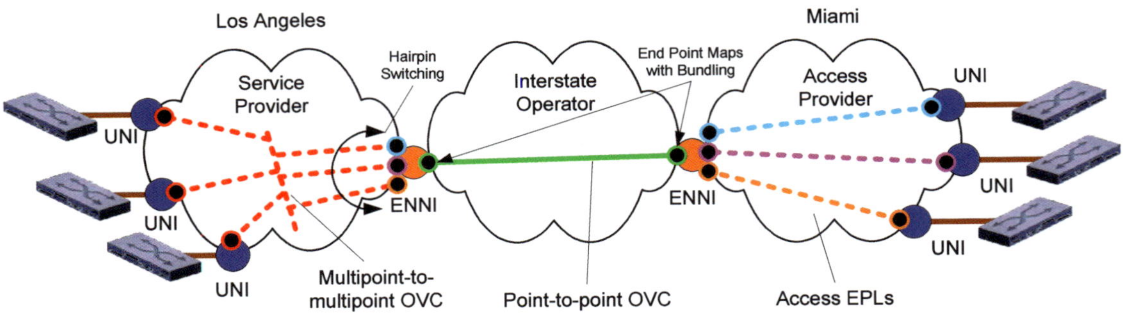

19. An Ethernet Mobile Backhaul Service provides:

- **a.** Radio access between cell phone users and cell tower base stations
- **b.** Temporary Ethernet services through a mobile WiMAX network
- **c.** Connectivity between two or more mobile phone users
- **d.** **Connectivity between RAN base stations and RAN network controllers**
- **e.** A gateway between the public mobile phone network and the Internet

Note: *Refer to* Mobile Backhaul Services.

20. What does MEF 18 contain?

- **a.** Test cases for compliance to functionality of Ethernet service traffic management
- **b.** **Tests cases for compliance of CESoETH functionality**
- **c.** Test cases for compliance of UNI Type 2 functionality
- **d.** Test cases for compliance to functionality of Ethernet services at the UNI

Note: *Refer to* Circuit Emulation Services over Ethernet.

21. A service provider uses Carrier Ethernet as the access network technology to deliver an IP service. What functionality does Ethernet access provide for the service provider?

- **a.** A Layer 3 (IP) VPN
- **b.** **Layer 2 connectivity to a Layer 3 service or an Internet Service Provider (ISP)**
- **c.** Connection to the Internet through a Layer 2 (Ethernet) firewall
- **d.** Layer 2 emulation of a Layer 3 (IP) access service
- **e.** A virtual private gateway to the Internet

Note: *Refer to* Ethernet Access to IP Services.

22. A company wants to connect two remote campuses to its corporate headquarters, extending all corporate VLANs across all three locations. Which two properties make the EP-LAN service a good choice for this application? (Choose two.)

    a. The service provider can forward frames without MAC address learning.

    b. The company does not have to coordinate CE-VLAN IDs with the service provider.

    c. The service can perform CE-VLAN ID translation if desired.

    d. The company can run STP across all locations if desired.

    e. The UNIs can support service multiplexing if desired.

    Note: *Refer to* EP-LAN Service *and* Building a Layer 2 VPN with Layer 2 Service.

23. Which two statements concerning Carrier Ethernet mobile backhaul are true? (Choose two.)

    a. A mobile backhaul service must support at least 2 classes of service.

    b. A mobile operator using SOAM uses the Subscriber MEG level.

    c. If LTE technology is used, the mobile backhaul service topology must be "hub and spoke."

    d. CoS Label H^+ must be used for packet-based synchronization traffic.

    Note: *(a) Refer to* Mobile Backhaul CoS Requirements. *(b) Refer to* MEG Levels in Mobile Backhaul. *(c) Refer to* Mobile Backhaul for LTE. *(d) Refer to* Mobile Backhaul CoS Requirements *(H^+ can be used, but it is not required).*

24. A mobile operator requests mobile backhaul services for a 4G LTE network that includes base stations (called eNodeBs) with Ethernet interfaces. Which of the following options would you recommend for interconnecting the eNodeBs to the Carrier Ethernet mobile backhaul service?

    a. Generic Inter-Working Function (GIWF)

    b. Circuit Emulation Services over Ethernet (CESoE)

    c. GIWF in combination with CESoE

    d. None of the above

    Note: *If base stations have Ethernet interfaces, GIWF and CESoE are not applicable. Refer to* Mobile Backhaul Use Cases.

25. An enterprise needs Carrier Ethernet services that give two branch offices access to critical business applications located at corporate headquarters, but not to each other. The critical business applications use VLAN IDs 10 and 20. CE devices do not support translation of CE VLAN IDs. All other VLANs should be

blocked. Both branch offices need access to both critical business applications. Which service arrangement will support this application?

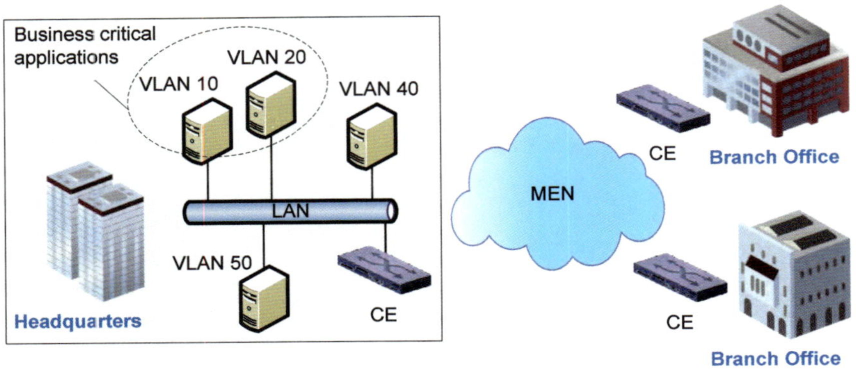

a. One Ethernet Private Line (EPL) service between each branch and the headquarters.

b. One Ethernet Virtual Private Line (EVPL) service between each branch and the headquarters. At the headquarters CE, both EVCs terminate on one UNI that is configured for service multiplexing.

c. One Ethernet Virtual Private Line (EVPL) service between each branch office and the headquarters. At the headquarters CE, the EVPLs terminate on different UNIs, which are both configured to support bundling.

d. One Ethernet Private Tree (EP-Tree) service with the headquarters UNI designated as the root and branch offices designated as leaves and with the root UNI configured to support bundling.

Note: *(a) Does not block VLANs. (b) Will not work because VLANs would be needed to separate branch offices, but VLANs are already used to identify the critical business application. This solution might be made to work if CE devices supported CE-VLAN ID translation, but that is not allowed. (c) Works. Bundling at the headquarters UNIs filters VLANs to allow access to the critical business applications but block others. (d) Bundling is not supported for port-based services.*

26. Which of the following is <u>not</u> required for a Carrier Ethernet service that carries CESoETH?

a. Bandwidth profiles include only CIR and CBS components (no EIR or EBS components).

b. The EVC must tunnel BPDUs.

c. The EVC is point-to-point.

d. All service frame delivery attributes should be set to "Deliver Unconditionally."

Note: *Refer to* MEN/CEN Requirements for CESoETH.

27. A subscriber has 10 sites with one UNI at each site. The 10 sites are interconnected by an EVP-LAN service that supports multiple VLANs between the sites. Each UNI also supports one EVPL service that connects the site to an Internet service provider (ISP). Which one of the following configuration options is <u>not</u> required?

- **a.** Configure all UNIs with *Bundling* set to *Yes*.
- **b.** **Configure *CE-VLAN ID preservation* for each EVPL service to *Yes*.**
- **c.** Configure *CE-VLAN ID preservation* for the EVP-LAN service to *Yes*.
- **d.** Configure all UNIs with *Service Multiplexing* set to *Yes*.

Note: *Refer to* Service Multiplexing, Bundling, and VLAN Tag Preservation Service Attributes.

14.13
Answers: MEF Certification

1. Which three of the following services can include CE 2.0 certified support for CoS Label H at Performance Tier 2? (Choose three.)

    a. **EPL**

    b. EP-LAN

    c. EP-Tree

    d. **EVPL**

    e. **Access EVPL**

    Note: *Multipoint services cannot include CE 2.0 certified support for CoS Label H at Performance Tier 2 because MEF 23.1 does not define CoS Performance Objective (CPO) values for multipoint services. Refer to* CPOs (per CoS Label and Performance Tier).

2. Which two of the following are requirements for an equipment vendor to participate in MEF certification testing? (Choose two.)

    a. **The equipment vendor must supply the equipment to Iometrix (the official MEF certification lab).**

    b. The equipment vendor must attend MEF quarterly member meetings.

    c. The equipment vendor must be ISO 9001 compliant.

    d. **The equipment vendor must be a member of the MEF.**

    Note: *Refer to* Equipment Certification.

3. Which two MEF test specifications are included in the CE 1.0 certification program? (Choose two.)

    a. **MEF 9, Abstract Test Suite for Ethernet Services at the UNI**

    b. **MEF 14, Abstract Test Suite for Traffic Management Phase 1**

    c. MEF 18, Abstract Test Suite for Circuit Emulation Services over Ethernet

    d. MEF 21, Abstract Test Suite for UNI Type 2 Part 1 Link OAM

    Note: *Refer to* Equipment Certification.

4. For MEF service certification, where does testing occur?

   a. At the Iometrix lab (the official MEF certification lab)

   b. On the service provider's production network

   c. On a representative network using MEF-certified networking equipment

   d. At an MEF testing facility

   Note: *Refer to* Service Certification.

5. Which one of the following is true for MEF CE 2.0 service certification?

   a. The service provider must provide on-site engineering support at the Iometrix lab during of testing.

   b. The service provider must be a member of the MEF.

   c. The service must terminate on MEF-certified equipment.

   d. Certification is lost if MEF membership is not renewed annually.

   Note: *Refer to* Service Certification.

6. Where is equipment tested for MEF equipment certification?

   a. At the Iometrix lab (the official MEF certification lab)

   b. On the service provider's production network

   c. On the MEF service certification network

   d. At the equipment vendor's facility

   Note: *Refer to* Equipment Certification.

7. CE 2.0 Certification allows equipment to be certified with respect to:

   a. Link OAM (per clause 57 of IEEE 802.3) at the UNI

   b. Any one or more of 8 Carrier Ethernet services: EPL, EVPL, EP-LAN, EVP-LAN, EP-Tree, EVP-Tree, Access EPL, and/or Access EVPL

   c. CESoETH (per ITU-T G.8261, G.823 and G.824)

   d. The following service only: EPL, EVPL, and/or E-LAN (EP-LAN together with EVP-LAN)

Note: *Refer to* Equipment Certification.

8. To participate in the MEF equipment or service certification program, a company must:

a. have a least one employee with MEF-CECP certification

b. be a member of the MEF

c. participate in MEF quarterly member meetings

d. have an in-house testing lab

Note: *Refer to* Equipment Certification *and* Service Certification.

A

About the Authors

Jon Kieffer

Jon Kieffer (MEF-CECP 2.0) is a technical writer working for Fujitsu Network Communications since 2001. In a previous academic career spanning 9 years, Jon taught software engineering at the University of Hull, U.K., and mechanical engineering at the Australian National University. Jon holds BS, MS, and PhD degrees in mechanical engineering and has authored 12 refereed journal articles and more than 30 conference papers in the fields of mechanical and controls engineering. In 2013, the Metro Ethernet Forum (MEF) recognized Jon as an outstanding contributor for his contributions to the MEF-CECP 2.0 Program.

Ralph Santitoro

Ralph Santitoro is a founding member and serves on the board of directors of the MEF where he has held leadership roles in the MEF Technical, Marketing, and Certification Committees. Ralph created the MEF's Carrier Ethernet Professional Certification Program and co-developed the original MEF-CECP exam and the new MEF-CECP 2.0 exam. Ralph was editor of MEF 6, the industry's first Ethernet Services Definitions specification, and authored the MEF's Ethernet Services primer with over 100,000 copies downloaded. Ralph has lectured at over 70 events and has published over 30 papers in the past 10 years on Carrier Ethernet, SDN, Network Virtualization, Cloud Computing, QoS, Network Security, and Mobile Backhaul. Ralph currently works with customers to identify and develop new architectures, APIs, and applications leveraging SDN and NFV.

Acknowledgements

The authors wish to thank the following individuals and organization for contributions that made this book possible: Harvey Clark, Jeffery Harris, Edward Hartman, Evan Olson, and the FNC (Fujitsu Network Communications) Services organization. This work benefits from reader feedback to our previous study guide[48], including comments from Manuel Campana, Mark Dempsky, Blake Hlavaty, Roger Marks, Ross Peterson, Froilan Pitagan, Raja Ramamurthy, and Erik Rodriguez.

This work was authored in DITA and produced using the DITA open toolkit.

48 *MEF-CECP Exam Study Guide for Carrier Ethernet Professionals*, Fujitsu Network Communications, 2012.

Index

D

E

F

G

H

I

J

K

L

M

S

T

U

V

W

Y

5214651R00279

Printed in Germany
by Amazon Distribution
GmbH, Leipzig